AF764646

GLOBAL ENERGY SECURITY

GLOBAL ENERGY SECURITY

ROGER H. ATTENBERG
EDITOR

Nova Science Publishers, Inc.
New York

Copyright © 2009 by Nova Science Publishers, Inc.

All rights reserved. No part of this book may be reproduced, stored in a retrieval system or transmitted in any form or by any means: electronic, electrostatic, magnetic, tape, mechanical photocopying, recording or otherwise without the written permission of the Publisher.

For permission to use material from this book please contact us:
Telephone 631-231-7269; Fax 631-231-8175
Web Site: http://www.novapublishers.com

NOTICE TO THE READER
The Publisher has taken reasonable care in the preparation of this book, but makes no expressed or implied warranty of any kind and assumes no responsibility for any errors or omissions. No liability is assumed for incidental or consequential damages in connection with or arising out of information contained in this book. The Publisher shall not be liable for any special, consequential, or exemplary damages resulting, in whole or in part, from the readers' use of, or reliance upon, this material.

Independent verification should be sought for any data, advice or recommendations contained in this book. In addition, no responsibility is assumed by the publisher for any injury and/or damage to persons or property arising from any methods, products, instructions, ideas or otherwise contained in this publication.

This publication is designed to provide accurate and authoritative information with regard to the subject matter covered herein. It is sold with the clear understanding that the Publisher is not engaged in rendering legal or any other professional services. If legal or any other expert assistance is required, the services of a competent person should be sought. FROM A DECLARATION OF PARTICIPANTS JOINTLY ADOPTED BY A COMMITTEE OF THE AMERICAN BAR ASSOCIATION AND A COMMITTEE OF PUBLISHERS.

LIBRARY OF CONGRESS CATALOGING-IN-PUBLICATION DATA

ISBN: 978-1-60692-087-9

Available upon request

Published by Nova Science Publishers, Inc. ✢ New York

CONTENTS

PREFACE

Access to cheap energy has become essential to the functioning of modern economies. However, the uneven distribution of energy supplies among countries and the critical need for energy has led to significant vulnerabilities. Threats to global energy security include the political instability of several energy producing countries, the manipulation of energy supplies, the competition over energy sources, attacks on supply infrastructure, as well as accidents and natural disasters. It is also the limited supplies of the most common forms of primary energy, i.e. Oil and Gas that changes perceptions on this topic. Although plenty of coal, up to 155 years worth, is readily available, coal is not the fossil fuel of choice for many more advanced countries because of its highly polluting nature. The potential need to change our primary energy sources in the foreseeable future is the crux of the energy security question, leading to higher prices, more limited access to sources of energy, competitions and political troubles, which in turn make the threat even larger.

Chapter 1 - Energy security is of increasing importance to the United States and its European allies, as some energy producers are using oil and gas for political leverage. Although most European allies believe that a market solution exists to ensure security of energy supplies, NATO has begun to discuss the issue as an allied concern. This report will be updated periodically. See also CRS Report RL33636, *The European Union's Energy Security Challenges*, by Paul Belkin.

Chapter 2 - Energy Independence and Security Act of 2007 excerpted from Public Law 110–140; 110th Congress; December 19, 2007.

Chapter 3 - Recent increases in energy prices and a steady escalation in global energy demand — expected to rise by nearly 60% over the next 20 years — have led U.S. policy-makers to engage in a wide ranging debate over how best to address the country's future energy requirements. Similarly, energy security has become a policy priority for the European Union (EU) and its 27 member states. The EU imports about 50% of its energy needs. Barring significant changes, the European Commission expects this figure to rise to 65% by 2030. About half of the EU's natural gas imports and 30% of its imported oil come from Russia. Europe's growing dependence on Russian energy, and long-term energy agreements between Russian firms and some European governments have fueled speculation that Moscow is using the "energy weapon" to try to influence European foreign and economic policy.

The EU has traditionally exerted little if any influence over individual member state energy policy. However, in March 2007, in the face of increasing concern about Europe's reliance on Russian energy, and growing public pressure to address global climate change,

EU member states agreed to forge an "Energy Policy for Europe." They have agreed on a set of EU-wide targets — some legally binding — to increase the use of renewable energy, and reduce carbon emissions. However, member states continue to pursue divergent external energy policies, particularly toward Russia, and some European countries remain reluctant to cede national control over energy markets.

The United States and Europe have steadily broadened the transatlantic energy dialogue to include joint promotion of collective energy security, energy efficiency and alternative energy sources. At the April 2007 U.S.-EU summit, leaders on both sides of the Atlantic agreed to advance cooperation to develop alternative and renewable energy technologies. However, U.S. officials have expressed concern at some European member states' unwillingness to exert more pressure on Russia to comply with EU market principles. On the other hand, European leaders appear increasingly frustrated with U.S. reluctance to pursue binding multilateral regulatory frameworks to reduce carbon emissions and promote energy efficiency.

Members of Congress have expressed an interest in efforts to increase European energy security, particularly vis-a-vis Russia. In the first session of the 110th Congress, committees held hearings that touched on the issue of European energy dependence on Russia, and on European efforts to increase energy efficiency and renewable energy use. The second session of the 1 10th Congress may also hold hearings and pursue legislation on various aspects of EU energy policy.

This report examines some of Europe's critical energy security challenges and EU efforts to coordinate a common European energy strategy. It also includes an overview of broader transatlantic energy security cooperation and will be updated as needed. For additional information, see CRS Report RL34261, *Russian Energy Policy Toward Neighboring Countries*, by Steven Woehrel, and CRS Report RS22409, *NATO and Energy Security,* by Paul Gallis.

Chapter 4 - Apart from China, there are two other large energy consumers in North-East Asia: Japan and South Korea. These economies, which belong to the OECD, are highly dependent on imports (which account for over 80% of domestic consumption in both cases), especially oil and natural gas. In recent years their energy security has been subject to serious threats. This chapter briefly addresses the position and energy forecasts for the two countries. It proceeds to analyse the strategic responses of Tokyo and Seoul to the deterioration (whether perceived or real) of their energy security, highlighting the strengths and weaknesses of each approach. Finally, the chapter details some of the lessons that other countries which are highly dependent on energy imports might learn from the experience of Japan and South Korea.

Chapter 5 - The Energy Independence and Security Act (P.L. 110-140, H.R. 6) is an omnibus energy policy law that consists mainly of provisions designed to increase energy efficiency and the availability of renewable energy. This report describes the key provisions of the enacted law, summarizes the legislative action on H.R. 6, and provides a summary of the provisions under each of the titles in the law.

The highlights of key provisions enacted into law are as follows:

- *Corporate Average Fuel Economy (CAFE).* The law sets a target of 35 miles per gallon for the combined fleet of cars and light trucks by model year 2020.

- *Renewable Fuels Standard (RFS)*. The law sets a modified standard that starts at 9.0 billion gallons in 2008 and rises to 36 billion gallons by 2022.
- *Energy Efficiency Equipment Standards.* The adopted bill includes a variety of new standards for lighting and for residential and commercial appliance equipment. The equipment includes residential refrigerators, freezers, refrigerator-freezers, metal halide lamps, and commercial walk-in coolers and freezers.
- *Repeal of Oil and Gas Tax Incentives.* The enacted law includes repeal of two tax subsidies in order to offset the estimated cost to implement the CAFE provision.

The two most controversial provisions of H.R. 6 that were not included in the enacted law were the proposed Renewable Energy Portfolio Standard (RPS) and most of the proposed tax provisions, which included repeal of tax subsidies for oil and gas and new incentives for energy efficiency and renewable energy.

In: Global Energy Security
Editor: Roger T. Attenberg

ISBN: 978-1-60692-087-9
©2009 Nova Science Publishers, Inc.

Chapter 1

NATO AND ENERGY SECURITY*

Paul Gallis

ABSTRACT

Energy security is of increasing importance to the United States and its European allies, as some energy producers are using oil and gas for political leverage. Although most European allies believe that a market solution exists to ensure security of energy supplies, NATO has begun to discuss the issue as an allied concern. This report[1] will be updated periodically. See also CRS Report RL33636, *The European Union's Energy Security Challenges*, by Paul Belkin.

ENERGY AS A SECURITY ISSUE

The growth of China and India as large consumers of energy, coupled with an inability to develop reliable and affordable alternatives to oil and natural gas, has led to the belief that the power to ensure access to international energy resources has shifted from energy consumers to energy producers.[2] In December 2005-January 2006, when Russia dramatically raised the price of natural gas that it was supplying to Ukraine, many saw an effort to squeeze Ukraine politically and economically to secure Kiev within Russia's orbit. Moscow's effort also underscored the shift towards the ability of energy producers to exert pressure on countries dependent upon them for supplies.[3]

* Excerpted from Congressional Research Service Report for Congress; order code RS22409.

1 This study was originally a memorandum for Senator Richard Lugar and is printed as a CRS report with his permission.

2 Paolo Scaroni, "Europeans Must Face the Threat to Energy Supplies," Financial Times, January 18, 2006, p. 15. Scaroni is head of ENI, the large Italian energy firm.

3 See CRS Report RS22378, Russia's Cutoff of Natural Gas to Ukraine: Context and Implications, by Bernard Gelb, Jim Nichol, and Steven Woehrel.

The United States and its European allies have begun to discuss the appropriate institutions and policies for ensuring energy security. The Bush Administration introduced a discussion of energy security at NATO in February 2006, with the support of key allies. At the same time, EU governments view energy security in a broad manner, and most believe that political and economic measures are the first steps to ensure access to energy resources. Most EU members are also members of NATO, and the two organizations may handle energy security in a complementary manner.

EUROPEAN VIEWS ON ENERGY SECURITY

Most European countries are heavily reliant upon imported energy. Today, EU countries as a whole import 50% of their energy needs, a figure expected to rise to 70% by 2030. Russia is a key supplier of oil and natural gas. Germany imports 32% of its energy from Russia. Poland imports two-thirds of its natural gas needs from Russia, and 97% of its oil. As a whole, EU countries import 25% of their energy needs from Russia. In one estimate, by 2030 EU countries will import 40% of their gas needs from Russia, and 45% of their oil from the Middle East.[4] In addition, oil in particular is found largely in unstable areas of the world such as the Middle East, a factor in U.S. and European concerns over energy security.

European governments view energy security primarily in an economic and political context. The EU floated a proposal meant to build interdependence between EU members and Russia to secure reliable energy supplies from Russia. The EU has discussed with Russia a structured arrangement in which Russia would sell energy not only to its principal customers in central and eastern Europe, but to more distant customers in western Europe. In return, the EU is asking Moscow to allow European companies to develop Russian energy reserves.[5] But Russia has rejected key elements of this proposal. Moscow for the most part has not allowed foreign ownership of its pipelines, and has squeezed out some foreign companies that have been developing its energy reserves. At the same time, it has secured access to some European markets, for example, through agreements to sell gas to Hungary and France.[6]

Russia has taken steps to build its leverage in European energy markets. In May 2007, Russia, Turkmenistan, and Kazakhstan agreed to build a new gas pipeline around the Caspian Sea. The new pipeline would send Central Asian natural gas to the Russian energy grid; Russia has repriced such gas, from another pipeline, twofold before selling it to European customers. The United States and some European governments have sought instead a trans-Caspian Sea pipeline that would bypass the Russian grid, and provide natural gas more cheaply to Europe, thereby diminishing as well greater potential Russian leverage tied to the supply of energy.[7] Russia has also discussed the linking its natural gas supply grid to that of Algeria, which

---

4 The United States also relies heavily on imported energy. In 1960, the United States imported 18% of its petroleum; in 2003, that figure reached 58%. "Schroeder and Putin Cementing Relationship," International Herald Tribune, September 8, 2005, p. 3; "Gas Crisis a Warning for Europe," International Herald Tribune, February 16, 2006, p. 3.

5 "Russia Insists It Is Reliable Gas Supplier," *Financial Times*, March 14, 2006, p. 5.

6 "Russia Defiant on Energy," *International Herald Tribune*, December 13, 2006, p. 1; and "Gazprom Gains Access to Hungary Market," *Financial Times*, July 14, 2006, p. 4. An exception to this policy is an agreement between Gazprom and Eni, an Italian company, to co-develop the large Shtokman gas field in eastern Russia.

7 "Russia Will Get Central Asian Pipeline," *New York Times*, May 13, 2007, p. 12.

also supplies gas to Europe. EU energy commissioner Andris Piebalgs has charged that the two governments may be planning to develop an energy cartel that would further weaken competitive pricing.[8]

To prevent impediments to competition and to improve energy security, the EU Commission is urging new infrastructure, including terminals for receiving liquified natural gas; the construction of new pipelines from the Caspian region and North Africa; and single European energy grids for both continental electricity and natural gas markets that would challenge the grip of national energy firms on their national markets.[9] The Commission has also recommended that companies that produce raw energy not be allowed to own distribution networks, a step intended to encourage competition. Some EU governments, such as France, have large public entities that own both the sources of energy and the distribution network, and oppose this proposal. Should the EU eventually adopt the Commission's proposal, Russian efforts to buy parts of the European energy grid might be set back.

Few observers believe that Moscow's pricing agreement for its gas exports to Ukraine indicates that the market process is working successfully. Some EU officials say that the agreement lacks transparency, and may mask involvement by criminal enterprises. They contend that Russia needs European (and other) firms' good will and continued investment in its decaying energy infrastructure to maintain existing production and develop its oil and gas reserves to sell energy products abroad.[10]

Some European and U.S. officials believe that Germany may become too reliant on Russian energy supplies and move away from its EU partners and the United States. East European states in particular, once in Moscow's sphere, believe that they could find themselves unable to ensure reliable and affordable energy supplies from Gazprom, the powerful state-controlled Russian energy company. They point to the former Schroeder government's deal with Gazprom to involve German companies in the development of a Russian-German gas pipeline under the Baltic Sea as a special arrangement that appears to promise a supply to Germany that other states might not enjoy.[11]

Some governments believe that Russia has little interest in market forces in the energy sector. In this view, Russia seeks high energy prices to maximize profits. These governments note that the Russian government has a prevailing control over Gazprom, hardly a model of capitalist entrepreneurship, and that Gazprom was behaving like a monopoly in ratcheting up the price of natural gas to Ukraine. Knowing that Ukraine had no reliable alternatives for gas supply, Gazprom raised prices threefold and threatened a sixfold rise. Gazprom also controls the transit of non-Russian energy supplies to Ukraine, and threatened rapid rises in transit fees as well. Russia has temporarily followed similar policies towards Georgia, Lithuania, and Belarus. Political motives seem apparent in such policies. In 2003, Putin himself said that Gazprom is a "powerful political and economic lever of influence over the rest of the world."[12]

---

[8] "EU Official Says Russia is Seeking Gas Cartel," *International Herald Tribune*, January 27, 2007, p. 1.

[9] "EU To Issue Warning on Foreign Energy Use," International Herald Tribune, March 7, 2006, p. 1.

[10] Interviews with European officials, 2006-2007; see also CRS Report RL33212, *Russian Oil and Gas Challenges*, by Robert Pirog.

[11] Interviews with U.S. and east European officials, 2006-2007.

[12] *Jamestown Eurasian Daily Monitor*, January 19, 2006; Putin cited in "EU's Barroso Demands Natural Gas Supply Assurances from Russia," *Bloomberg News*, March 16, 2006.

NATO AND ENERGY SECURITY

Some U.S. officials believe that NATO could play a role in building international political solidarity in the event of a deliberate disruption of energy flows. In this view, NATO might coordinate policies among member states and with non-member partner governments to share resources and to bring an end to an energy disruption. NATO might also provide security for infrastructure in energy-producing states facing unrest.

Iran has threatened to use its energy reserves to attain political objectives. In response to possible sanctions due to its refusal to comply with requirements by the International Atomic Energy Agency on its nuclear program, Iran has threatened to cut off or limit its energy supplies to buyers. Beyond deliberate policies affecting energy security, there are many countries in Central Asia and the Middle East that are unstable, have a need for new energy infrastructure investment, and have insecure transportation systems due to political unrest. Some of these countries are in NATO's Partnership for Peace program, or desire a closer association with NATO.

NATO member states increasingly believe that the alliance must be a global player with global partners. This trend is evident in Afghanistan, for example, where Australia, New Zealand, and Japan are expending resources to bring stability through NATO's International Security Assistance Force, even though the three countries are not NATO members. NATO's role in energy security could be complementary to the EU's effort to strengthen market forces and interdependence in the international energy sector. U.S. officials agree with their EU counterparts that market forces can lead to greater energy security. Diversification of supply, for example, through building more pipelines that are secure, is one course of action. Joint investment efforts to build such pipelines in and with energy producers such as Kazakhstan and Azerbaijan could be an important step in this direction. Both countries are members of NATO's Partnership for Peace program, and are seeking closer relations with the United States and its allies.

Development of more liquified natural gas (LNG) transport and reception facilities from distant suppliers, such as Nigeria, into Europe could be another course of action. Coupled with the development of new oil and gas pipelines could be an offer from NATO (and/or EU) members to provide security for energy infrastructure in periods of unrest or conflict in supplier and transit countries.[13]

The Military Option

NATO governments (although not NATO as a whole) have already been involved in military efforts to secure energy resources. The first Gulf War, while not a NATO operation, involved key member states such as the United States, France, Britain, and Italy that sought not only to liberate Kuwait but also to ensure that Iraq did not control Kuwaiti oil and threaten Saudi Arabia and other Gulf producers.

NATO governments also took part in a military operation in the 1980s explicitly designed to secure the supply of oil. *Operation Earnest Will* was an effort, primarily by NATO states, to

[13] Discussions at the Organization for Economic Cooperation and Development, February 2006.

protect tanker traffic in the Gulf during the Iran-Iraq War (1980-1988). Beginning in 1984, Iran first, and then Iraq, attacked neutral oil tankers to cut off the other's means of financial support. Iran attacked Kuwaiti and Saudi tankers in those two countries' own waters to ensure that all Gulf states understood that none was secure. The Soviet Union, followed shortly thereafter by the United States, made offers to the Kuwaitis, who lost the most tankers, to reflag their vessels under the USSR and the U.S. flags, an offer that was accepted. After Iraqi aircraft attacked the USS Stark in 1987, killing 37 sailors, the Reagan Administration formed a coalition of like-minded states, above all from NATO, to protect tanker traffic in the Gulf. Britain, France, and the Netherlands were important participants in *Operation Earnest Will*. The allies captured Iranian vessels mining shipping lanes in the Gulf, and engaged in firefights with Iranian troops using oil platforms to fire on ships.

In February 2006, NATO governments discussed a range of potential actions in the event of future disruption of oil supplies caused by military action. Some member states reportedly raised the possibility of protecting tanker traffic and oil platforms in periods of conflict, and using satellites to monitor developments in areas where energy resources come under threat.[14]

Congressional Action

There was increasing congressional interest in energy security during the 109th Congress. The Senate passed S.Res. 456, which urged the Administration to raise the issue of energy security in the North Atlantic Council, and to report the results of such a meeting to Congress. Senator Lugar gave a speech during the NATO Riga summit in November 2006 in which he urged that energy security be raised to an Article V, or mutual security, issue. Several House and Senate committees held hearings on the issue of energy security, with attention to NATO's possible role. Similar hearings and legislation are possible in the 110th Congress.

Conclusion

NATO is attempting to become a global security organization, still concentrating on protection of the interests of the United States and its Canadian and European partners, but engaging non-member states as global partners. NATO's role in energy security remains uncertain, however, as some individual members may prefer a greater role for the EU.

A political role in energy security for NATO seems most likely in the near future. Under NATO's Istanbul Cooperation Initiative of 2004, the allies have begun discussions with Bahrain, Qatar, Kuwait, and the United Arab Emirates to build practical cooperation in the security field, including the fight against terrorism.[15] Some Middle Eastern governments are concerned about terrorist attacks on their oil facilities, but it is not publicly known whether NATO has discussed this issue with the four governments.

Partnership for Peace countries, such as Kazakhstan and Azerbaijan, that are important energy producers often seek ways to associate themselves more closely with NATO, in part to

[14] Discussions with officials from several NATO governments, February 2006.

diminish Russian influence on their soil, in part to develop reliable partners in an unstable region. It is possible that NATO will seek ways to provide security for the energy infrastructure of such countries. At the same time, the EU may encourage its member states to invest more heavily in that infrastructure.

There is division in the EU over management of the Union's growing dependence on Russian oil and gas. Several states, led by Poland, wish to engage NATO more fully in ensuring energy security in this relationship. While in the early stages of discussion, Poland is exploring a role for NATO and the United States, perhaps only diplomatically, in which U.S. leverage on Moscow could be an element for encouraging responsible Russian behavior and deflecting any Russian attempt to divide the Europeans.

A NATO role in energy security could prove to be premature. Most EU governments clearly prefer that market forces secure access to energy. A well-structured commercial partnership with Russia might be one mark of such a policy. Another would be the effort of the EU3 (Germany, France, and Britain) and the United States to curtail Iran's nuclear program. The EU3 desire completion of that effort in the UN before there is any discussion of a military organization like NATO assuming responsibility for a broader policy of energy security.

Some EU governments also believe that discussion of energy security at NATO sends the wrong signal to other governments, which might assume that the allies are contemplating military action to ensure the flow of oil and gas. Some of these governments propose instead that there first be a high-level "seminar" that includes the United States, representatives of key EU countries, Russia, and such countries as Uzbekistan, Kazakhstan, and Azerbaijan. There, some participants would reiterate the importance of market forces and the interdependence of producers and suppliers, and the need to protect and maintain energy infrastructure.

In addition, some NATO partner governments in Central Asia and the Middle East might be reluctant to accept allied assistance in securing the resource that is central to their survival. The belief is widespread in the Middle East that the United States invaded Iraq in part to secure access to its oil. There might be popular opposition to any NATO effort to secure energy infrastructure in some of these countries. Moreover, the United States has been unable to provide full security to pipelines in Iraq, and NATO might have similar difficulties in partnership countries. Russia is also a factor. Turkmenistan and Kazakhstan depend upon Russia as a transit country for their pipeline shipments to the west, and could be subject to Moscow's pressure to spurn NATO proposals of assistance.

[15] Istanbul Cooperation Initiative, NATO, Brussels, January 2006.

In: Global Energy Security
Editor: Roger T. Attenberg
ISBN: 978-1-60692-087-9
©2009 Nova Science Publishers, Inc.

Chapter 2

ENERGY INDEPENDENCE AND SECURITY ACT OF 2007*

PUBLIC LAW 110–140 110TH CONGRESS

An Act

Dec. 19, 2007 [H.R. 6]

To move the United States toward greater energy independence and security, to increase the production of clean renewable fuels, to protect consumers, to increase the efficiency of products, buildings, and vehicles, to promote research on and deploy greenhouse gas capture and storage options, and to improve the energy performance of the Federal Government, and for other purposes.

42 USC 17001 note

Be it enacted by the Senate and House of Representatives of the United States of America in Congress assembled,

SECTION 1. SHORT TITLE; TABLE OF CONTENTS

Short Title — This Act may be cited as the "Energy Independence and Security Act of 2007".

Table of Contents — The table of contents of this Act is as follows:

* Excerpted from Public Law 110–140; 110th Congress; December 19, 2007

TITLE I — ENERGY SECURITY THROUGH IMPROVED VEHICLE FUEL ECONOMY

Subtitle A—Increased Corporate Average Fuel Economy Standards

Subtitle B—Improved Vehicle Technology

Subtitle C—Federal Vehicle Fleets

TITLE II — ENERGY SECURITY THROUGH INCREASED PRODUCTION OF BIOFUELS

Subtitle A—Renewable Fuel Standard

Subtitle B—Biofuels Research and Development

Subtitle C—Biofuels Infrastructure

Subtitle D—Environmental Safeguards

TITLE III — ENERGY SAVINGS THROUGH IMPROVED STANDARDS FOR APPLIANCE AND LIGHTING

Subtitle A — Appliance Energy Efficiency

Subtitle B—Lighting Energy Efficiency

TITLE IV — ENERGY SAVINGS IN BUILDINGS AND INDUSTRY

Subtitle A—Residential Building Efficiency

Subtitle B—High-Performance Commercial Buildings

Subtitle C—High-Performance Federal Buildings

Subtitle D—Industrial Energy Efficiency

Subtitle E—Healthy High-Performance Schools

Subtitle F—Institutional Entities

Subtitle G—Public and Assisted Housing

Subtitle H—General Provisions

TITLE V—ENERGY SAVINGS IN GOVERNMENT AND PUBLIC INSTITUTIONS

Subtitle A—United States Capitol Complex

Subtitle B—Energy Savings Performance Contracting

Subtitle C—Energy Efficiency in Federal Agencies

Subtitle D—Energy Efficiency of Public Institutions

Subtitle E—Energy Efficiency and Conservation Block Grants

TITLE VI — ACCELERATED RESEARCH AND DEVELOPMENT

Subtitle A—Solar Energy

Subtitle B—Geothermal Energy

Subtitle C—Marine and Hydrokinetic Renewable Energy Technologies

Subtitle D—Energy Storage for Transportation and Electric Power

Subtitle E—Miscellaneous Provisions

TITLE VII—CARBON CAPTURE AND SEQUESTRATION

Subtitle A—Carbon Capture and Sequestration Research, Development, and Demonstration

Subtitle B—Carbon Capture and Sequestration Assessment and Framework

TITLE VIII—IMPROVED MANAGEMENT OF ENERGY POLICY

Subtitle A—Management Improvements

Subtitle B—Prohibitions on Market Manipulation and False Information

TITLE IX—INTERNATIONAL ENERGY PROGRAMS

Subtitle A—Assistance to Promote Clean and Efficient Energy Technologies in Foreign Countries

Subtitle B—International Clean Energy Foundation

Subtitle C—Miscellaneous Provisions

TITLE X — GREEN JOBS

TITLE XI — ENERGY TRANSPORTATION AND INFRASTRUCTURE

Subtitle A—Department of Transportation

Subtitle B—Railroads

Subtitle C—Marine Transportation

Subtitle D—Highways

TITLE XII — SMALL BUSINESS ENERGY PROGRAMS

TITLE XIII — SMART GRID

TITLE XIV—POOL AND SPA SAFETY

TITLE XV — REVENUE PROVISIONS

TITLE XVI — EFFECTIVE DATE

42 USC 17001. SEC. 2. DEFINITIONS

In this Act:

Department— The term "Department" means the Department of Energy.

Institution of Higher Education — The term "institution of higher education" has the meaning given the term in section 10 1(a) of the Higher Education Act of 1965 (20 U.S.C. 1001(a)).

Secretary — The term "Secretary" means the Secretary of Energy.

42 USC 17002. SEC. 3. RELATIONSHIP TO OTHER LAW

Except to the extent expressly provided in this Act or an amendment made by this Act, nothing in this Act or an amendment made by this Act supersedes, limits the authority provided or responsibility conferred by, or authorizes any violation of any provision of law (including a regulation), including any energy or environmental law or regulation.

TITLE I — ENERGY SECURITY THROUGH IMPROVED VEHICLE FUEL ECONOMY

Ten-in-Ten Fuel Economy Act.

Subtitle A—Increased Corporate Average Fuel Economy Standards

42 USC 30101 note.

Sec. 101. Short Title.

This subtitle may be cited as the "Ten-in-Ten Fuel Economy Act".

Sec. 102. Average fuel economy standards for automobiles and certain other vehicles.

(a) INCREASED STANDARDS.—Section 32902 of title 49, United States Code, is amended—

(1) in subsection (a)—

(A) by striking "NON-PASSENGER AUTOMOBILES.—" and inserting "PRESCRIPTION OF STANDARDS BY REGULATION.—";

(B) by striking "(except passenger automobiles)" in subsection (a); and

(C) by striking the last sentence;

(2) by striking subsection (b) and inserting the following:

"(b) Standards For Automobiles And Certain Other Vehicles.—

"(1) IN GENERAL.—The Secretary of Transportation, after consultation with the Secretary of Energy and the Administrator of the Environmental Protection Agency, shall prescribe separate average fuel economy standards for—

(A) passenger automobiles manufactured by manufacturers in each model year beginning with model year 2011 in accordance with this subsection;

(B) non-passenger automobiles manufactured by manufacturers in each model year beginning with model year 2011 in accordance with this subsection; and

(C) work trucks and commercial medium-duty or heavy-duty on-highway vehicles in accordance with subsection (k).

"(2) FUEL ECONOMY STANDARDS FOR AUTOMOBILES.—

(A) AUTOMOBILE FUEL ECONOMY AVERAGE FOR MODEL YEARS 2011 THROUGH 2020.—The Secretary shall prescribe a separate average fuel economy standard for passenger automobiles and a separate average fuel economy standard for non-passenger automobiles for each model year beginning with model year 2011 to achieve a combined fuel economy average for model year 2020 of at least 35 miles per gallon for the total fleet of passenger and non-passenger automobiles manufactured for sale in the United States for that model year.

(B) AUTOMOBILE FUEL ECONOMY AVERAGE FOR MODEL YEARS 2021 THROUGH 2030.—For model years 2021 through 2030, the average fuel economy required to be attained by each fleet of passenger and non-passenger automobiles manufactured for sale in the United States shall be the maximum feasible average fuel economy standard for each fleet for that model year.

(C) PROGRESS TOWARD STANDARD REQUIRED.—In prescribing average fuel economy standards under subparagraph (A), the Secretary shall prescribe annual fuel economy standard increases that increase the applicable average fuel economy standard ratably beginning with model year 2011 and ending with model year 2020.

"(3) AUTHORITY OF THE SECRETARY.—The Secretary shall—

(A) prescribe by regulation separate average fuel economy standards for passenger and non-passenger automobiles based on 1 or more vehicle attributes related to fuel economy and express each standard in the form of a mathematical function; and

(B) issue regulations under this title prescribing average fuel economy standards for at least 1, but not more than 5, model years.

"(4) MINIMUM STANDARD.—In addition to any standard prescribed pursuant to paragraph (3), each manufacturer shall also meet the minimum standard for domestically manufactured passenger automobiles, which shall be the greater of—

(A) 27.5 miles per gallon; or

(B) 92 percent of the average fuel economy projected by the Secretary for the combined domestic and non- domestic passenger automobile fleets manufactured for sale in the United States by all manufacturers in the model year, which projection shall be published in the Federal Register when the standard for that model year is promulgated in accordance with this section."; and

(3) in subsection (c)—

(A) by striking "(1) Subject to paragraph (2) of this subsection, the" and inserting "The"; and

(B) by striking paragraph (2).

(b) FUEL ECONOMY STANDARD FOR COMMERCIAL MEDIUM-DUTY AND HEAVY-DUTY ON-HIGHWAY VEHICLES AND WORK TRUCKS.—

Section 32902 of title 49, United States Code, is amended by adding at the end the following:

"(k) COMMERCIAL MEDIUM- AND HEAVY-DUTY ON-HIGHWAY VEHICLES AND WORK TRUCKS.—

"(1) STUDY.—Not later than 1 year after the National Academy of Sciences publishes the results of its study under section 108 of the Ten-in-Ten Fuel Economy Act, the Secretary of Transportation, in consultation with the Secretary of Energy and the Administrator of the Environmental Protection Agency, shall examine the fuel efficiency of commercial medium- and heavy-duty on-highway vehicles and work trucks and determine—

(A) the appropriate test procedures and methodologies for measuring the fuel efficiency of such vehicles and work trucks;

(B) the appropriate metric for measuring and expressing commercial medium- and heavy-duty on-highway vehicle and work truck fuel efficiency performance, taking into consideration, among other things, the work performed by such on-highway vehicles and work trucks and types of operations in which they are used;

(C) the range of factors, including, without limitation, design, functionality, use, duty cycle, infrastructure, and total overall energy consumption and operating costs that affect commercial medium- and heavy-duty on-highway vehicle and work truck fuel efficiency; and

(D) such other factors and conditions that could have an impact on a program to improve commercial medium- and heavy-duty on-highway vehicle and work truck fuel efficiency.

"(2) RULEMAKING.—Not later than 24 months after completion of the study required under paragraph (1), the Secretary, in consultation with the Secretary of Energy and the Administrator of the Environmental Protection Agency, by regulation, shall determine in a rulemaking proceeding how to implement a commercial medium- and heavy-duty on-highway vehicle and work truck fuel efficiency improvement program designed to achieve the maximum feasible improvement, and shall adopt and implement appropriate test methods, measurement metrics, fuel economy standards, and compliance and enforcement protocols that are appropriate, cost-effective, and technologically feasible for commercial medium- and heavy-duty on- highway vehicles and work trucks. The Secretary may prescribe separate standards for different classes of vehicles under this subsection.

"(3) LEAD-TIME; REGULATORY STABILITY.—The commercial medium- and heavy-duty on-highway vehicle and work truck fuel economy standard adopted pursuant to this subsection shall provide not less than—

(A) 4 full model years of regulatory lead-time; and

(B) 3 full model years of regulatory stability.".

Sec. 103. Definitions.

(a) IN GENERAL.—Section 3290 1(a) of title 49, United States Code, is amended—

(1) by striking paragraph (3) and inserting the following:

"(3) except as provided in section 32908 of this title, 'automobile' means a 4-wheeled vehicle that is propelled by fuel, or by alternative fuel, manufactured primarily for use on public streets, roads, and highways and rated at less than 10,000 pounds gross vehicle weight, except—

(A) a vehicle operated only on a rail line;

(B) a vehicle manufactured in different stages by 2 or more manufacturers, if no intermediate or final-stage manufacturer of that vehicle manufactures more than 10,000 multi-stage vehicles per year; or

(C) a work truck.";

(2) by redesignating paragraphs (7) through (16) as paragraphs (8) through (17), respectively;

(3) by inserting after paragraph (6) the following:

"(7) 'commercial medium- and heavy-duty on-highway vehicle' means an on-highway vehicle with a gross vehicle weight rating of 10,000 pounds or more.";

(4) in paragraph (9)(A), as redesignated, by inserting "or a mixture of biodiesel and diesel fuel meeting the standard established by the American Society for Testing and Materials or under section 211(u) of the Clean Air Act (42 U.S.C. 7545(u)) for fuel containing 20 percent biodiesel (commonly known as 'B20')" after "alternative fuel";

(5) by redesignating paragraph (17), as redesignated, as paragraph (18);

(6) by inserting after paragraph (16), as redesignated, the following:

"(17) 'non-passenger automobile' means an automobile that is not a passenger automobile or a work truck."; and

(7) by adding at the end the following:

"(19) 'work truck' means a vehicle that—

(A) is rated at between 8,500 and 10,000 pounds gross vehicle weight; and

(B) is not a medium-duty passenger vehicle (as defined in section 86.1803–01 of title 40, Code of Federal Regulations, as in effect on the date of the enactment of the Ten-in-Ten Fuel Economy Act).".

Sec. 104. Credit trading program

(a) IN GENERAL.—Section 32903 of title 49, United States Code, is amended—

(1) by striking "section 32902(b)–(d) of this title" each place it appears and inserting "subsections (a) through (d) of section 32902";

(2) in subsection (a)(2)—

(A) by striking "3 consecutive model years" and inserting "5 consecutive model years";

(B) by striking "clause (1) of this subsection," and inserting "paragraph (1)";

(3) by redesignating subsection (f) as subsection (h); and

(4) by inserting after subsection (e) the following:

"(f) CREDIT TRADING AMONG MANUFACTURERS.—

(1) IN GENERAL.—The Secretary of Transportation may establish, by regulation, a fuel economy credit trading program to allow manufacturers whose automobiles exceed the average fuel economy standards prescribed under section 32902 to earn credits to be sold to manufacturers whose automobiles fail to achieve the prescribed standards such that the total oil savings associated with manufacturers that exceed the prescribed standards are preserved when trading credits to manufacturers that fail to achieve the prescribed standards.

(2) LIMITATION.—The trading of credits by a manufacturer to the category of passenger automobiles manufactured domestically is limited to the extent that the fuel economy level of such automobiles shall comply with the requirements of section 32902(b)(4), without regard to any trading of credits from other manufacturers.

"(g) CREDIT TRANSFERRING WITHIN A MANUFACTURER'S FLEET.—

"(1) IN GENERAL.—The Secretary of Transportation shall establish by regulation a fuel economy credit transferring program to allow any manufacturer whose automobiles exceed any of the average fuel economy standards prescribed under section 32902 to transfer the credits earned under this section and to apply such credits within that manufacturer's fleet to a compliance category of automobiles that fails to achieve the prescribed standards.

"(2) YEARS FOR WHICH USED.—Credits transferred under this subsection are available to be used in the same model years that the manufacturer could have applied such credits under subsections (a), (b), (d), and (e), as well as for the model year in which the manufacturer earned such credits.

"(3) MAXIMUM INCREASE.—The maximum increase in any compliance category attributable to transferred credits is—

(A) for model years 2011 through 2013, 1.0 mile per gallon;

(B) for model years 2014 through 2017, 1.5 miles per gallon; and

(C) for model year 2018 and subsequent model years, 2.0 miles per gallon.

"(4) LIMITATION.—The transfer of credits by a manufacturer to the category of passenger automobiles manufactured domestically is limited to the extent that the fuel economy level of such automobiles shall comply with the requirements under section 32904(b)(4), without regard to any transfer of credits from other categories of automobiles described in paragraph (6)(B).

"(5) YEARS AVAILABLE.—A credit may be transferred under this subsection only if it is earned after model year 2010.

"(6) DEFINITIONS.—In this subsection:

(A) FLEET.—The term 'fleet' means all automobiles manufactured by a manufacturer in a particular model year.

(B) COMPLIANCE CATEGORY OF AUTOMOBILES.—The term 'compliance category of automobiles' means any of the following 3 categories of automobiles for which compliance is separately calculated under this chapter:

(i) Passenger automobiles manufactured domestically.

(ii) Passenger automobiles not manufactured domestically.

(iii) Non-passenger automobiles.".

(b) CONFORMING AMENDMENTS.—

(1) LIMITATIONS.—Section 32902(h) of title 49, United States Code, is amended—

(A) in paragraph (1), by striking "and" at the end;

(B) in paragraph (2), by striking the period at the end and inserting "; and"; and

(C) by adding at the end the following:

"(3) may not consider, when prescribing a fuel economy standard, the trading, transferring, or availability of credits under section 32903.".

(2) SEPARATE CALCULATIONS.—Section 32904(b)(1)(B) is amended by striking "chapter." and inserting "chapter, except for the purposes of section 32903."

Sec. 105. Consumer information

Section 32908 of title 49, United States Code, is amended by adding at the end the following:

"(g) CONSUMER INFORMATION.—

"(1) PROGRAM.—The Secretary of Transportation, in consultation with the Secretary of Energy and the Administrator of the Environmental Protection Agency, shall develop and implement by rule a program to require manufacturers—

"(A) to label new automobiles sold in the United States with—

''(i) information reflecting an automobile's performance on the basis of criteria that the Administrator shall develop, not later than 18 months after the date of the enactment of the Ten-in-Ten Fuel Economy Act, to reflect fuel economy and greenhouse gas and other emissions over the useful life of the automobile;

"(ii) a rating system that would make it easy for consumers to compare the fuel economy and greenhouse gas and other emissions of automobiles at the point of purchase, including a designation of automobiles—

(A) with the lowest greenhouse gas emissions over the useful life of the vehicles; and

(B) the highest fuel economy; and

"(iii) a permanent and prominent display that an automobile is capable of operating on an alternative fuel; and

"(B) to include in the owner's manual for vehicles capable of operating on alternative fuels information that describes that capability and the benefits of using alternative fuels, including the renewable nature and environmental benefits of using alternative fuels.

"(2) CONSUMER EDUCATION.—

"(A) IN GENERAL.—The Secretary of Transportation, in consultation with the Secretary of Energy and the Administrator of the Environmental Protection Agency, shall develop and implement by rule a consumer education program to improve consumer understanding of automobile performance described in paragraph (1)(A)(i) and to inform consumers of the benefits of using alternative fuel in automobiles and the location of stations with alternative fuel capacity.

"(B) FUEL SAVINGS EDUCATION CAMPAIGN.—The Secretary of Transportation shall establish a consumer education campaign on the fuel savings that would be recognized from the purchase of vehicles equipped with thermal management technologies, including energy efficient air conditioning systems and glass.

(3) FUEL TANK LABELS FOR ALTERNATIVE FUEL AUTOMOBILES.—The Secretary of Transportation shall by rule require a label to be attached to the fuel compartment of vehicles capable of operating on alternative fuels, with the form of alternative fuel stated on the label. A label attached in compliance with the requirements of section 32905(h) is deemed to meet the requirements of this paragraph.

(4) RULEMAKING DEADLINE.—The Secretary of Transportation shall issue a final rule under this subsection not later than 42 months after the date of the enactment of the Ten- in-Ten Fuel Economy Act.".

Sec. 106. Continued Applicability of Existing Standards

Nothing in this subtitle, or the amendments made by this subtitle, shall be construed to affect the application of section 32902 of title 49, United States Code, to passenger automobiles or non- passenger automobiles manufactured before model year 2011.

Sec. 107. National Academy of Sciences Studies.

(a) IN GENERAL.—As soon as practicable after the date of enactment of this Act, the Secretary of Transportation shall execute an agreement with the National Academy of Sciences to develop a report evaluating vehicle fuel economy standards, including—

(A) an assessment of automotive technologies and costs to reflect developments since the Academy's 2002 report evaluating the corporate average fuel economy standards was conducted;

(B) an analysis of existing and potential technologies that may be used practically to improve automobile and medium- duty and heavy-duty truck fuel economy;

(C) an analysis of how such technologies may be practically integrated into the automotive and medium-duty and heavy- duty truck manufacturing process; and

(D) an assessment of how such technologies may be used to meet the new fuel economy standards under chapter 329 of title 49, United States Code, as amended by this subtitle.

(b) REPORT.—The Academy shall submit the report to the Secretary, the Committee on Commerce, Science, and Transportation of the Senate, and the Committee on Energy and Commerce of the House of Representatives, with its findings and recommendations not later than 5 years after the date on which the Secretary executes the agreement with the Academy.

(c) QUINQUENNIAL UPDATES.—After submitting the initial report, the Academy shall update the report at 5 year intervals thereafter through 2025.

Sec. 108. National Academy Of Sciences Study Of Medium-Duty And Heavy-Duty Truck Fuel Economy

(a) IN GENERAL.—As soon as practicable after the date of enact Contracts. ment of this Act, the Secretary of Transportation shall execute

(A) an agreement with the National Academy of Sciences to develop a report evaluating medium-duty and heavy-duty truck fuel economy standards, including—

(B) an assessment of technologies and costs to evaluate fuel economy for medium-duty and heavy-duty trucks;

(C) an analysis of existing and potential technologies that may be used practically to improve medium-duty and heavy- duty truck fuel economy;

(D) an analysis of how such technologies may be practically integrated into the medium-duty and heavy-duty truck manufacturing process;

(E) an assessment of how such technologies may be used to meet fuel economy standards to be prescribed under section 32902(k) of title 49, United States Code, as amended by this subtitle; and

(F) associated costs and other impacts on the operation of medium-duty and heavy-duty trucks, including congestion.

(b) REPORT.—The Academy shall submit the report to the Secretary, the Committee on Commerce, Science, and Transportation of the Senate, and the Committee on Energy and Commerce of the House of Representatives, with its findings and recommendations not later than 1 year after the date on which the Secretary executes the agreement with the Academy.

Sec. 109. Extension Of Flexible Fuel Vehicle Credit Program

(a) IN GENERAL.—Section 32906 of title 49, United States Code, is amended to read as follows:

"§ 32906. Maximum fuel economy increase for alternative fuel automobiles

"(a) IN GENERAL.—For each of model years 1993 through 2019 for each category of automobile (except an electric automobile), the maximum increase in average fuel economy for a manufacturer attributable to dual fueled automobiles is—

"(1) 1.2 miles a gallon for each of model years 1993 through 2014;

"(2) 1.0 miles per gallon for model year 2015;

"(3) 0.8 miles per gallon for model year 2016;

"(4) 0.6 miles per gallon for model year 2017;

"(5) 0.4 miles per gallon for model year 2018;

"(6) 0.2 miles per gallon for model year 2019; and

"(7) 0 miles per gallon for model years after 2019.

"(b) CALCULATION.—In applying subsection (a), the Administrator of the Environmental Protection Agency shall determine the increase in a manufacturer's average fuel economy attributable to dual fueled automobiles by subtracting from the manufacturer's average fuel economy calculated under section 32905(e) the number equal to what the manufacturer's average fuel economy would be if it were calculated by the formula under section 32904(a)(1) by including as the denominator for each model of dual fueled automobiles the fuel economy when the automobiles are operated on gasoline or diesel fuel.".

(b) CONFORMING AMENDMENTS.—Section 32905 of title 49, United States Code, is amended—

(A) in subsection (b), by striking "1993–2010," and inserting "1993 through 2019,";

(B) in subsection (d), by striking "1993–20 10," and inserting "1993 through 2019,";

(C) by striking subsections (f) and (g); and

(D) by redesignating subsection (h) as subsection (f).

(c) B20 BIODIESEL FLEXIBLE FUEL CREDIT.—Section 32905(b)(2) of title 49, United States Code, is amended to read as follows: "(2) .5 divided by the fuel economy—

(A) measured under subsection (a) when operating the model on alternative fuel; or

(B) measured based on the fuel content of B20 when operating the model on B20, which is deemed to contain 0.15 gallon of fuel.".

Sec. 110. Periodic Review Of Accuracy Of Fuel Economy Labeling Procedures

Beginning in December 2009, and not less often than every 5 years thereafter, the Administrator of the Environmental Protection Agency, in consultation with the Secretary of Transportation, shall—

(A) reevaluate the fuel economy labeling procedures described in the final rule published in the Federal Register on December 27, 2006 (71 Fed. Reg. 77,872; 40 CFR parts 86 and 600) to determine whether changes in the factors used to establish the labeling procedures warrant a revision of that process; and

(B) submit a report to the Committee on Commerce, Science, and Transportation of the Senate and the Committee on Energy and Commerce of the House of Representatives that describes the results of the reevaluation process.

Sec. 111. Consumer Tire Information

(a) IN GENERAL.—Chapter 323 of title 49, United States Code, is amended by inserting after section 32304 the following:

"§ 32304A. Consumer tire information

"(a) RULEMAKING.—

(1) IN GENERAL.—Not later than 24 months after the date of enactment of the Ten-in-Ten Fuel Economy Act, the Secretary of Transportation shall, after notice and opportunity for comment, promulgate rules establishing a national tire fuel efficiency consumer information program for replacement tires designed for use on motor vehicles to educate consumers about the effect of tires on automobile fuel efficiency, safety, and durability.

(2) ITEMS INCLUDED IN RULE.—The rulemaking shall include—

(A) a national tire fuel efficiency rating system for motor vehicle replacement tires to assist consumers in making more educated tire purchasing decisions;

(B) requirements for providing information to consumers, including information at the point of sale and other potential information dissemination methods, including the Internet;

(C) specifications for test methods for manufacturers to use in assessing and rating tires to avoid variation among test equipment and manufacturers; and

(D) a national tire maintenance consumer education program including, information on tire inflation pressure, alignment, rotation, and tread wear to maximize fuel efficiency, safety, and durability of replacement tires.

"(3) APPLICABILITY.—This section shall apply only to replacement tires covered under section 575.104(c) of title 49, Code of Federal Regulations, in effect on the date of the enactment of the Ten-in-Ten Fuel Economy Act.

(a) CONSULTATION.—The Secretary shall consult with the Secretary of Energy and the Administrator of the Environmental Protection Agency on the means of conveying tire fuel efficiency consumer information.

(b) REPORT TO CONGRESS.—The Secretary shall conduct periodic assessments of the rules promulgated under this section to determine the utility of such rules to consumers, the level of cooperation by industry, and the contribution to national goals pertaining to energy consumption. The Secretary shall transmit periodic reports detailing the findings of such assessments to the Senate Committee on Commerce, Science, and Transportation and the House of Representatives Committee on Energy and Commerce.

(c) TIRE MARKING.—The Secretary shall not require permanent labeling of any kind on a tire for the purpose of tire fuel efficiency information.

(d) APPLICATION WITH STATE AND LOCAL LAWS AND REGULATIONS.—Nothing in this section prohibits a State or political subdivision thereof from enforcing a law or regulation on tire fuel efficiency consumer information that was in effect on January 1, 2006. After a requirement promulgated under this section is in effect, a State or political subdivision thereof may adopt or enforce a law or regulation on tire fuel efficiency consumer information enacted or promulgated after January 1, 2006, if the requirements of that law or regulation are identical to the requirement promulgated under this section. Nothing in this section shall be construed to preempt a State or political subdivision thereof from regulating the fuel efficiency of tires (including establishing testing methods for determining compliance with such standards) not otherwise preempted under this chapter.".

(b) ENFORCEMENT.—Section 32308 of title 49, United States Code, is amended—

(1) by redesignating subsections (c) and (d) as subsections (d) and (e), respectively; and

(2) by inserting after subsection (b) the following:

"(c) SECTION 32304A.—Any person who fails to comply with Penalties. the national tire fuel efficiency information program under section 32304A is liable to the United States Government for a civil penalty of not more than $50,000 for each violation.".

(c) CONFORMING AMENDMENT.—The chapter analysis for chapter 323 of title 49, United States Code, is amended by inserting after the item relating to section 32304 the following:

"32304A. Consumer tire information".

Sec. 112. Use Of Civil Penalties For Research And Development

Section 32912 of title 49, United States Code, is amended by adding at the end the following:

"(e) USE OF CIVIL PENALTIES.—For fiscal year 2008 and each fiscal year thereafter, from the total amount deposited in the general fund of the Treasury during the preceding fiscal year from fines, penalties, and other funds obtained through enforcement actions conducted pursuant to this section (including funds obtained under consent decrees), the Secretary of the Treasury, subject to the availability of appropriations, shall—

(1) transfer 50 percent of such total amount to the account providing appropriations to the Secretary of Transportation for the administration of this chapter, which shall be used by the Secretary to support rulemaking under this chapter; and

(2) transfer 50 percent of such total amount to the account providing appropriations to the Secretary of Transportation for the administration of this chapter, which shall be used by the Secretary to carry out a program to make grants to manufacturers for retooling, reequipping, or expanding existing manufacturing facilities in the United States to produce advanced technology vehicles and components.".

Sec. 113. Exemption From Separate Calculation Requirement

(a) REPEAL.—Paragraphs (6), (7), and (8) of section 32904(b) of title 49, United States Code, are repealed.

(b) EFFECT OF REPEAL ON EXISTING EXEMPTIONS.—Any exemption granted under section 32904(b)(6) of title 49, United States Code, prior to the date of the enactment of this Act shall remain in effect subject to its terms through model year 2013.

(c) ACCRUAL AND USE OF CREDITS.—Any manufacturer holding an exemption under section 32904(b)(6) of title 49, United States Code, prior to the date of the enactment of this Act may accrue and use credits under sections 32903 and 32905 of such title beginning with model year 2011.

Subtitle B—Improved Vehicle Technology

Sec. 131. Transportation Electrification.

(a) DEFINITIONS.—In this section:

(1) ADMINISTRATOR.—The term "Administrator" means the Administrator of the Environmental Protection Agency.

(2) BATTERY.—The term "battery" means an electrochemical energy storage system powered directly by electrical current.

(3) ELECTRIC TRANSPORTATION TECHNOLOGY.—The term "electric transportation technology" means—

(A) technology used in vehicles that use an electric motor for all or part of the motive power of the vehicles, including battery electric, hybrid electric, plug-in hybrid electric, fuel cell, and plug-in fuel cell vehicles, or rail transportation; or

(B) equipment relating to transportation or mobile sources of air pollution that use an electric motor to replace an internal combustion engine for all or part of the work of the equipment, including—

(C) corded electric equipment linked to transportation or mobile sources of air pollution; and

(D) electrification technologies at airports, ports, truck stops, and material-handling facilities.

(4) NONROAD VEHICLE.—The term "nonroad vehicle" means a vehicle—

"(A) powered—

(1) by a nonroad engine, as that term is defined in section 216 of the Clean Air Act (42 U.S.C. 7550); or

(2) fully or partially by an electric motor powered by a fuel cell, a battery, or an off-board source of electricity; and

"(B) that is not a motor vehicle or a vehicle used solely for competition.

(5) PLUG-IN ELECTRIC DRIVE VEHICLE.—The term "plug- in electric drive vehicle" means a vehicle that—

(1) draws motive power from a battery with a capacity of at least 4 kilowatt-hours;

(2) can be recharged from an external source of electricity for motive power; and

(3)is a light-, medium-, or heavy-duty motor vehicle or nonroad vehicle (as those terms are defined in section 216 of the Clean Air Act (42 U.S.C. 7550)).

(6) QUALIFIED ELECTRIC TRANSPORTATION PROJECT.—The

(A) term "qualified electric transportation project" means an electric transportation technology project that would significantly reduce emissions of criteria pollutants, greenhouse gas emissions, and petroleum, including—
(B) shipside or shoreside electrification for vessels;
(C) truck-stop electrification;
(D) electric truck refrigeration units;
(E) battery-powered auxiliary power units for trucks;
(F) electric airport ground support equipment;
(G) electric material and cargo handling equipment;
(H) electric or dual-mode electric rail;
(I) any distribution upgrades needed to supply electricity to the project; and
(J) any ancillary infrastructure, including panel upgrades, battery chargers, in-situ transformers, and trenching.

(b) PLUG-IN ELECTRIC DRIVE VEHICLE PROGRAM.—

(1) ESTABLISHMENT.—The Secretary shall establish a competitive program to provide grants on a cost-shared basis to State governments, local governments, metropolitan transportation authorities, air pollution control districts, private or nonprofit entities, or combinations of those governments, authorities, districts, and entities, to carry out one or more projects to encourage the use of plug-in electric drive vehicles or other emerging electric vehicle technologies, as determined by the Secretary.

(2) ADMINISTRATION.—The Secretary shall, in consultation Grants. with the Secretary of Transportation and the Administrator, establish requirements for applications for grants under this section, including reporting of data to be summarized for dissemination to grantees and the public, including safety, vehicle, and component performance, and vehicle and component life cycle costs.

(3) PRIORITY.—In making awards under this subsection, the Secretary shall—

(A) give priority consideration to applications that—

(i) encourage early widespread use of vehicles described in paragraph (1); and

(ii) are likely to make a significant contribution to the advancement of the production of the vehicles in the United States; and

(B) ensure, to the maximum extent practicable, that the program established under this subsection includes a variety of applications, manufacturers, and end-uses.

(4) REPORTING.—The Secretary shall require a grant recipient under this subsection to submit to the Secretary, on an annual basis, data relating to safety, vehicle performance, life cycle costs, and emissions of vehicles demonstrated under the grant, including emissions of greenhouse gases.

(5) COST SHARING.—Section 988 of the Energy Policy Act of 2005 (42 U.S.C. 16352) shall apply to a grant made under this subsection.

(6) AUTHORIZATION OF APPROPRIATIONS.—There is authorized to be appropriated to carry out this subsection $90,000,000 for each of fiscal years 2008 through 2012, of which not less than 1/3 of the total amount appropriated shall be available each fiscal year to make grants to local and municipal governments.

(c) NEAR-TERM TRANSPORTATION SECTOR ELECTRIFICATION PROGRAM.—

(1) IN GENERAL.—Not later than 1 year after the date of enactment of this Act, the Secretary, in consultation with the Secretary of Transportation and the Administrator, shall establish a program to provide grants for the conduct of qualified electric transportation projects.

(2) PRIORITY.—In providing grants under this subsection, the Secretary shall give priority to large-scale projects and large-scale aggregators of projects.

(3) COST SHARING.—Section 988 of the Energy Policy Act of 2005 (42 U.S.C. 16352) shall apply to a grant made under this subsection.

(4) AUTHORIZATION OF APPROPRIATIONS.—There is authorized to be appropriated to carry out this subsection $95,000,000 for each of fiscal years 2008 through 2013.

(d) EDUCATION PROGRAM.—

(1) IN GENERAL.—The Secretary shall develop a nationwide electric drive transportation technology education program under which the Secretary shall provide teaching materials to secondary schools and high schools; and

(i) assistance for programs relating to electric drive system and component engineering to institutions of higher education.

(2) ELECTRIC VEHICLE COMPETITION.—The program established under paragraph (1) shall include a plug-in hybrid electric vehicle competition for institutions of higher education, which shall be known as the "Dr. Andrew Frank Plug-In Electric Vehicle Competition".

(3) ENGINEERS.—In carrying out the program established under paragraph (1), the Secretary shall provide financial assistance to institutions of higher education to create new, or support existing, degree programs to ensure the availability of trained electrical and mechanical engineers with the skills necessary for the advancement of—

(i) plug-in electric drive vehicles; and

(ii) other forms of electric drive transportation technology vehicles.

(4) AUTHORIZATION OF APPROPRIATIONS.—There are authorized to be appropriated such sums as may be necessary to carry out this subsection.

Sec. 132. Domestic Manufacturing Conversion Grant Program.

Section 712 of the Energy Policy Act of 2005 (42 U.S.C. 16062) is amended to read as follows:

"Sec. 712. Domestic Manufacturing Conversion Grant Program.

"(a) PROGRAM.—

(1) IN GENERAL.—The Secretary shall establish a program to encourage domestic production and sales of efficient hybrid and advanced diesel vehicles and components of those vehicles.

(2) INCLUSIONS.—The program shall include grants to automobile manufacturers and suppliers and hybrid component manufacturers to encourage domestic production of efficient hybrid, plug-in electric hybrid, plug-in electric drive, and advanced diesel vehicles.

(3) PRIORITY.—Priority shall be given to the refurbishment or retooling of manufacturing facilities that have recently ceased operation or will cease operation in the near future.

"(b) COORDINATION WITH STATE AND LOCAL PROGRAMS.—The Secretary may coordinate implementation of this section with State and local programs designed to accomplish similar goals, including the retention and retraining of skilled workers from the manufacturing facilities, including by establishing matching grant arrangements.

"(c) AUTHORIZATION OF APPROPRIATIONS.—There are authorized to be appropriated to the Secretary such sums as may be necessary to carry out this section.".

Sec. 133. Inclusion Of Electric Drive In Energy Policy Act Of 1992.

Section 508 of the Energy Policy Act of 1992 (42 U.S.C. 13258) is amended—

(1) by redesignating subsections (a) through (d) as subsections (b) through (e), respectively;

(2) by inserting before subsection (b) the following: "(a) DEFINITIONS.—In this section:

"(1) FUEL CELL ELECTRIC VEHICLE.—The term 'fuel cell electric vehicle' means an on-road or non-road vehicle that uses a fuel cell (as defined in section 803 of the Spark M. Matsunaga Hydrogen Act of 2005 (42 U.S.C. 16152)).

"(2) HYBRID ELECTRIC VEHICLE.—The term 'hybrid electric vehicle' means a new qualified hybrid motor vehicle (as defined in section 30B(d)(3) of the Internal Revenue Code of 1986).

"(3) MEDIUM- OR HEAVY-DUTY ELECTRIC VEHICLE.—The term 'medium- or heavy-duty electric vehicle' means an electric, hybrid electric, or plug-in hybrid electric vehicle with a gross vehicle weight of more than 8,501 pounds.

"(4) NEIGHBORHOOD ELECTRIC VEHICLE.—The term 'neighborhood electric vehicle' means a 4-wheeled on-road or nonroad vehicle that—

"(A) has a top attainable speed in 1 mile of more than 20 mph and not more than 25 mph on a paved level surface; and

"(B) is propelled by an electric motor and on-board, rechargeable energy storage system that is rechargeable using an off-board source of electricity.

"(5) PLUG-IN ELECTRIC DRIVE VEHICLE.—The term 'plug- in electric drive vehicle' means a vehicle that—

"(C) draws motive power from a battery with a capacity of at least 4 kilowatt-hours;

"(D) can be recharged from an external source of electricity for motive power; and

"(E) is a light-, medium-, or heavy duty motor vehicle or nonroad vehicle (as those terms are defined in section 216 of the Clean Air Act (42 U.S.C. 7550)).";

(3) in subsection (b) (as redesignated by paragraph (1))—

(A) by striking "The Secretary" and inserting the following:

"(1) ALLOCATION.—The Secretary"; and

(B) by adding at the end the following:

"(2) ELECTRIC VEHICLES.—Not later than January 31, 2009, the Secretary shall—

"(A) allocate credit in an amount to be determined by the Secretary for—

"(i) acquisition of—
"(I) a hybrid electric vehicle;
"(II) a plug-in electric drive vehicle;
"(III) a fuel cell electric vehicle;
"(IV) a neighborhood electric vehicle; or
"(V) a medium- or heavy-duty electric vehicle; and
"(ii) investment in qualified alternative fuel infrastructure or nonroad equipment, as determined by the Secretary; and

"(B) allocate more than 1, but not to exceed 5, credits for investment in an emerging technology relating to any vehicle described in subparagraph (A) to encourage—
(i) a reduction in petroleum demand;
(ii) technological advancement; and
(iii) a reduction in vehicle emissions.";

(4) in subsection (c) (as redesignated by paragraph (1)), by striking "subsection (a)" and inserting "subsection (b)"; and

(5) by adding at the end the following:

"(f) AUTHORIZATION OF APPROPRIATIONS.—There are authorized to be appropriated such sums as are necessary to carry out this section for each of fiscal years 2008 through 2013.".

Sec. 134. Loan Guarantees For Fuel-Efficient Automobile Parts Manufacturers

(a) IN GENERAL.—Section 712(a)(2) of the Energy Policy Act of 2005 (42 U.S.C. 16062(a)(2)) (as amended by section 132) is amended by inserting "and loan guarantees under section 1703" after "grants".

(b) CONFORMING AMENDMENT.—Section 1703(b) of the Energy Policy Act of 2005 (42 U.S.C. 16513(b)) is amended by striking paragraph (8) and inserting the following:

"(8) Production facilities for the manufacture of fuel efficient vehicles or parts of those vehicles, including electric drive vehicles and advanced diesel vehicles.".

Sec. 135. Advanced Battery Loan Guarantee Program

(a) ESTABLISHMENT OF PROGRAM.—The Secretary shall establish a program to provide guarantees of loans by private institutions for the construction of facilities for the manufacture of advanced vehicle batteries and battery systems that are developed and produced in the United States, including advanced lithium ion batteries and hybrid electrical system and component manufacturers and software designers.

(b) REQUIREMENTS.—The Secretary may provide a loan guarantee under subsection (a) to an applicant if—

(1) without a loan guarantee, credit is not available to the applicant under reasonable terms or conditions sufficient to finance the construction of a facility described in subsection (a);

(2) the prospective earning power of the applicant and the character and value of the security pledged provide a reasonable assurance of repayment of the loan to be guaranteed in accordance with the terms of the loan; and

(3) the loan bears interest at a rate determined by the Secretary to be reasonable, taking into account the current average yield on outstanding obligations of the

United States with remaining periods of maturity comparable to the maturity of the loan.

(c) CRITERIA.—In selecting recipients of loan guarantees from among applicants, the Secretary shall give preference to proposals that—

(1) meet all applicable Federal and State permitting requirements;

(2) are most likely to be successful; and

(3) are located in local markets that have the greatest need for the facility.

(d) MATURITY.—A loan guaranteed under subsection (a) shall have a maturity of not more than 20 years.

(e) TERMS AND CONDITIONS.—The loan agreement for a loan guaranteed under subsection (a) shall provide that no provision of the loan agreement may be amended or waived without the consent of the Secretary.

(f) ASSURANCE OF REPAYMENT.—The Secretary shall require that an applicant for a loan guarantee under subsection (a) provide an assurance of repayment in the form of a performance bond, insurance, collateral, or other means acceptable to the Secretary in an amount equal to not less than 20 percent of the amount of the loan.

(g) GUARANTEE FEE.—The recipient of a loan guarantee under subsection (a) shall pay the Secretary an amount determined by the Secretary to be sufficient to cover the administrative costs of the Secretary relating to the loan guarantee.

(h) FULL FAITH AND CREDIT.—The full faith and credit of the United States is pledged to the payment of all guarantees made under this section. Any such guarantee made by the Secretary shall be conclusive evidence of the eligibility of the loan for the guarantee with respect to principal and interest. The validity of the guarantee shall be incontestable in the hands of a holder of the guaranteed loan.

(i) REPORTS.—Until each guaranteed loan under this section has been repaid in full, the Secretary shall annually submit to Congress a report on the activities of the Secretary under this section.

(j) AUTHORIZATION OF APPROPRIATIONS.—There are authorized to be appropriated such sums as are necessary to carry out this section.

(k) TERMINATION OF AUTHORITY.—The authority of the Secretary to issue a loan guarantee under subsection (a) terminates on the date that is 10 years after the date of enactment of this Act.

Sec. 136. Advanced Technology Vehicles Manufacturing Incentive Program

(a) DEFINITIONS.—In this section:

(1) ADVANCED TECHNOLOGY VEHICLE.—The term “advanced technology vehicle” means a light duty vehicle that meets—

(A) the Bin 5 Tier II emission standard established in regulations issued by the Administrator of the Environmental Protection Agency under section 202(i) of the Clean Air Act (42 U.S.C. 7521(i)), or a lower-numbered Bin emission standard;

(B) any new emission standard in effect for fine particulate matter prescribed by the Administrator under that Act (42 U.S.C. 7401 et seq.); and

(C) at least 125 percent of the average base year combined fuel economy for vehicles with substantially similar attributes.

(2) COMBINED FUEL ECONOMY.—The term “combined fuel economy” means—

(A) the combined city/highway miles per gallon values, as reported in accordance with section 32904 of title 49, United States Code; and

(B) in the case of an electric drive vehicle with the ability to recharge from an off-board source, the reported mileage, as determined in a manner consistent with the Society of Automotive Engineers recommended practice for that configuration or a similar practice recommended by the Secretary.

(3) ENGINEERING INTEGRATION COSTS.—The term "engineering integration costs" includes the cost of engineering tasks relating to—

(A) incorporating qualifying components into the design of advanced technology vehicles; and

(B) designing tooling and equipment and developing manufacturing processes and material suppliers for production facilities that produce qualifying components or advanced technology vehicles.

(4) QUALIFYING COMPONENTS.—The term "qualifying components" means components that the Secretary determines to be—

(A) designed for advanced technology vehicles; and

(B) installed for the purpose of meeting the performance requirements of advanced technology vehicles.

(b) ADVANCED VEHICLES MANUFACTURING FACILITY.—The Secretary shall provide facility funding awards under this section to automobile manufacturers and component suppliers to pay not more than 30 percent of the cost of—

(1) reequipping, expanding, or establishing a manufacturing facility in the United States to produce—

(A) qualifying advanced technology vehicles; or

(B) qualifying components; and

(2) engineering integration performed in the United States of qualifying vehicles and qualifying components.

(c) PERIOD OF AVAILABILITY.—An award under subsection (b) shall apply to—

(1) facilities and equipment placed in service before December 30, 2020; and

(2) engineering integration costs incurred during the period beginning on the date of enactment of this Act and ending on December 30, 2020.

(d) DIRECT LOANPROGRAM.—

(1) IN GENERAL.—Not later than 1 year after the date of enactment of this Act, and subject to the availability of appropriated funds, the Secretary shall carry out a program to provide a total of not more than $25,000,000,000 in loans to eligible individuals and entities (as determined by the Secretary) for the costs of activities described in subsection (b).

(2) APPLICATION.—An applicant for a loan under this subsection shall submit to the Secretary an application at such time, in such manner, and containing such information as the Secretary may require, including a written assurance that—

(A) all laborers and mechanics employed by contractors or subcontractors during construction, alteration, or repair that is financed, in whole or in part, by a loan under this section shall be paid wages at rates not less than those prevailing on similar construction in the locality, as determined by the Secretary of Labor in accordance with sections 3141–3144, 3146, and 3147 of title 40, United States Code; and

(B) the Secretary of Labor shall, with respect to the labor standards described in this paragraph, have the authority and functions set forth in Reorganization Plan Numbered 14 of 1950 (5 U.S.C. App.) and section 3145 of title 40,

United States Code.

(3) SELECTION OF ELIGIBLE PROJECTS.—The Secretary shall select eligible projects to receive loans under this subsection in cases in which, as determined by the Secretary, the award recipient—

(A) is financially viable without the receipt of additional Federal funding associated with the proposed project;

(B) will provide sufficient information to the Secretary for the Secretary to ensure that the qualified investment is expended efficiently and effectively; and

(C) has met such other criteria as may be established and published by the Secretary.

(4) RATES, TERMS, AND REPAYMENT OF LOANS.—A loan provided under this subsection—

(A) shall have an interest rate that, as of the date on which the loan is made, is equal to the cost of funds to the Department of the Treasury for obligations of comparable maturity;

(B) shall have a term equal to the lesser of—

(i) the projected life, in years, of the eligible project to be carried out using funds from the loan, as determined by the Secretary; and

(ii) 25 years;

(C) may be subject to a deferral in repayment for not more than 5 years after the date on which the eligible project carried out using funds from the loan first begins operations, as determined by the Secretary; and

(D) shall be made by the Federal Financing Bank.

(e) IMPROVEMENT.—The Secretary shall issue regulations that require that, in order for an automobile manufacturer to be eligible for an award or loan under this section during a particular year, the adjusted average fuel economy of the manufacturer for light duty vehicles produced by the manufacturer during the most recent year for which data are available shall be not less than the average fuel economy for all light duty vehicles of the manufacturer for model year 2005. In order to determine fuel economy baselines for eligibility of a new manufacturer or a manufacturer that has not produced previously produced equivalent vehicles, the Secretary may substitute industry averages.

(f) FEES.—Administrative costs shall be no more than $100,000 or 10 basis point of the loan.

(g)PRIORITY.—The Secretary shall, in making awards or loans to those manufacturers that have existing facilities, give priority to those facilities that are oldest or have been in existence for at least 20 years. Such facilities can currently be sitting idle.

(h) SET ASIDE FOR SMALL AUTOMOBILE MANUFACTURERS AND COMPONENT SUPPLIERS.—

(1) DEFINITION OF COVERED FIRM.—In this subsection, the term "covered firm" means a firm that—

(A) employs less than 500 individuals; and

(B) manufactures automobiles or components of automobiles.

(2) SET ASIDE.—Of the amount of funds that are used to provide awards for each fiscal year under subsection (b), the Secretary shall use not less than 10 percent to provide awards to covered firms or consortia led by a covered firm.

(i) AUTHORIZATION OF APPROPRIATIONS.—There are authorized to be appropriated such sums as are necessary to carry out this section for each of fiscal years 2008 through 2012.

Subtitle C—Federal Vehicle Fleets

Sec. 141. Federal Vehicle Fleets

Section 303 of the Energy Policy Act of 1992 (42 U.S.C. 13212) is amended—

(1) by redesignating subsection (f) as subsection (g); and

(2) by inserting after subsection (e) the following new subsection:

"(f) VEHICLE EMISSION REQUIREMENTS.—

"(1) DEFINITIONS.—In this subsection:

"(A) FEDERAL AGENCY.—The term 'Federal agency' does not include any office of the legislative branch, except that it does include the House of Representatives with respect to an acquisition described in paragraph (2)(C).

"(B) MEDIUM DUTY PASSENGER VEHICLE.—The term 'medium duty passenger vehicle' has the meaning given that term section 523.2 of title 49 of the Code of Federal Regulations, as in effect on the date of enactment of this paragraph.

"(C) MEMBER'S REPRESENTATIONAL ALLOWANCE — The term 'Member's Representational Allowance' means the allowance described in section 10 1(a) of the House of Representatives Administrative Reform Technical Corrections Act (2 U.S.C. 57b(a)).

"(2) PROHIBITION.—

"(A) IN GENERAL.—Except as provided in subparagraph (B), no Federal agency shall acquire a light duty motor vehicle or medium duty passenger vehicle that is not a low greenhouse gas emitting vehicle.

"(B) EXCEPTION.—The prohibition in subparagraph (A) shall not apply to acquisition of a vehicle if the head of the agency certifies in writing, in a separate certification for each individual vehicle purchased, either—

(i) that no low greenhouse gas emitting vehicle is available to meet the functional needs of the agency and details in writing the functional needs that could not be met with a low greenhouse gas emitting vehicle; or

(ii) that the agency has taken specific alternative more cost-effective measures to reduce petroleum consumption that—

"(I) have reduced a measured and verified quantity of greenhouse gas emissions equal to or greater than the quantity of greenhouse gas reductions that would have been achieved through acquisition of a low greenhouse gas emitting vehicle over the lifetime of the vehicle; or

"(II) will reduce each year a measured and verified quantity of greenhouse gas emissions equal to or greater than the quantity of greenhouse gas reductions that would have been achieved each year through acquisition of a low greenhouse gas emitting vehicle.

"(C) SPECIAL RULE FOR VEHICLES PROVIDED BY FUNDS Applicability. CONTAINED IN MEMBERS' REPRESENTATIONAL ALLOWANCE.—

This paragraph shall apply to the acquisition of a light duty motor vehicle or medium duty passenger vehicle using any portion of a Member's Representational Allowance, including an acquisition under a long-term lease.

"(3) GUIDANCE.—

"(A) IN GENERAL.—Each year, the Administrator of the Environmental Protection Agency shall issue guidance identifying the makes and model numbers of vehicles that are low greenhouse gas emitting vehicles.

"(B) CONSIDERATION.—In identifying vehicles under subparagraph (A), the Administrator shall take into account the most stringent standards for vehicle greenhouse gas emissions applicable to and enforceable against motor vehicle manufacturers for vehicles sold anywhere in the United States.

"(C) REQUIREMENT.—The Administrator shall not identify any vehicle as a low greenhouse gas emitting vehicle if the vehicle emits greenhouse gases at a higher rate than such standards allow for the manufacturer's fleet average grams per mile of carbon dioxide-equivalent emissions for that class of vehicle, taking into account any emissions allowances and adjustment factors such standards provide.".

Sec. 142. Federal Fleet Conservation Requirements

Part J of title III of the Energy Policy and Conservation Act (42 U.S.C. 6374 et seq.) is amended by adding at the end the following:

"Sec. 400ff. Federal Fleet Conservation Requirements

"(a) MANDATORY REDUCTION IN PETROLEUM CONSUMPTION.—

"(1) IN GENERAL.—Not later than 18 months after the date of enactment of this section, the Secretary shall issue regulations for Federal fleets subject to section 400AA to require that, beginning in fiscal year 2010, each Federal agency shall reduce petroleum consumption and increase alternative fuel consumption each year by an amount necessary to meet the goals described in paragraph (2).

"(2) GOALS.—The goals of the requirements under paragraph (1) are that not later than October 1, 2015, and for each year thereafter, each Federal agency shall achieve at least a 20 percent reduction in annual petroleum consumption and a 10 percent increase in annual alternative fuel consumption, as calculated from the baseline established by the Secretary for fiscal year 2005.

"(3) MILESTONES.—The Secretary shall include in the regulations described in paragraph (1)—

"(A) interim numeric milestones to assess annual agency progress towards accomplishing the goals described in that paragraph; and

"(B) a requirement that agencies annually report on progress towards meeting each of the milestones and the 2015 goals.

"(b) PLAN.—

"(1) REQUIREMENT.—

"(A) IN GENERAL.—The regulations under subsection (a) shall require each Federal agency to develop a plan, and implement the measures specified in the plan by dates specified in the plan, to meet the required petroleum reduction levels and the alternative fuel consumption increases, including the milestones specified by the Secretary.

"(B) "(B) INCLUSIONS.—The plan shall—

"(i) identify the specific measures the agency will use to meet the requirements of subsection (a)(2); and

"(ii) quantify the reductions in petroleum consumption or increases in alternative fuel consumption projected to be achieved by each measure each year.

"(2) MEASURES.—The plan may allow an agency to meet the required petroleum reduction level through—

"(A) the use of alternative fuels;

"(B) the acquisition of vehicles with higher fuel economy, including hybrid vehicles, neighborhood electric vehicles, electric vehicles, and plug-in hybrid vehicles if the vehicles are commercially available;

"(C) the substitution of cars for light trucks;

"(D) an increase in vehicle load factors;

"(E) a decrease in vehicle miles traveled;

"(F) a decrease in fleet size; and

"(G) other measures.".

TITLE II—ENERGY SECURITY THROUGH INCREASED PRODUCTION OF BIOFUELS

Subtitle A—Renewable Fuel Standard

Sec. 201. Definitions

Section 211(o)(1) of the Clean Air Act (42 U.S.C. 7545(o)) is amended to read as follows:

"(1) DEFINITIONS.—In this section:

"(A) ADDITIONAL RENEWABLE FUEL.—The term 'additional renewable fuel' means fuel that is produced from renewable biomass and that is used to replace or reduce the quantity of fossil fuel present in home heating oil or jet fuel.

"(B) ADVANCED BIOFUEL.—

"(iii) IN GENERAL.—The term 'advanced biofuel' means renewable fuel, other than ethanol derived from corn starch, that has lifecycle greenhouse gas emissions, as determined by the Administrator, after notice and opportunity for comment, that are at least 50 percent less than baseline lifecycle greenhouse gas emissions.

"(iv) INCLUSIONS.—The types of fuels eligible for consideration as 'advanced biofuel' may include any of the following:

"(I) Ethanol derived from cellulose, hemicellulose, or lignin.

"(II) Ethanol derived from sugar or starch (other than corn starch).

"(III) Ethanol derived from waste material, including crop residue, other vegetative waste material, animal waste, and food waste and yard waste.

"(IV) Biomass-based diesel.

"(V) Biogas (including landfill gas and sewage waste treatment gas) produced through the conversion of organic matter from renewable biomass.

"(VI) Butanol or other alcohols produced through the conversion of organic matter from renewable biomass.

"(VII) Other fuel derived from cellulosic biomass.

"(C) BASELINE LIFECYCLE GREENHOUSE GAS EMISSIONS.—The term 'baseline lifecycle greenhouse gas emissions' means the average lifecycle greenhouse gas emissions, as determined by the Administrator, after notice and opportunity for comment, for gasoline or diesel (whichever is being replaced by the renewable fuel) sold or distributed as transportation fuel in 2005.

"(D) BIOMASS-BASED DIESEL.—The term 'biomass-based diesel' means renewable fuel that is biodiesel as defined in section 312(f) of the Energy Policy Act of 1992 (42 U.S.C. 13220(f)) and that has lifecycle greenhouse gas emissions, as determined by the Administrator, after notice and opportunity for comment, that are at least 50 percent less than the baseline lifecycle greenhouse gas emissions. Notwithstanding the preceding sentence, renewable fuel derived from co-processing biomass with a petroleum feedstock shall be advanced biofuel if it meets the requirements of subparagraph (B), but is not biomass-based diesel.

"(E) CELLULOSIC BIOFUEL.—The term 'cellulosic biofuel' means renewable fuel derived from any cellulose, hemicellulose, or lignin that is derived from renewable biomass and that has lifecycle greenhouse gas emissions, as determined by the Administrator, that are at least 60 percent less than the baseline lifecycle greenhouse gas emissions.

"(F) CONVENTIONAL BIOFUEL.—The term 'conventional biofuel' means renewable fuel that is ethanol derived from corn starch.

"(G) GREENHOUSE GAS.—The term 'greenhouse gas' means carbon dioxide, hydrofluorocarbons, methane, nitrous oxide, perfluorocarbons, sulfur hexafluoride. The Administrator may include any other anthropogenically-emitted gas that is determined by the Administrator, after notice and comment, to contribute to global warming.

"(H) LIFECYCLE GREENHOUSE GAS EMISSIONS.—The term 'lifecycle greenhouse gas emissions' means the aggregate quantity of greenhouse gas emissions (including direct emissions and significant indirect emissions such as significant emissions from land use changes), as determined by the Administrator, related to the full fuel lifecycle, including all stages of fuel and feedstock production and distribution, from feedstock generation or extraction through the distribution and delivery and use of the finished fuel to the ultimate consumer, where the mass values for all greenhouse gases are adjusted to account for their relative global warming potential.

"(I) RENEWABLE BIOMASS.—The term 'renewable biomass' means each of the following:

"(i) Planted crops and crop residue harvested from agricultural land cleared or cultivated at any time prior to the enactment of this sentence that is either actively managed or fallow, and nonforested.

"(ii) Planted trees and tree residue from actively managed tree plantations on non-federal land cleared at any time prior to enactment of this sentence, including land belonging to an Indian tribe or an Indian individual, that is held in trust by the United States or subject to a restriction against alienation imposed by the United States.

"(iii) Animal waste material and animal byproducts.

"(iv) Slash and pre-commercial thinnings that are from non-federal forestlands, including forestlands belonging to an Indian tribe or an Indian individual, that are held in trust by the United States or subject to a restriction against alienation imposed by the United States, but not forests or forestlands that are ecological communities with a global or State ranking of critically imperiled, imperiled, or rare pursuant to a State Natural Heritage Program, old growth forest, or late successional forest.

"(v) Biomass obtained from the immediate vicinity of buildings and other areas regularly occupied by people, or of public infrastructure, at risk from wildfire.

"(vi) Algae.

"(vii) Separated yard waste or food waste, including recycled cooking and trap grease.

"(J) RENEWABLE FUEL.—The term 'renewable fuel' means fuel that is produced from renewable biomass and that is used to replace or reduce the quantity of fossil fuel present in a transportation fuel.

"(K) SMALL REFINERY.—The term 'small refinery' means a refinery for which the average aggregate daily crude oil throughput for a calendar year (as determined by dividing the aggregate throughput for the calendar year by the number of days in the calendar year) does not exceed 75,000 barrels.

"(L) TRANSPORTATION FUEL.—The term 'transportation fuel' means fuel for use in motor vehicles, motor vehicle engines, nonroad vehicles, or nonroad engines (except for ocean-going vessels).".

Sec. 202. Renewable Fuel Standard

(a) RENEWABLE FUEL PROGRAM.—Paragraph (2) of section 211(o) (42 U.S.C. 7545(o)(2)) of the Clean Air Act is amended as follows:

(1) REGULATIONS.—Clause (i) of subparagraph (A) is amended by adding the following at the end thereof: "Not later than 1 year after the date of enactment of this sentence, the Administrator shall revise the regulations under this paragraph to ensure that transportation fuel sold or introduced into commerce in the United States (except in noncontiguous States or territories), on an annual average basis, contains at least the applicable volume of renewable fuel, advanced biofuel, cellulosic biofuel, and biomass-based diesel, determined in accordance with subparagraph (B) and, in the case of any such renewable fuel produced from new facilities that commence construction after the date of enactment of this sentence, achieves at least a 20 percent reduction in lifecycle greenhouse gas emissions compared to baseline lifecycle greenhouse gas emissions.".

(2) APPLICABLE VOLUMES OF RENEWABLE FUEL.—Subparagraph (B) is amended to read as follows:

"(B) APPLICABLE VOLUMES.—

"(i) CALENDAR YEARS AFTER 2005.—

(I) RENEWABLE FUEL.—For the purpose of subparagraph (A), the applicable volume of renewable fuel for the calendar years 2006 through 2022 shall be determined in accordance with the following table:

“Calendar year:	Applicable volume of renewable fuel (in billions of gallons):
2006	4.0
2007	4.7
2008	9.0
2009	11.1
2010	12.95
2011	13.95
2012	15.2
2013	16.55
2014	18.15
2015	20.5
2016	22.25
2017	24.0
2018	26.0
2019	28.0
2020	30.0
2021	33.0
2022	36.0

(II) ADVANCED BIOFUEL.—For the purpose of subparagraph (A), of the volume of renewable fuel required under subclause (I), the applicable volume of advanced biofuel for the calendar years 2009 through 2022 shall be determined in accordance with the following table:

“Calendar year:	Applicable volume of Advanced biofuel (in billions of gallons):
2009	0.6
2010	0.95
2011	1.35
2012	2.0
2013	2.75
2014	3.75
2015	5.5
2016	7.25
2017	9.0
2018	11.0
2019	13.0
2020	15.0
2021	18.0
2022	21.0

(III) CELLULOSIC BIOFUEL.—For the purpose of subparagraph (A), of the volume of advanced biofuel required under subclause (II), the applicable volume of cellulosic biofuel for the calendar years 2010 through 2022 shall be determined in accordance with the following table:

“Calendar year:	Applicable volume of cellulosic biofuel (in billions of gallons):
2010	0.1
2011	0.25
2012	0.5
2013	1.0
2014	1.75
2015	3.0
2016	4.25
2017	5.5
2018	7.0
2019	8.5
2020	10.5
2021	13.5
2022	16.0

(IV) BIOMASS-BASED DIESEL.—For the purpose of subparagraph (A), of the volume of advanced biofuel required under subclause (II), the applicable volume of biomass-based diesel for the calendar years 2009 through 2012 shall be determined in accordance with the following table:

“Calendar year:	Applicable volume of biomass-based diesel (in billions of gallons):
2009	0.5
2010	0.65
2011	0.80
2012	1.0

“(ii) OTHER CALENDAR YEARS.—For the purposes of subparagraph (A), the applicable volumes of each fuel specified in the tables in clause (i) for calendar years after the calendar years specified in the tables shall be determined by the Administrator, in coordination with the Secretary of Energy and the Secretary of Agriculture, based on a review of the implementation of the program during calendar years specified in the tables, and an analysis of—

“(I) the impact of the production and use of renewable fuels on the environment, including on air quality, climate change, conversion of wetlands, ecosystems, wildlife habitat, water quality, and water supply;

“(II) the impact of renewable fuels on the energy security of the United States;

“(III) the expected annual rate of future commercial production of renewable fuels, including advanced biofuels in each category (cellulosic biofuel and biomass-based diesel);

“(IV) the impact of renewable fuels on the infrastructure of the United States, including deliverability of materials, goods, and products other than renewable fuel, and the sufficiency of infrastructure to deliver and use renewable fuel;

“(V) the impact of the use of renewable fuels on the cost to consumers of transportation fuel and on the cost to transport goods; and

"(VI) the impact of the use of renewable fuels on other factors, including job creation, the price and supply of agricultural commodities, rural economic development, and food prices.

The Administrator shall promulgate rules establishing the applicable volumes under this clause no later than 14 months before the first year for which such applicable volume will apply.

"(iii) APPLICABLE VOLUME OF ADVANCED BIOFUEL.— For the purpose of making the determinations in clause (ii), for each calendar year, the applicable volume of advanced biofuel shall be at least the same percentage of the applicable volume of renewable fuel as in calendar year 2022.

"(iv) APPLICABLE VOLUME OF CELLULOSIC BIOFUEL.—For the purpose of making the determinations in clause (ii), for each calendar year, the applicable volume of cellulosic biofuel established by the Administrator shall be based on the assumption that the Administrator will not need to issue a waiver for such years under paragraph (7)(D).

"(v) MINIMUM APPLICABLE VOLUME OF BIOMASS- BASED DIESEL.—For the purpose of making the determinations in clause (ii), the applicable volume of biomass-based diesel shall not be less than the applicable volume listed in clause (i)(IV) for calendar year 2012.".

(b) APPLICABLE PERCENTAGES.—Paragraph (3) of section 211(o) of the Clean Air Act (42 U.S.C. 7545(o)(3)) is amended as follows:

(1) In subparagraph (A), by striking "2011" and inserting "2021".

(2) In subparagraph (A), by striking "gasoline" and inserting "transportation fuel, biomass-based diesel, and cellulosic biofuel".

(3) In subparagraph (B), by striking "2012" and inserting "2021" in clause (i).

(4) In subparagraph (B), by striking "gasoline" and inserting "transportation fuel" in clause (ii)(II).

(c) MODIFICATION OF GREENHOUSE GAS PERCENTAGES.—Paragraph (4) of section 211(o) of the Clean Air Act (42 U.S.C. 7545(o)(4)) is amended to read as follows:

"(4) MODIFICATION OF GREENHOUSE GAS REDUCTION PERCENTAGES.—

"(A) "(A) IN GENERAL.—The Administrator may, in the regulations under the last sentence of paragraph (2)(A)(i), adjust the 20 percent, 50 percent, and 60 percent reductions in lifecycle greenhouse gas emissions specified in paragraphs (2)(A)(i) (relating to renewable fuel), (1)(D) (relating to biomass-based diesel), (1)(B)(i) (relating to advanced biofuel), and (1)(E) (relating to cellulosic biofuel) to a lower percentage. For the 50 and 60 percent reductions, the Administrator may make such an adjustment only if he determines that generally such reduction is not commercially feasible for fuels made using a variety of feedstocks, technologies, and processes to meet the applicable reduction.

"(B) AMOUNT OF ADJUSTMENT.—In promulgating regulations under this paragraph, the specified 50 percent reduction in greenhouse gas emissions from advanced biofuel and in biomass-based diesel may not be reduced below 40 percent. The specified 20 percent reduction in greenhouse gas emissions from renewable fuel may not be reduced below 10 percent, and the specified 60 percent reduction in greenhouse gas emissions from cellulosic biofuel may not be reduced below 50 percent.

"(C) ADJUSTED REDUCTION LEVELS.—An adjustment under this paragraph to a percent less than the specified 20 percent greenhouse gas reduction for renewable fuel shall be the minimum possible adjustment, and the adjusted greenhouse gas reduction shall be established by the Administrator at the maximum achievable level, taking cost in consideration, for natural gas fired corn-based ethanol plants, allowing for the use of a variety of technologies and processes. An adjustment in the 50 or 60 percent greenhouse gas levels shall be the minimum possible adjustment for the fuel or fuels concerned, and the adjusted greenhouse gas reduction shall be established at the maximum achievable level, taking cost in consideration, allowing for the use of a variety of feedstocks, technologies, and processes.

"(D) 5-YEARREVIEW.—Whenever the Administrator Deadline. makes any adjustment under this paragraph, not later than 5 years thereafter he shall review and revise (based upon the same criteria and standards as required for the initial adjustment) the regulations establishing the adjusted level.

"(E) SUBSEQUENT ADJUSTMENTS.—After the Administrator has promulgated a final rule under the last sentence of paragraph (2)(A)(i) with respect to the method of determining lifecycle greenhouse gas emissions, except as provided in subparagraph (D), the Administrator may not adjust the percent greenhouse gas reduction levels unless he determines that there has been a significant change in the analytical methodology used for determining the lifecycle greenhouse gas emissions. If he makes such determination, he may adjust the 20, 50, or 60 percent reduction levels through rulemaking using the criteria and standards set forth in this paragraph.

"(F) LIMIT ON UPWARD ADJUSTMENTS.—If, under subparagraph (D) or (E), the Administrator revises a percent level adjusted as provided in subparagraphs (A), (B), and (C) to a higher percent, such higher percent may not exceed the applicable percent specified in paragraph (2)(A)(i), (1)(D), (1)(B)(i), or (1)(E).

"(G) APPLICABILITY OF ADJUSTMENTS.—If the Administrator adjusts, or revises, a percent level referred to in this paragraph or makes a change in the analytical methodology used for determining the lifecycle greenhouse gas emissions, such adjustment, revision, or change (or any combination thereof) shall only apply to renewable fuel from new facilities that commence construction after the effective date of such adjustment, revision, or change.".

(d) CREDITS FOR ADDITIONAL RENEWABLE FUEL.—Paragraph (5) of section 211(o) of the Clean Air Act (42 U.S.C. 7545(o)(5)) is amended by adding the following new subparagraph at the end thereof:

"(E) CREDITS FOR ADDITIONAL RENEWABLE FUEL.—The Administrator may issue regulations providing: (i) for the generation of an appropriate amount of credits by any person that refines, blends, or imports additional renewable fuels specified by the Administrator; and (ii) for the use of such credits by the generator, or the transfer of all or a portion of the credits to another person, for the purpose of complying with paragraph (2).".

(e) WAIVERS.—

(1) IN GENERAL.—Paragraph (7)(A) of section 211(o) of the Clean Air Act (42 U.S.C. 7545(o)(7)(A)) is amended by inserting ", by any person subject to the requirements of this subsection, or by the Administrator on his own motion" after "one or more States" in subparagraph (A) and by striking out "State" in subparagraph (B).

(2) CELLULOSIC BIOFUEL.—Paragraph (7) of section 211(o) of the Clean Air Act (42 U.S.C. 7545(o)(7)) is amended by adding the following at the end thereof:

"(D) CELLULOSIC BIOFUEL.—(i) For any calendar year for which the projected volume of cellulosic biofuel production is less than the minimum applicable volume established under paragraph (2)(B), as determined by the Administrator based on the estimate provided under paragraph (3)(A), not later than November 30 of the preceding calendar year, the Administrator shall reduce the applicable volume of cellulosic biofuel required under paragraph (2)(B) to the projected volume available during that calendar year. For any calendar year in which the Administrator makes such a reduction, the Administrator may also reduce the applicable volume of renewable fuel and advanced biofuels requirement established under paragraph (2)(B) by the same or a lesser volume.

"(ii) Whenever the Administrator reduces the minimum cellulosic biofuel volume under this subparagraph, the Administrator shall make available for sale cellulosic biofuel credits at the higher of $0.25 per gallon or the amount by which $3.00 per gallon exceeds the average wholesale price of a gallon of gasoline in the United States. Such amounts shall be adjusted for inflation by the Administrator for years after 2008.

"(iii) Eighteen months after the date of enactment of this subparagraph, the Administrator shall promulgate regulations to govern the issuance of credits under this subparagraph. The regulations shall set forth the method for determining the exact price of credits in the event of a waiver. The price of such credits shall not be changed more frequently than once each quarter. These regulations shall include such provisions, including limiting the credits' uses and useful life, as the Administrator deems appropriate to assist market liquidity and transparency, to provide appropriate certainty for regulated entities and renewable fuel producers, and to limit any potential misuse of cellulosic biofuel credits to reduce the use of other renewable fuels, and for such other purposes as the Administrator determines will help achieve the goals of this subsection. The regulations shall limit the number of cellulosic biofuel credits for any calendar year to the minimum applicable volume (as reduced under this subparagraph) of cellulosic biofuel for that year.".

(3) BIOMASS-BASED DIESEL.—Paragraph (7) of section 211(o) of the Clean Air Act (42 U.S.C. 7545(o)(7)) is amended by adding the following at the end thereof:

"(E) BIOMASS-BASED DIESEL.—

"(i) MARKET EVALUATION.—The Administrator, in consultation with the Secretary of Energy and the Secretary of Agriculture, shall periodically evaluate the impact of the biomass-based diesel requirements established under this paragraph on the price of diesel fuel.

"(ii) WAIVER.—If the Administrator determines that there is a significant renewable feedstock disruption or other market circumstances that would make the price of biomass-based diesel fuel increase significantly, the Administrator, in consultation with the Secretary of Energy and the Secretary of Agriculture, shall issue an order to reduce, for up to a 60-day period, the quantity of biomass-based diesel required under subparagraph (A) by an appropriate quantity that does not exceed 15 percent of the applicable annual requirement for biomass-based diesel. For any calendar year in which the Administrator makes a reduction under this subparagraph, the Administrator may also reduce the applicable volume of renewable fuel and advanced biofuels requirement established under paragraph (2)(B) by the same or a lesser volume.

"(iii) EXTENSIONS.—If the Administrator determines that the feedstock disruption or circumstances described in clause (ii) is continuing beyond the 60- day period described in clause (ii) or this clause, the Administrator, in consultation with the Secretary of Energy and the Secretary of Agriculture, may issue an order to reduce, for up to an additional 60-day period, the quantity of biomass-based diesel required under subparagraph (A) by an appropriate quantity that does not exceed an additional 15 percent of the applicable annual requirement for biomass-based diesel.

"(F) MODIFICATION OF APPLICABLE VOLUMES.—For any of the tables in paragraph (2)(B), if the Administrator waives—

"(i) at least 20 percent of the applicable volume requirement set forth in any such table for 2 consecutive years; or

"(ii) at least 50 percent of such volume requirement for a single year, the Administrator shall promulgate a rule (within 1 year after issuing such waiver) that modifies the applicable volumes set forth in the table concerned for all years following the final year to which the waiver applies, except that no such modification in applicable volumes shall be made for any year before 2016. In promulgating such a rule, the Administrator shall comply with the processes, criteria, and standards set forth in paragraph (2)(B)(ii).".

Sec. 203. Study Of Impact Of Renewable Fuel Standard

(a) IN GENERAL.—The Secretary of Energy, in consultation with the Secretary of Agriculture and the Administrator of the Environmental Protection Agency, shall enter into an arrangement with the National Academy of Sciences under which the Academy shall conduct a study to assess the impact of the requirements described in section 211(o) of the Clean Air Act on each industry relating to the production of feed grains, livestock, food, forest products, and energy.

(b) PARTICIPATION.—In conducting the study under this section, the National Academy of Sciences shall seek the participation, and consider the input, of—

(1) producers of feed grains;

(2) producers of livestock, poultry, and pork products;

(3) producers of food and food products;
(4) producers of energy;
(5) individuals and entities interested in issues relating to conservation, the environment, and nutrition;
(6) users and consumers of renewable fuels;
(7) producers and users of biomass feedstocks; and
(8) land grant universities.

(c) CONSIDERATIONS.—In conducting the study, the National Academy of Sciences shall consider—
(1) the likely impact on domestic animal agriculture feedstocks that, in any crop year, are significantly below current projections;
(2) policy options to alleviate the impact on domestic animal agriculture feedstocks that are significantly below current projections; and
(3) policy options to maintain regional agricultural and silvicultural capability.

(d) COMPONENTS.—The study shall include—
(1) a description of the conditions under which the requirements described in section 211(o) of the Clean Air Act should be suspended or reduced to prevent adverse impacts to domestic animal agriculture feedstocks described in subsection (c)(2) or regional agricultural and silvicultural capability described in subsection (c)(3); and
(2) recommendations for the means by which the Federal Government could prevent or minimize adverse economic hardships and impacts.

(e) DEADLINE FOR COMPLETION OF STUDY.—Not later than 18 months after the date of enactment of this Act, the Secretary shall submit to Congress a report that describes the results of the study under this section.

(f) PERIODIC REVIEWS.—Section 2 11(o) of the Clean Air Act is amended by adding the following at the end thereof:
"(11) PERIODIC REVIEWS.—To allow for the appropriate adjustment of the requirements described in subparagraph (B) of paragraph (2), the Administrator shall conduct periodic reviews of—
"(A) existing technologies;
"(B) the feasibility of achieving compliance with the requirements; and
"(C) the impacts of the requirements described in subsection (a)(2) on each individual and entity described in paragraph (2).".

Sec. 204. Environmental and Resource Conservation Impacts.

(a) IN GENERAL.—Not later than 3 years after the enactment of this section and every 3 years thereafter, the Administrator of the Environmental Protection Agency, in consultation with the Secretary of Agriculture and the Secretary of Energy, shall assess and report to Congress on the impacts to date and likely future impacts of the requirements of section 211(o) of the Clean Air Act on the following:
(1) Environmental issues, including air quality, effects on hypoxia, pesticides, sediment, nutrient and pathogen levels in waters, acreage and function of waters, and soil environmental quality.
(2) Resource conservation issues, including soil conservation, water availability, and ecosystem health and biodiversity, including impacts on forests, grasslands, and wetlands.
(3) The growth and use of cultivated invasive or noxious plants and their impacts on the environment and agriculture. In advance of preparing the report required by

this subsection, the Administrator may seek the views of the National Academy of Sciences or another appropriate independent research institute. The report shall include the annual volume of imported renewable fuels and feedstocks for renewable fuels, and the environmental impacts outside the United States of producing such fuels and feedstocks. The report required by this subsection shall include recommendations for actions to address any adverse impacts found.

(d) EFFECT ON AIR QUALITY AND OTHER ENVIRONMENTAL REQUIREMENTS.—Except as provided in section 211(o)(12) of the Clean Air Act, nothing in the amendments made by this title to section 211(o) of the Clean Air Act shall be construed as superseding, or limiting, any more environmentally protective requirement under the Clean Air Act, or under any other provision of State or Federal law or regulation, including any environmental law or regulation.

Sec. 205. Biomass-Based Diesel And Biodiesel Labeling.

(a) IN GENERAL.—Each retail diesel fuel pump shall be labeled in a manner that informs consumers of the percent of biomass- based diesel or biodiesel that is contained in the biomass-based diesel blend or biodiesel blend that is offered for sale, as determined by the Federal Trade Commission.

(b) LABELING REQUIREMENTS.—Not later than 180 days after the date of enactment of this section, the Federal Trade Commission shall promulgate biodiesel labeling requirements as follows:

(1) Biomass-based diesel blends or biodiesel blends that contain less than or equal to 5 percent biomass-based diesel or biodiesel by volume and that meet ASTM D975 diesel specifications shall not require any additional labels.

(2) Biomass-based diesel blends or biodiesel blends that contain more than 5 percent biomass-based diesel or biodiesel by volume but not more than 20 percent by volume shall be labeled "contains biomass-based diesel or biodiesel in quantities between 5 percent and 20 percent".

(3) Biomass-based diesel or biodiesel blends that contain more than 20 percent biomass based or biodiesel by volume shall be labeled "contains more than 20 percent biomass-based diesel or biodiesel".

(c) DEFINITIONS.—In this section:

(1) ASTM.—The term "ASTM" means the American Society of Testing and Materials.

(2) BIOMASS-BASED DIESEL.—The term "biomass-based diesel" means biodiesel as defined in section 312(f) of the Energy Policy Act of 1992 (42 U.S.C. 13220(f)).

(3) BIODIESEL.—The term "biodiesel" means the monoalkyl esters of long chain fatty acids derived from plant or animal matter that meet—

(4) the registration requirements for fuels and fuel additives under this section; and

(A) the requirements of ASTM standard D6751.

(1) BIOMASS-BASED DIESEL AND BIODIESEL BLENDS.—The terms "biomass-based diesel blend" and "biodiesel blend" means a blend of "biomass-based diesel" or "biodiesel" fuel that is blended with petroleum-based diesel fuel.

Sec. 206. Study Of Credits For Use Of Renewable Electricity In Electric Vehicles

(a) DEFINITION OF ELECTRIC VEHICLE.—In this section, the term "electric vehicle" means an electric motor vehicle (as defined in section 601 of the Energy Policy Act of 1992 (42 U.S.C. 13271)) for which the rechargeable storage battery—

(1) receives a charge directly from a source of electric current that is external to the vehicle; and

(2) provides a minimum of 80 percent of the motive power of the vehicle.

(b) STUDY.—The Administrator of the Environmental Protection Agency shall conduct a study on the feasibility of issuing credits under the program established under section 211(o) of the Clean Air Act to electric vehicles powered by electricity produced from renewable energy sources.

(c) REPORT.—Not later than 18 months after the date of enactment of this Act, the Administrator shall submit to the Committee on Energy and Natural Resources of the United States Senate and the Committee on Energy and Commerce of the United States House of Representatives a report that describes the results of the study, including a description of—

(1) existing programs and studies on the use of renewable electricity as a means of powering electric vehicles; and

(2) alternatives for—

(A) designing a pilot program to determine the feasibility of using renewable electricity to power electric vehicles as an adjunct to a renewable fuels mandate;

(B) allowing the use, under the pilot program designed under subparagraph (A), of electricity generated from nuclear energy as an additional source of supply;

(C) identifying the source of electricity used to power electric vehicles; and

(D) equating specific quantities of electricity to quantities of renewable fuel under section 211(o) of the Clean Air Act.

Sec. 207. Grants For Production Of Advanced Biofuels.

(a) IN GENERAL.—The Secretary of Energy shall establish a grant program to encourage the production of advanced biofuels.

(b) REQUIREMENTS AND PRIORITY.—In making grants under this section, the Secretary—

(1) shall make awards to the proposals for advanced biofuels with the greatest reduction in lifecycle greenhouse gas emissions compared to the comparable motor vehicle fuel lifecycle emissions during calendar year 2005; and

(2) shall not make an award to a project that does not achieve at least an 80 percent reduction in such lifecycle greenhouse gas emissions.

(a) AUTHORIZATION OF APPROPRIATIONS.—There is authorized to be appropriated to carry out this section $500,000,000 for the period of fiscal years 2008 through 2015.

Sec. 208. Integrated Consideration Of Water Quality In Determinations On Fuels And Fuel Additives

Section 211(c)(1) of the Clean Air Act (42 U.S.C. 7545(c)(1)) is amended as follows:

(1) By striking "nonroad vehicle (A) if in the judgment of the Administrator" and inserting "nonroad vehicle if, in the judgment of the Administrator, any fuel or

fuel additive or"; and

(2) In subparagraph (A), by striking "air pollution which" and inserting "air pollution or water pollution (including any degradation in the quality of groundwater) that".

Sec. 209. Anti-Backsliding

Section 211 of the Clean Air Act (42 U.S.C. 7545) is amended by adding at the end the following:

"(v) PREVENTION OF AIR QUALITY DETERIORATION.—

"(1) STUDY.—

"(A) IN GENERAL.—Not later than 18 months after the date of enactment of this subsection, the Administrator shall complete a study to determine whether the renewable fuel volumes required by this section will adversely impact air quality as a result of changes in vehicle and engine emissions of air pollutants regulated under this Act.

"(B) CONSIDERATIONS.—The study shall include consideration of—

"(i) different blend levels, types of renewable fuels, and available vehicle technologies; and

"(ii) appropriate national, regional, and local air quality control measures.

"(2) REGULATIONS.—Not later than 3 years after the date Deadline. of enactment of this subsection, the Administrator shall—

"(A) promulgate fuel regulations to implement appropriate measures to mitigate, to the greatest extent achievable, considering the results of the study under paragraph (1), any adverse impacts on air quality, as the result of the renewable volumes required by this section; or

"(B) make a determination that no such measures are necessary.".

Sec. 210. Effective Date, Savings Provision, And Transition Rules

(a) TRANSITION RULES.—(1) For calendar year 2008, transportation fuel sold or introduced into commerce in the United States (except in noncontiguous States or territories), that is produced from facilities that commence construction after the date of enactment of this Act shall be treated as renewable fuel within the meaning of section 211(o) of the Clean Air Act only if it achieves at least a 20 percent reduction in lifecycle greenhouse gas emissions compared to baseline lifecycle greenhouse gas emissions. For calendar years 2008 and 2009, any ethanol plant that is fired with natural gas, biomass, or any combination thereof is deemed to be in compliance with such 20 percent reduction requirement and with the 20 percent reduction requirement of section 211(o)(1) of the Clean Air Act. The terms used in this subsection shall have the same meaning as provided in the amendment made by this Act to section 211(o) of the Clean Air Act.

(2) Until January 1, 2009, the Administrator of the Environmental Protection Agency shall implement section 211(o) of the Clean Air Act and the rules promulgated under that section in accordance with the provisions of that section as in effect before the enactment of this Act and in accordance with the rules promulgated before the enactment of this Act, except that for calendar year 2008, the number "9.0" shall be substituted for the number "5.4" in the table in section 211(o)(2)(B) and in the corresponding rules promulgated to carry out those

provisions. The Administrator is authorized to take such other actions as may be necessary to carry out this paragraph notwithstanding any other provision of law.

(b) SAVINGS CLAUSE.—Section 211(o) of the Clean Air Act (42 U.S.C. 7545(o)) is amended by adding the following new paragraph at the end thereof:

"(12) EFFECT ON OTHER PROVISIONS.—Nothing in this subsection, or regulations issued pursuant to this subsection, shall affect or be construed to affect the regulatory status of carbon dioxide or any other greenhouse gas, or to expand or limit regulatory authority regarding carbon dioxide or any other greenhouse gas, for purposes of other provisions (including section 165) of this Act. The previous sentence shall not affect implementation and enforcement of this subsection.".

(c) EFFECTIVE DATE.—The amendments made by this title to section 211(o) of the Clean Air Act shall take effect January 1, 2009, except that the Administrator shall promulgate regulations to carry out such amendments not later than 1 year after the enactment of this Act.

SUBTITLE B—BIOFUELS RESEARCH AND DEVELOPMENT

Sec. 221. Biodiesel.

(a) BIODIESEL STUDY.—Not later than 180 days after the date of enactment of this Act, the Secretary, in consultation with the Administrator of the Environmental Protection Agency, shall submit to Congress a report on any research and development challenges inherent in increasing the proportion of diesel fuel sold in the United States that is biodiesel.

(b) MATERIAL FOR THE ESTABLISHMENT OF STANDARDS.—The Director of the National Institute of Standards and Technology, in consultation with the Secretary, shall make publicly available the physical property data and characterization of biodiesel and other biofuels as appropriate.

Sec. 222. Biogas.

Not later than 180 days after the date of enactment of this Act, the Secretary, in consultation with the Administrator of the Environmental Protection Agency, shall submit to Congress a report on any research and development challenges inherent in increasing the amount of transportation fuels sold in the United States that are fuel with biogas or a blend of biogas and natural gas.

Sec. 223. Grants For Biofuel Production Research And Development In Certain States.

(a) IN GENERAL.—The Secretary shall provide grants to eligible entities for research, development, demonstration, and commercial application of biofuel production technologies in States with low rates of ethanol production, including low rates of production of cellulosic biomass ethanol, as determined by the Secretary.

(b) ELIGIBILITY.—To be eligible to receive a grant under this section, an entity shall—

(1) (A) be an institution of higher education (as defined in section 2 of the Energy Policy Act of 2005 (42 U.S.C. 15801)), including tribally controlled colleges or universities, located in a State described in subsection (a); or

(B) be a consortium including at least 1 such institution of higher education and industry, State agencies, Indian tribal agencies, National Laboratories, or local government agencies located in the State; and

(2) have proven experience and capabilities with relevant technologies.

(a) AUTHORIZATION OF APPROPRIATIONS.—There are authorized to be appropriated to the Secretary to carry out this section $25,000,000 for each of fiscal years 2008 through 2010.

Sec. 224. Biorefinery Energy Efficiency.

Section 932 of the Energy Policy Act of 2005 (42 U.S.C. 16232) is amended by adding at the end the following new subsections:

"(g) BIOREFINERY ENERGY EFFICIENCY.—The Secretary shall establish a program of research, development, demonstration, and commercial application for increasing energy efficiency and reducing energy consumption in the operation of biorefinery facilities.

"(h) RETROFIT TECHNOLOGIES FOR THE DEVELOPMENT OF ETHANOL FROM CELLULOSIC MATERIALS.—The Secretary shall establish a program of research, development, demonstration, and commercial application on technologies and processes to enable biorefineries that exclusively use corn grain or corn starch as a feedstock to produce ethanol to be retrofitted to accept a range of biomass, including lignocellulosic feedstocks.".

Sec. 225. Study Of Optimization Of Flexible Fueled Vehicles To Use E–85 Fuel.

(a) IN GENERAL.—The Secretary, in consultation with the Secretary of Transportation and the Administrator of the Environmental Protection Agency, shall conduct a study of whether optimizing flexible fueled vehicles to operate using E–85 fuel would increase the fuel efficiency of flexible fueled vehicles.

(b) REPORT.—Not later than 180 days after the date of enactment of this Act, the Secretary shall submit to the Committee on Science and Technology and the Committee on Energy and Commerce of the House of Representatives, and to the Committee on Energy and Natural Resources, the Committee on Environment and Public Works, and the Committee on Commerce, Science, and Transportation of the Senate, a report that describes the results of the study under this section, including any recommendations of the Secretary.

Sec. 226. Study Of Engine Durability And Performance Associated With The Use Of Biodiesel.

(a) IN GENERAL.—Not later than 30 days after the date of enactment of this Act, the Secretary, in consultation with the Administrator of the Environmental Protection Agency, shall initiate a study on the effects of the use of biodiesel on the performance and durability of engines and engine systems.

(b) COMPONENTS.—The study under this section shall include—

(1) an assessment of whether the use of biodiesel lessens the durability and performance of conventional diesel engines and engine systems; and

(2) an assessment of the effects referred to in subsection (a) with respect to biodiesel blends at varying concentrations, including the following percentage concentrations of biodiesel:

(A) 5 percent biodiesel.

(B) 10 percent biodiesel.

(C) 20 percent biodiesel.

(D) 30 percent biodiesel.
(E) 100 percent biodiesel.

(c) REPORT.—Not later than 24 months after the date of enactment of this Act, the Secretary shall submit to the Committee on Science and Technology and the Committee on Energy and Commerce of the House of Representatives, and to the Committee on Energy and Natural Resources and the Committee on Environment and Public Works of the Senate, a report that describes the results of the study under this section, including any recommendations of the Secretary.

Sec. 227. Study Of Optimization Of Biogas Used In Natural Gas Vehicles.

(a) IN GENERAL.—The Secretary, in consultation with the Administrator of the Environmental Protection Agency and the Secretary of Transportation, shall conduct a study of methods of increasing the fuel efficiency of vehicles using biogas by optimizing natural gas vehicle systems that can operate on biogas, including the advancement of vehicle fuel systems and the combination of hybrid-electric and plug-in hybrid electric drive platforms with natural gas vehicle systems using biogas.

(b) REPORT.—Not later than 180 days after the date of enactment of this Act, the Secretary shall submit to the Committee on Energy and Natural Resources, the Committee on Environment and Public Works, and the Committee on Commerce, Science, and Transportation of the Senate, and to the Committee on Science and Technology and the Committee on Energy and Commerce of the House of Representatives, a report that describes the results of the study, including any recommendations of the Secretary.

Sec. 228. Algal Biomass.

(a) IN GENERAL.—Not later than 90 days after the date of enactment of this Act, the Secretary shall submit to the Committee on Science and Technology of the House of Representatives and the Committee on Energy and Natural Resources of the Senate, a report on the progress of the research and development that is being conducted on the use of algae as a feedstock for the production of biofuels.

(b) CONTENTS.—The report shall identify continuing research and development challenges and any regulatory or other barriers found by the Secretary that hinder the use of this resource, as well as recommendations on how to encourage and further its development as a viable transportation fuel.

Sec. 229. Biofuels And Biorefinery Information Center.

(a) IN GENERAL.—The Secretary, in cooperation with the Secretary of Agriculture, shall establish a biofuels and biorefinery information center to make available to interested parties information on—

(1) renewable fuel feedstocks, including the varieties of fuel capable of being produced from various feedstocks;
(2) biorefinery processing techniques related to various renewable fuel feedstocks;
(3) the distribution, blending, storage, and retail dispensing infrastructure necessary for the transport and use of renewable fuels;
(4) Federal and State laws and incentives related to renewable fuel production and use;
(5) renewable fuel research and development advancements;
(6) renewable fuel development and biorefinery processes and technologies;
(7) renewable fuel resources, including information on programs and incentives for renewable fuels;

(8) renewable fuel producers;
(9) renewable fuel users; and
(10) potential renewable fuel users.

(b) ADMINISTRATION.—In administering the biofuels and biorefinery information center, the Secretary shall—
(1) continually update information provided by the center;
(2) make information available relating to processes and technologies for renewable fuel production;
(3) make information available to interested parties on the process for establishing a biorefinery; and
(4) make information and assistance provided by the center available through a toll-free telephone number and website.

(c) COORDINATION AND NONDUPLICATION.—To the maximum extent practicable, the Secretary shall ensure that the activities under this section are coordinated with, and do not duplicate the efforts of, centers at other government agencies.

(d) AUTHORIZATION OF APPROPRIATIONS.—There are authorized to be appropriated such sums as are necessary to carry out this section.

Sec. 230. Cellulosic Ethanol And Biofuels Research.

(a) DEFINITION OF ELIGIBLE ENTITY.—In this section, the term "eligible entity" means—
(1) an 1890 Institution (as defined in section 2 of the Agricultural Research, Extension, and Education Reform Act of 1998 (7 U.S.C. 7061));
(2) a part B institution (as defined in section 322 of the Higher Education Act of 1965 (20 U.S.C. 1061)) (commonly referred to as "Historically Black Colleges and Universities");
(3) a tribal college or university (as defined in section 316(b) of the Higher Education Act of 1965 (20 U.S.C. 1059c(b))); or
(4) a Hispanic-serving institution (as defined in section 502(a) of the Higher Education Act of 1965 (20 U.S.C. 1101a(a))).

(b) GRANTS.—The Secretary shall make cellulosic ethanol and biofuels research and development grants to 10 eligible entities selected by the Secretary to receive a grant under this section through a peer-reviewed competitive process.

(c) COLLABORATION.—An eligible entity that is selected to receive a grant under subsection (b) shall collaborate with 1 of the Bioenergy Research Centers of the Office of Science of the Department.

(d) AUTHORIZATION OF APPROPRIATIONS.—There is authorized to be appropriated to the Secretary to make grants described in subsection (b) $50,000,000 for fiscal year 2008, to remain available until expended.

Sec. 231. Bioenergy Research And Development, Authorization Of Appropriation.

Section 931 of the Energy Policy Act of 2005 (42 U.S.C. 16231) is amended—
(1) in subsection (b)—
(A) in paragraph (2), by striking "and" at the end;
(B) in paragraph (3), by striking the period at the end and inserting "; and"; and
(C) by adding at the end the following:
"(4)$963,000,000 for fiscal year 2010."; and
(2) in subsection (c)—
(A) in paragraph (2)—
(i) by striking "$251,000,000" and inserting "$377,000,000"; and
(ii) by striking "and" at the end;

(B) in paragraph (3)—
(i) by striking “$274,000,000” and inserting “$398,000,000”; and
(ii) by striking the period at the end and inserting “; and”; and
(C) by adding at the end the following:
"(4)$419,000,000 for fiscal year 2010, of which $150,000,000 shall be for section 932(d).”.

Sec. 232. Environmental Research And Development.

(a) IN GENERAL.—Section 977 of the Energy Policy Act of 2005 (42 U.S.C. 16317) is amended—
(1) in subsection (a)(1), by striking “and computational biology” and inserting “computational biology, and environmental science”; and
(2) in subsection (b)—
(A) in paragraph (1), by inserting “in sustainable production systems that reduce greenhouse gas emissions” after “hydrogen”;
(B) in paragraph (3), by striking “and” at the end;
(C) by redesignating paragraph (4) as paragraph (5); and
(D) by inserting after paragraph (3) the following:
“(4) develop cellulosic and other feedstocks that are less resource and land intensive and that promote sustainable use of resources, including soil, water, energy, forests, and land, and ensure protection of air, water, and soil quality; and”.
(b) TOOLS AND EVALUATION.—Section 307(d) of the Biomass Research and Development Act of 2000 (7 U.S.C. 8606(d)) is amended—
(1) in paragraph (3)(E), by striking “and” at the end;
(2) in paragraph (4), by striking the period at the end and inserting a semicolon; and
(3) by adding at the end the following:
“(5) the improvement and development of analytical tools to facilitate the analysis of life-cycle energy and greenhouse gas emissions, including emissions related to direct and indirect land use changes, attributable to all potential biofuel feedstocks and production processes; and
“(6) the systematic evaluation of the impact of expanded biofuel production on the environment, including forest lands, and on the food supply for humans and animals.”.
(c) SMALL-SCALE PRODUCTION AND USE OF BIOFUELS.—Section 307(e) of the Biomass Research and Development Act of 2000 (7 U.S.C. 8606(e)) is amended—
(1) in paragraph (2), by striking “and” at the end;
(2) in paragraph (3), by striking the period at the end and inserting “; and”; and
(3) by adding at the end the following:
"(4)to facilitate small-scale production, local, and on-farm use of biofuels, including the development of small-scale gasification technologies for production of biofuel from cellulosic feedstocks.”.

Sec. 233. Bioenergy Research Centers.

Section 977 of the Energy Policy Act of 2005 (42 U.S.C. 16317) is amended by adding at the end the following:
“(f) BIOENERGY RESEARCH CENTERS.—
“(1) ESTABLISHMENT OF CENTERS.—In carrying out the program under subsection (a), the Secretary shall establish at least 7 bioenergy research centers, which may be of varying size.

Administration for Defense District or Subdistrict of a Petroleum Administration for Defense District.

"(3) GOALS.—The goals of the centers established under this subsection shall be to accelerate basic transformational research and development of biofuels, including biological processes.

"(4) SELECTION AND DURATION.—

"(A) IN GENERAL.—A center under this subsection shall be selected on a competitive basis for a period of 5 years.

"(B) REAPPLICATION.—After the end of the period described in subparagraph (A), a grantee may reapply for selection on a competitive basis.

"(5) INCLUSION.—A center that is in existence on the date of enactment of this subsection—

"(A) shall be counted towards the requirement for establishment of at least 7 bioenergy research centers; and

"(B) may continue to receive support for a period of 5 years beginning on the date of establishment of the center.".

Sec. 234. University Based Research And Development Grant Program.

(a) ESTABLISHMENT.—The Secretary shall establish a competitive grant program, in a geographically diverse manner, for projects submitted for consideration by institutions of higher education to conduct research and development of renewable energy technologies. Each grant made shall not exceed $2,000,000.

(b) ELIGIBILITY.—Priority shall be given to institutions of higher education with—

(1) established programs of research in renewable energy;

(2) locations that are low income or outside of an urbanized area;

(3) a joint venture with an Indian tribe; and

(4) proximity to trees dying of disease or insect infestation as a source of woody biomass.

(c) AUTHORIZATION OF APPROPRIATIONS.—There are authorized to be appropriated to the Secretary $25,000,000 for carrying out this section.

(d) DEFINITIONS.—In this section:

(1) INDIAN TRIBE.—The term "Indian tribe" has the meaning as defined in section 126(c) of the Energy Policy Act of 2005.

(2) RENEWABLE ENERGY.—The term "renewable energy" has the meaning as defined in section 902 of the Energy Policy Act of 2005.

(3) URBANIZED AREA.—The term "urbanized area" has the meaning as defined by the U.S. Bureau of the Census.

SUBTITLE C—BIOFUELS INFRASTRUCTURE

Sec. 241. Prohibition On Franchise Agreement Restrictions Related To Renewable Fuel Infrastructure.

(a) IN GENERAL.—Title I of the Petroleum Marketing Practices Act (15 U.S.C. 2801 et seq.) is amended by adding at the end the following:

"SEC. 107. PROHIBITION ON RESTRICTION OF INSTALLATION OF RENEWABLE FUEL PUMPS.

"(a) DEFINITION.—In this section:

"(1) RENEWABLE FUEL.—The term 'renewable fuel' means any fuel—

"(A) at least 85 percent of the volume of which consists of ethanol; or

"(B) any mixture of biodiesel and diesel or renewable diesel (as defined in regulations adopted pursuant to section 211(o) of the Clean Air Act (40 CFR, part 80)), determined without regard to any use of kerosene and containing at least 20 percent biodiesel or renewable diesel.

"(2) FRANCHISE-RELATED DOCUMENT.—The term 'franchise- related document' means—

"(A) a franchise under this Act; and

"(B) any other contract or directive of a franchisor relating to terms or conditions of the sale of fuel by a franchisee.

"(b) PROHIBITIONS.—

"(1) IN GENERAL.—No franchise-related document entered into or renewed on or after the date of enactment of this section shall contain any provision allowing a franchisor to restrict the franchisee or any affiliate of the franchisee from—

"(A) installing on the marketing premises of the franchisee a renewable fuel pump or tank, except that the franchisee's franchisor may restrict the installation of a tank on leased marketing premises of such franchisor;

"(B) converting an existing tank or pump on the marketing premises of the franchisee for renewable fuel use, so long as such tank or pump and the piping connecting them are either warranted by the manufacturer or certified by a recognized standards setting organization to be suitable for use with such renewable fuel;

"(C) advertising (including through the use of signage) the sale of any renewable fuel;

"(D) selling renewable fuel in any specified area on the marketing premises of the franchisee (including any area in which a name or logo of a franchisor or any other entity appears);

"(E) purchasing renewable fuel from sources other than the franchisor if the franchisor does not offer its own renewable fuel for sale by the franchisee;

"(F) listing renewable fuel availability or prices, including on service station signs, fuel dispensers, or light poles; or

"(G) allowing for payment of renewable fuel with a credit card, so long as such activities described in subparagraphs (A) through (G) do not constitute mislabeling, misbranding, willful adulteration, or other trademark violations by the franchisee.

"(2) EFFECT OF PROVISION.—Nothing in this section shall be construed to preclude a franchisor from requiring the franchisee to obtain reasonable indemnification and insurance policies.

"(c) EXCEPTION TO 3-GRADE REQUIREMENT.—No franchise- related document that requires that 3 grades of gasoline be sold by the applicable franchisee shall prevent the franchisee from selling a renewable fuel in lieu of 1, and only 1, grade of gasoline.".

(b) ENFORCEMENT.—Section 105 of the Petroleum Marketing Practices Act (15 U.S.C. 2805) is amended by striking "102 or 103" each place it appears and inserting "102, 103, or 107".

(c) CONFORMING AMENDMENTS.—

(1) IN GENERAL.—Section 10 1(13) of the Petroleum Marketing Practices Act (15 U.S.C. 2801(13)) is amended by aligning the margin of subparagraph (C) with subparagraph (B).

(2) TABLE OF CONTENTS.—The table of contents of the Petroleum Marketing

Practices Act (15 U.S.C. 2801 note) is amended—

(A) by inserting after the item relating to section 106 the following:

(B) "Sec. 107. Prohibition on restriction of installation of renewable fuel pumps."; and

(C) by striking the item relating to section 202 and inserting the following:

(D) "Sec. 202. Automotive fuel rating testing and disclosure requirements.".

Sec. 242. Renewable Fuel Dispenser Requirements.

(a) MARKET PENETRATION REPORTS.—The Secretary, in consultation with the Secretary of Transportation, shall determine and report to Congress annually on the market penetration for flexible-fuel vehicles in use within geographic regions to be established by the Secretary.

(b) DISPENSER FEASIBILITY STUDY.—Not later than 24 months after the date of enactment of this Act, the Secretary, in consultation with the Department of Transportation, shall report to the Congress on the feasibility of requiring motor fuel retailers to install E– 85 compatible dispensers and related systems at retail fuel facilities in regions where flexible-fuel vehicle market penetration has reached 15 percent of motor vehicles. In conducting such study, the Secretary shall consider and report on the following factors:

(1) The commercial availability of E–85 fuel and the number of competing E–85 wholesale suppliers in a given region.

(2) The level of financial assistance provided on an annual basis by the Federal Government, State governments, and nonprofit entities for the installation of E–85 compatible infrastructure.

(3) The number of retailers whose retail locations are unable to support more than 2 underground storage tank dispensers.

(4) The expense incurred by retailers in the installation and sale of E–85 compatible dispensers and related systems and any potential effects on the price of motor vehicle fuel.

Sec. 243. Ethanol Pipeline Feasibility Study.

(a) IN GENERAL.—The Secretary, in coordination with the Secretary of Transportation, shall conduct a study of the feasibility of the construction of pipelines dedicated to the transportation of ethanol.

(b) FACTORS FOR CONSIDERATION.—In conducting the study under subsection (a), the Secretary shall take into consideration—

(1) the quantity of ethanol production that would make dedicated pipelines economically viable;

(2) existing or potential barriers to the construction of pipelines dedicated to the transportation of ethanol, including technical, siting, financing, and regulatory barriers;

(3) market risk (including throughput risk) and means of mitigating the risk;

(4) regulatory, financing, and siting options that would mitigate the risk and help ensure the construction of 1 or more pipelines dedicated to the transportation of ethanol;

(5) financial incentives that may be necessary for the construction of pipelines dedicated to the transportation of ethanol, including the return on equity that sponsors of the initial dedicated ethanol pipelines will require to invest in the

pipelines;

(6) technical factors that may compromise the safe transportation of ethanol in pipelines, including identification of remedial and preventive measures to ensure pipeline integrity; and

(7) such other factors as the Secretary considers to be appropriate.

(c) REPORT.—Not later than 15 months after the date of enactment of this Act, the Secretary shall submit to Congress a report describing the results of the study conducted under this section.

(d) AUTHORIZATION OF APPROPRIATIONS.—There is authorized to be appropriated to the Secretary to carry out this section $1,000,000 for each of fiscal years 2008 and 2009, to remain available until expended.

Sec. 244. Renewable Fuel Infrastructure Grants.

(a) DEFINITION OF RENEWABLE FUEL BLEND.—For purposes of this section, the term "renewable fuel blend" means a gasoline blend that contains not less than 11 percent, and not more than 85 percent, renewable fuel or diesel fuel that contains at least 10 percent renewable fuel.

(b) INFRASTRUCTURE DEVELOPMENT GRANTS.—

(1) ESTABLISHMENT.—The Secretary shall establish a program for making grants for providing assistance to retail and wholesale motor fuel dealers or other entities for the installation, replacement, or conversion of motor fuel storage and dispensing infrastructure to be used exclusively to store and dispense renewable fuel blends.

(2) SELECTION CRITERIA.—Not later than 12 months after the date of enactment of this Act, the Secretary shall establish criteria for evaluating applications for grants under this subsection that will maximize the availability and use of renewable fuel blends, and that will ensure that renewable fuel blends are available across the country. Such criteria shall provide for—

(A) consideration of the public demand for each renewable fuel blend in a particular geographic area based on State registration records showing the number of flexible- fuel vehicles;

(B) consideration of the opportunity to create or expand corridors of renewable fuel blend stations along interstate or State highways;

(C) consideration of the experience of each applicant with previous, similar projects;

(D) consideration of population, number of flexible- fuel vehicles, number of retail fuel outlets, and saturation of flexible-fuel vehicles; and

(E) priority consideration to applications that—

(i) are most likely to maximize displacement of petroleum consumption, measured as a total quantity and a percentage;

(ii) are best able to incorporate existing infrastructure while maximizing, to the extent practicable, the use of renewable fuel blends; and

(iii) demonstrate the greatest commitment on the part of the applicant to ensure funding for the proposed project and the greatest likelihood that the project will be maintained or expanded after Federal assistance under this subsection is completed.

(3) LIMITATIONS.—Assistance provided under this subsection shall not exceed—

(A) 33 percent of the estimated cost of the installation, replacement, or conversion of motor fuel storage and dispensing infrastructure; or

(B) $180,000 for a combination of equipment at any one retail outlet location.

(4) OPERATION OF RENEWABLE FUEL BLEND STATIONS.—The Secretary shall establish rules that set forth requirements for grant recipients under this section that include providing to the public the renewable fuel blends, establishing a marketing plan that informs consumers of the price and availability of the renewable fuel blends, clearly labeling the dispensers and related equipment, and providing periodic reports on the status of the renewable fuel blend sales, the type and amount of the renewable fuel blends dispensed at each location, and the average price of such fuel.

(5) NOTIFICATION REQUIREMENTS.—Not later than the date on which each renewable fuel blend station begins to offer renewable fuel blends to the public, the grant recipient that used grant funds to construct or upgrade such station shall notify the Secretary of such opening. The Secretary shall add each new renewable fuel blend station to the renewable fuel blend station locator on its Website when it receives notification under this subsection.

(6) DOUBLE COUNTING.—No person that receives a credit under section 30C of the Internal Revenue Code of 1986 may receive assistance under this section.

(7) RESERVATION OF FUNDS.—The Secretary shall reserve funds appropriated for the renewable fuel blends infrastructure development grant program for technical and marketing assistance described in subsection (c).

(c) RETAIL TECHNICAL AND MARKETING ASSISTANCE.—The Secretary shall enter into contracts with entities with demonstrated experience in assisting retail fueling stations in installing refueling systems and marketing renewable fuel blends nationally, for the provision of technical and marketing assistance to recipients of grants under this section. Such assistance shall include—

(1) technical advice for compliance with applicable Federal

(2) and State environmental requirements;

(3) help in identifying supply sources and securing long- term contracts; and

(4) provision of public outreach, education, and labeling materials.

(d) REFUELING INFRASTRUCTURE CORRIDORS.—

(1) IN GENERAL.—The Secretary shall establish a competitive grant pilot program (referred to in this subsection as the "pilot program"), to be administered through the Vehicle Technology Deployment Program of the Department, to provide not more than 10 geographically-dispersed project grants to State governments, Indian tribal governments, local governments, metropolitan transportation authorities, or partnerships of those entities to carry out 1 or more projects for the purposes described in paragraph (2).

(2) GRANT PURPOSES.—A grant under this subsection shall be used for the establishment of refueling infrastructure corridors, as designated by the Secretary, for renewable fuel blends, including—

(A) installation of infrastructure and equipment necessary to ensure adequate distribution of renewable fuel blends within the corridor;

(B) installation of infrastructure and equipment necessary to directly support vehicles powered by renewable fuel blends; and

(C) operation and maintenance of infrastructure and equipment installed as part of a project funded by the grant.

(3) APPLICATIONS.—

(A) REQUIREMENTS.—

(i) IN GENERAL.—Subject to clause (ii), not later Deadline. than 90 days after the date of enactment of this Act, the Secretary shall issue requirements for use in applying for grants under the pilot program.

(ii) MINIMUM REQUIREMENTS.—At a minimum, the Secretary shall require that an application for a grant under this subsection—

(I) be submitted by—

(aa) the head of a State, tribal, or local government or a metropolitan transportation authority, or any combination of those entities; and

(bb) a registered participant in the Vehicle Technology Deployment Program of the Department; and

(II) include—

(aa) a description of the project proposed in the application, including the ways in which the project meets the requirements of this subsection;

(bb) an estimate of the degree of use of the project, including the estimated size of fleet of vehicles operated with renewable fuels blend available within the geographic region of the corridor, measured as a total quantity and a percentage;

(cc) an estimate of the potential petroleum displaced as a result of the project (measured as a total quantity and a percentage), and a plan to collect and disseminate petroleum displacement and other relevant data relating to the project to be funded under the grant, over the expected life of the project;

(dd) a description of the means by which the project will be sustainable without Federal assistance after the completion of the term of the grant;

(ee) a complete description of the costs of the project, including acquisition, construction, operation, and maintenance costs over the expected life of the project; and

(ff) a description of which costs of the project will be supported by Federal assistance under this subsection.

(B) PARTNERS.—An applicant under subparagraph (A) may carry out a project under the pilot program in partnership with public and private entities.

(4) SELECTION CRITERIA.—In evaluating applications under the pilot program, the Secretary shall—

(A) consider the experience of each applicant with previous, similar projects; and

(B) give priority consideration to applications that—

(i) are most likely to maximize displacement of petroleum consumption, measured as a total quantity and a percentage;

(ii) are best able to incorporate existing infrastructure while maximizing, to the extent practicable, the use of advanced biofuels;

(iii) demonstrate the greatest commitment on the part of the applicant to ensure funding for the proposed project and the greatest likelihood that the project will be maintained or expanded after Federal assistance under this subsection is completed;

(iv) represent a partnership of public and private entities; and

(v) exceed the minimum requirements of paragraph (3)(A)(ii).

(5) PILOT PROJECT REQUIREMENTS.—
(A) MAXIMUM AMOUNT.—The Secretary shall provide not more than $20,000,000 in Federal assistance under the pilot program to any applicant.
(B) COST SHARING.—The non-Federal share of the cost of any activity relating to renewable fuel blend infrastructure development carried out using funds from a grant under this subsection shall be not less than 20 percent.
(C) MAXIMUM PERIOD OF GRANTS.—The Secretary shall not provide funds to any applicant under the pilot program for more than 2 years.
(D) DEPLOYMENT AND DISTRIBUTION.—The Secretary shall seek, to the maximum extent practicable, to ensure a broad geographic distribution of project sites funded by grants under this subsection.
(E) TRANSFER OF INFORMATION AND KNOWLEDGE.—The Secretary shall establish mechanisms to ensure that the information and knowledge gained by participants in the pilot program are transferred among the pilot program participants and to other interested parties, including other applicants that submitted applications.
(6) SCHEDULE.—
(A) INITIAL GRANTS.—
(i) IN GENERAL.—Not later than 90 days after the date of enactment of this Act, the Secretary shall publish in the Federal Register, Commerce Business Daily, and such other publications as the Secretary considers to be appropriate, a notice and request for applications to carry out projects under the pilot program.
(ii) DEADLINE.—An application described in clause (i) shall be submitted to the Secretary by not later than 180 days after the date of publication of the notice under that clause.
(iii) INITIAL SELECTION.—Not later than 90 days after the date by which applications for grants are due under clause (ii), the Secretary shall select by competitive, peer-reviewed proposal up to 5 applications for projects to be awarded a grant under the pilot program.
(B) ADDITIONAL GRANTS.—
(i) IN GENERAL.—Not later than 2 years after the date of enactment of this Act, the Secretary shall publish in the Federal Register, Commerce Business Daily, and such other publications as the Secretary considers to be appropriate, a notice and request for additional applications to carry out projects under the pilot program that incorporate the information and knowledge obtained through the implementation of the first round of projects authorized under the pilot program.
(ii) DEADLINE.—An application described in clause (i) shall be submitted to the Secretary by not later than 180 days after the date of publication of the notice under that clause.
(iii) INITIAL SELECTION.—Not later than 90 days after the date by which applications for grants are due under clause (ii), the Secretary shall select by competitive, peer-reviewed proposal such additional applications for projects to be awarded a grant under the pilot program as the Secretary determines to be appropriate.
(7) REPORTS TO CONGRESS.—
(A) INITIAL REPORT.—Not later than 60 days after the date on which grants are awarded under this subsection, the Secretary shall submit to Congress a report containing—

(i) an identification of the grant recipients and a description of the projects to be funded under the pilot program;
(ii) an identification of other applicants that submitted applications for the pilot program but to which funding was not provided; and
(iii)a description of the mechanisms used by the Secretary to ensure that the information and knowledge gained by participants in the pilot program are transferred among the pilot program participants and to other interested parties, including other applicants that submitted applications.

(B) EVALUATION.—Not later than 2 years after the date of enactment of this Act, and annually thereafter until the termination of the pilot program, the Secretary shall submit to Congress a report containing an evaluation of the effectiveness of the pilot program, including an assessment of the petroleum displacement and benefits to the environment derived from the projects included in the pilot program.

(f) RESTRICTION.—No grant shall be provided under subsection (b) or (c) to a large, vertically integrated oil company.

(g) AUTHORIZATION OF APPROPRIATIONS.—There are authorized to be appropriated to the Secretary for carrying out this section $200,000,000 for each of the fiscal years 2008 through 2014.

Sec. 245. Study Of The Adequacy Of Transportation Of Domestically-Produced Renewable Fuel By Railroads And Other Modes Of Transportation.

(a) STUDY.—

(1) IN GENERAL.—The Secretary, in coordination with the Secretary of Transportation, shall jointly conduct a study of the adequacy of transportation of domestically-produced renewable fuels by railroad and other modes of transportation as designated by the Secretaries.

(2) COMPONENTS.—In conducting the study under paragraph (1), the Secretaries shall—

(A) consider the adequacy of existing railroad and other transportation and distribution infrastructure, equipment, service and capacity to move the necessary quantities of domestically-produced renewable fuel within the time- frames;

(B) (i) consider the projected costs of moving the domestically-produced renewable fuel by railroad and other modes of transportation; and
(ii) consider the impact of the projected costs on the marketability of the domestically-produced renewable fuel;

(C) identify current and potential impediments to the reliable transportation and distribution of adequate supplies of domestically-produced renewable fuel at reasonable prices, including practices currently utilized by domestic producers, shippers, and receivers of renewable fuels;

(D) consider whether adequate competition exists within and between modes of transportation for the transportation and distribution of domestically-produced renewable fuel and, whether inadequate competition leads to an unfair price for the transportation and distribution of domestically-produced renewable fuel or unacceptable service for transportation of domestically-produced renewable fuel;

(E) consider whether Federal agencies have adequate legal authority to address instances of inadequate competition when inadequate competition is found to prevent domestic producers for renewable fuels from obtaining a fair and reasonable transportation price or acceptable service for the transportation and distribution of domestically-produced renewable fuels;

(F) consider whether Federal agencies have adequate legal authority to address railroad and transportation service problems that may be resulting in inadequate supplies of domestically-produced renewable fuel in any area of the United States;

(G) consider what transportation infrastructure capital expenditures may be necessary to ensure the reliable transportation of adequate supplies of domestically-produced renewable fuel at reasonable prices within the United States and which public and private entities should be responsible for making such expenditures; and

(H) provide recommendations on ways to facilitate the reliable transportation of adequate supplies of domestically- produced renewable fuel at reasonable prices.

(e) REPORT.—Not later than 180 days after the date of enactment of this Act, the Secretaries shall jointly submit to the Committee on Commerce, Science and Transportation, the Committee on Energy and Natural Resources, and the Committee on Environment and Public Works of the Senate and the Committee on Transportation and Infrastructure and the Committee on Energy and Commerce of the House of Representatives a report that describes the results of the study conducted under subsection (a).

Sec. 246. Federal Fleet Fueling Centers.

(a) IN GENERAL.—Not later than January 1, 2010, the head of each Federal agency shall install at least 1 renewable fuel pump at each Federal fleet fueling center in the United States under the jurisdiction of the head of the Federal agency.

(b) REPORT.—Not later than October 31 of the first calendar year beginning after the date of the enactment of this Act, and each October 31 thereafter, the President shall submit to Congress a report that describes the progress toward complying with subsection (a), including identifying—

(1) the number of Federal fleet fueling centers that contain at least 1 renewable fuel pump; and

(2) the number of Federal fleet fueling centers that do not contain any renewable fuel pumps.

(c) DEPARTMENT OF DEFENSE FACILITY.—This section shall not apply to a Department of Defense fueling center with a fuel turnover rate of less than 100,000 gallons of fuel per year.

(d) AUTHORIZATION OF APPROPRIATIONS.—There are authorized to be appropriated such sums as are necessary to carry out this section.

Sec. 247. Standard Specifications For Biodiesel.

Section 211 of the Clean Air Act (42 U.S.C. 7545) is amended by redesignating subsection (s) as subsection (t), redesignating subsection (r) (relating to conversion assistance for cellulosic biomass, waste-derived ethanol, approved renewable fuels) as subsection (s) and by adding the following new subsection at the end thereof:

"(u) STANDARD SPECIFICATIONS FOR BIODIESEL.—(1) Unless the American Society for Testing and Materials has adopted a standard for diesel fuel containing 20 percent biodiesel (commonly known as 'B20') within 1 year after the date of enactment of this subsection, the Administrator shall initiate a rulemaking to establish a uniform per gallon fuel standard for such fuel and designate an identification number so that vehicle manufacturers are able to design engines to use fuel meeting such standard.

"(2) Unless the American Society for Testing and Materials has adopted a standard for diesel fuel containing 5 percent biodiesel (commonly known as 'B5') within 1 year after the date of enactment of this subsection, the Administrator shall initiate a rulemaking to establish a uniform per gallon fuel standard for such fuel and designate an identification so that vehicle manufacturers are able to design engines to use fuel meeting such standard.

"(3) Whenever the Administrator is required to initiate a rule- making under paragraph (1) or (2), the Administrator shall promulgate a final rule within 18 months after the date of the enactment of this subsection.

"(4) Not later than 180 days after the enactment of this subsection, the Administrator shall establish an annual inspection and enforcement program to ensure that diesel fuel containing biodiesel sold or distributed in interstate commerce meets the standards established under regulations under this section, including testing and certification for compliance with applicable standards of the American Society for Testing and Materials. There are authorized to be appropriated to carry out the inspection and enforcement program under this paragraph $3,000,000 for each of fiscal years 2008 through 2010.

"(5) For purposes of this subsection, the term 'biodiesel' has the meaning provided by section 312(f) of Energy Policy Act of 1992 (42 U.S.C. 13220(f)).".

Sec. 248. Biofuels Distribution And Advanced Biofuels Infrastructure.

(a) IN GENERAL.—The Secretary, in coordination with the Secretary of Transportation and in consultation with the Administrator of the Environmental Protection Agency, shall carry out a program of research, development, and demonstration relating to existing transportation fuel distribution infrastructure and new alternative distribution infrastructure.

(b) FOCUS.—The program described in subsection (a) shall focus on the physical and chemical properties of biofuels and efforts to prevent or mitigate against adverse impacts of those properties in the areas of—

(1) corrosion of metal, plastic, rubber, cork, fiberglass, glues, or any other material used in pipes and storage tanks;

(2) dissolving of storage tank sediments;

(3) clogging of filters;

(4) contamination from water or other adulterants or pollutants;

(5) poor flow properties related to low temperatures;

(6) oxidative and thermal instability in long-term storage and uses;

(7) microbial contamination;

(8) problems associated with electrical conductivity; and

(9) such other areas as the Secretary considers appropriate.

Subtitle D—Environmental Safeguards

Sec. 251. Waiver For Fuel Or Fuel Additives.

Section 211(f)(4) of the Clean Air Act (42 U.S.C. 7545(f)) is amended to read as follows:

"(4) The Administrator, upon application of any manufacturer of any fuel or fuel additive, may waive the prohibitions established under paragraph (1) or (3) of this subsection or the limitation specified in paragraph (2) of this subsection, if he determines that the applicant has established that such fuel or fuel additive or a specified concentration thereof, and the emission products of such fuel or fuel additive or specified concentration thereof, will not cause or contribute to a failure of any emission control device or system (over the useful life of the motor vehicle, motor vehicle engine, nonroad engine or nonroad vehicle in which such device or system is used) to achieve compliance by the vehicle or engine with the emission standards with respect to which it has been certified pursuant to sections 206 and 2 13(a). The Administrator shall take final action to grant or deny an application submitted under this paragraph, after public notice and comment, within 270 days of the receipt of such an application.".

TITLE III — ENERGY SAVINGS THROUGH IMPROVED STANDARDS FOR APPLIANCE AND LIGHTING

Subtitle A—Appliance Energy Efficiency

Sec. 301. External Power Supply Efficiency Standards.

(a) DEFINITIONS.—Section 321 of the Energy Policy and Conservation Act (42 U.S.C. 6291) is amended—

(1) in paragraph (36)—

(A) by striking "(36) The" and inserting the following: "(36) EXTERNAL POWER SUPPLY.—

"(A) IN GENERAL.—The"; and

(B) by adding at the end the following:

"(B) ACTIVE MODE.—The term 'active mode' means the mode of operation when an external power supply is connected to the main electricity supply and the output is connected to a load.

"(C) CLASS A EXTERNAL POWER SUPPLY.—

"(i) " IN GENERAL.—The term 'class A external power supply' means a device that—

"(I) is designed to convert line voltage AC input into lower voltage AC or DC output;

"(II) is able to convert to only 1 AC or DC output voltage at a time;

"(III) is sold with, or intended to be used with, a separate end-use product that constitutes the primary load;

"(IV) is contained in a separate physical enclosure from the end-use product;

"(V) is connected to the end-use product via a removable or hard-wired male/female electrical connection, cable, cord, or other wiring; and

"(VI) has nameplate output power that is less than or equal to 250 watts.

i. EXCLUSIONS.—The term 'class A external power supply' does not include any device that—

"(I) requires Federal Food and Drug Administration listing and approval as a medical device in accordance with section 513 of the Federal Food, Drug, and Cosmetic Act (21 U.S.C. 360c); or

"(II) powers the charger of a detachable battery pack or charges the battery of a product that is fully or primarily motor operated.

"(D) NO-LOAD MODE.—The term 'no-load mode' means the mode of operation when an external power supply is connected to the main electricity supply and the output is not connected to a load."; and

(2) by adding at the end the following:

"(52) DETACHABLE BATTERY.—The term 'detachable battery' means a battery that is—

"(A) contained in a separate enclosure from the product; and

"(B) intended to be removed or disconnected from the product for recharging.".

(b) TEST PROCEDURES.—Section 323(b) of the Energy Policy and Conservation Act (42 U.S.C. 6293(b)) is amended by adding at the end the following:

"(6)CLASS A EXTERNAL POWER SUPPLIES.—Test procedures for class A external power supplies shall be based on the 'Test Method for Calculating the Energy Efficiency of Single- Voltage External AC–DC and AC–AC Power Supplies' published by the Environmental Protection Agency on August 11, 2004, except that the test voltage specified in section 4(d) of that test method shall be only 115 volts, 60 Hz.".

(c) EFFICIENCY STANDARDS FOR CLASS A EXTERNAL POWER SUPPLIES.—Section 325(u) of the Energy Policy and Conservation Act (42 U.S.C. 6295(u)) is amended by adding at the end the following:

"(6) EFFICIENCY STANDARDS FOR CLASS A EXTERNAL POWER SUPPLIES.—

"(A) IN GENERAL.—Subject to subparagraphs (B) through (D), a class A external power supply manufactured on or after the later of July 1, 2008, or the date of enactment of this paragraph shall meet the following standards:

"(B) NONCOVERED SUPPLIES.—A class A external power supply shall not be subject to subparagraph (A) if the class A external power supply is—

"(i) manufactured during the period beginning on July 1, 2008, and ending on June 30, 2015; and

"(ii) made available by the manufacturer as a service part or a spare part for an end-use product—

"(I) that constitutes the primary load; and

"(II) was manufactured before July 1, 2008.

“Active Mode	
“Nameplate Output	**Required Efficiency (decimal equivalent of a percent-age)**
Less than 1 watt	0.5 times the Nameplate Output
From 1 watt to not more than 51 watts	The sum of 0.09 times the Natural Logarithm of the Nameplate Output and 0.5
Greater than 51 watts	0.85
“No-Load	**Mode**
‘‘Nameplate Output	**Maximum Consumption**
Not more than 250 watts	0.5 watts

"(C) MARKING.—Any class A external power supply manufactured on or after the later of July 1, 2008 or the date of enactment of this paragraph shall be clearly and permanently marked in accordance with the External Power Supply International Efficiency Marking Protocol, as referenced in the ‘Energy Star Program Requirements for Single Voltage External AC–DC and AC–AC Power Supplies, version 1.1’ published by the Environmental Protection Agency.

"(D) AMENDMENT OF STANDARDS.—

"(i) FINAL RULE BY JULY 1, 2011.—

"(I)“ IN GENERAL.—Not later than July 1, 2011, Publication. the Secretary shall publish a final rule to determine whether the standards established under subparagraph (A) should be amended.

"(II) ADMINISTRATION.—The final rule shall—

“(aa) contain any amended standards; and

“(bb) apply to products manufactured on or after July 1, 2013.

"(ii) FINAL RULE BY JULY 1, 2015.—

"(I)“ IN GENERAL.—Not later than July 1, 2015 Publication. the Secretary shall publish a final rule to determine whether the standards then in effect should be amended.

"(II) ADMINISTRATION.—The final rule shall—

“(aa) contain any amended standards; and

“(bb) apply to products manufactured on or after July 1, 2017.

"(7) END-USE PRODUCTS.—An energy conservation standard for external power supplies shall not constitute an energy conservation standard for the separate end-use product to which the external power supplies is connected.”.

Sec. 302. Updating Appliance Test Procedures.

Policy and Conservation Act (42 U.S.C. 6293(b)(1)) is amended

(a) CONSUMER APPLIANCES.—Section 323(b)(1) of the Energy by striking “(1)” and all that follows through the end of the paragraph and inserting the following:

“(1) TEST PROCEDURES.—

“(A) AMENDMENT.—At least once every 7 years, the Secretary shall review test procedures for all covered products and—

"(i) amend test procedures with respect to any covered product, if the Secretary determines that amended test procedures would more accurately or fully comply with the requirements of paragraph (3); or

"(ii) publish notice in the Federal Register of any determination not to amend a test procedure.".

(b) INDUSTRIAL EQUIPMENT.—Section 343(a) of the Energy Policy and Conservation Act (42 U.S.C. 6313(a)) is amended by striking "(a)" and all that follows through the end of paragraph (1) and inserting the following:

"(a) PRESCRIPTION BY SECRETARY; REQUIREMENTS.—

"(1) TEST PROCEDURES.—

"(A) AMENDMENT.—At least once every 7 years, the Secretary shall conduct an evaluation of each class of covered equipment and—

"(i) if the Secretary determines that amended test procedures would more accurately or fully comply with the requirements of paragraphs (2) and (3), shall prescribe test procedures for the class in accordance with this section; or

"(ii) shall publish notice in the Federal Register of any determination not to amend a test procedure.".

Sec. 303. Residential Boilers.

Section 325(f) of the Energy Policy and Conservation Act (42 U.S.C. 6295(f)) is amended—

(1) in the subsection heading, by inserting "AND BOILERS" after "FURNACES";

(2) by redesignating paragraph (3) as paragraph (4); and

(3) by inserting after paragraph (2) the following:

"(3) BOILERS.—

"(A) IN GENERAL.—Subject to subparagraphs (B) and (C), boilers manufactured on or after September 1, 2012, shall meet the following requirements:

Boiler Type	Minimum Annual Fuel Utilization Efficiency	Design Requirements
Gas Hot Water................	82%	No Constant Burning Pilot, Automatic Means for Adjusting Water Temperature
Gas Steam	80%	No Constant Burning Pilot
Oil Hot Water	84%	Automatic Means for Adjusting Temperature
Oil Steam.......................	82%	None
Electric Hot Water	None	Automatic Means for Adjusting Temperature
Electric Steam	None	None

"(B) AUTOMATIC MEANS FOR ADJUSTING WATER TEMPERATURE.—

"(i) " IN GENERAL.—The manufacturer shall equip each gas, oil, and electric hot water boiler (other than a boiler equipped with a tankless domestic water heating coil) with automatic means for adjusting the temperature of the water supplied by the boiler to ensure that an incremental change in inferred heat load produces a corresponding incremental change in the temperature of water supplied.

"(ii) SINGLE INPUT RATE.—For a boiler that fires at 1 input rate, the requirements of this subparagraph may be satisfied by providing an automatic means that allows the burner or heating element to fire only when the means has determined that the inferred heat load cannot be met by the residual heat of the water in the system.

"(iii) NO INFERRED HEAT LOAD.—When there is no inferred heat load with respect to a hot water boiler, the automatic means described in clauses (i) and (ii) shall limit the temperature of the water in the boiler to not more than 140 degrees Fahrenheit.

"(iv) OPERATION.—A boiler described in clause (i) or (ii) shall be operable only when the automatic means described in clauses (i), (ii), and (iii) is installed.

"(B) EXCEPTION.—A boiler that is manufactured to operate without any need for electricity or any electric connection, electric gauges, electric pumps, electric wires, or electric devices shall not be required to meet the requirements of this paragraph.".

Sec. 304. Furnace Fan Standard Process.

Paragraph (4)(D) of section 325(f) of the Energy Policy and Conservation Act (42 U.S.C. 6295(f)) (as redesignated by section 303(4)) is amended by striking "the Secretary may" and inserting "not later than December 31, 2013, the Secretary shall".

Sec. 305. Improving Schedule For Standards Updating And Clarifying State Authority.

(a) CONSUMER APPLIANCES.—Section 325 of the Energy Policy and Conservation Act (42 U.S.C. 6295) is amended by striking subsection (m) and inserting the following:

"(m) AMENDMENT OF STANDARDS.—

"(1) IN GENERAL.—Not later than 6 years after issuance of any final rule establishing or amending a standard, as required for a product under this part, the Secretary shall publish—

"(A) a notice of the determination of the Secretary that standards for the product do not need to be amended, based on the criteria established under subsection (n)(2); or

"(B) a notice of proposed rulemaking including new proposed standards based on the criteria established under subsection (o) and the procedures established under subsection (p).

"(2) NOTICE.—If the Secretary publishes a notice under paragraph (1), the Secretary shall—

"(A) publish a notice stating that the analysis of the Department is publicly available; and

"(B) provide an opportunity for written comment.

"(3) AMENDMENT OF STANDARD; NEW DETERMINATION.—

"(A) AMENDMENT OF STANDARD.—Not later than 2 years after a notice is issued under paragraph (1)(B), the Secretary shall publish a final rule amending the standard for the product.

"(B) NEW DETERMINATION.—Not later than 3 years after a determination under paragraph (1)(A), the Secretary shall make a new determination and publication under subparagraph (A) or (B) of paragraph (1).

"(4) APPLICATION TO PRODUCTS.—

"(A) IN GENERAL.—Except as provided in subparagraph (B), an amendment prescribed under this subsection shall apply to—

"(i) with respect to refrigerators, refrigerator- freezers, freezers, room air conditioners, dishwashers, clothes washers, clothes dryers, fluorescent lamp ballasts, and kitchen ranges and ovens, such a product that is manufactured after the date that is 3 years after publication of the final rule establishing an applicable standard; and

"(ii) with respect to central air conditioners, heat pumps, water heaters, pool heaters, direct heating equipment, and furnaces, such a product that is manufactured after the date that is 5 years after publication of the final rule establishing an applicable standard.

"(B) OTHER NEW STANDARDS.—A manufacturer shall not be required to apply new standards to a product with respect to which other new standards have been required during the prior 6-year period.

"(5) REPORTS.—The Secretary shall promptly submit to the Committee on Energy and Commerce of the House of Representatives and the Committee on Energy and Natural Resources of the Senate—

"(A) a progress report every 180 days on compliance with this section, including a specific plan to remedy any failures to comply with deadlines for action established under this section; and

"(B) all required reports to the Court or to any party to the Consent Decree in State of New York v Bodman, Consolidated Civil Actions No. 05 Civ. 7807 and No. 05 Civ. 7808.".

(b) INDUSTRIAL EQUIPMENT.—Section 342(a)(6) of the Energy Policy and Conservation Act (42 U.S.C. 6313(a)(6)) is amended—

(1) by redesignating subparagraph (C) as subparagraph (D); and

(2) by striking "(6)(A)(i)" and all that follows through the end of subparagraph (B) and inserting the following:

"(6) AMENDED ENERGY EFFICIENCY STANDARDS.—

"(A) IN GENERAL.—

"(i) " ANALYSIS OF POTENTIAL ENERGY SAVINGS.—If ASHRAE/IES Standard 90.1 is amended with respect to any small commercial package air conditioning and heating equipment, large commercial package air conditioning and heating equipment, very large commercial package air conditioning and heating equipment, packaged terminal air conditioners, packaged terminal heat pumps, warm-air furnaces, packaged boilers, storage water heaters, instantaneous water heaters, or unfired hot water storage tanks, not later than 180 days after the amendment of the standard, the Secretary shall publish in the Federal Register for public comment an analysis of the energy savings potential of amended energy efficiency standards.

"(ii) AMENDED UNIFORM NATIONAL STANDARD FOR PRODUCTS.—

"(I) IN GENERAL.—Except as provided in sub- clause (II), not later than 18 months after the date of publication of the amendment to the ASHRAE/IES Standard 90.1 for a product described in clause (i), the Secretary shall establish an amended uniform national standard for the product at the minimum level specified in the amended ASHRAE/IES Standard 90.1.

"(II) MORE STRINGENT STANDARD.—Subclause (I) shall not apply if the Secretary determines, by rule published in the Federal Register, and supported by clear and convincing evidence, that adoption of a uniform national standard more stringent than the amended ASHRAE/IES Standard 90.1 for the product would result in significant additional conservation of energy and is technologically feasible and economically justified.

"(B) RULE.—If the Secretary makes a determination described in clause (ii)(II) for a product described in clause (i), not later than 30 months after the date of publication of the amendment to the ASHRAE/IES Standard 90.1 for the product, the Secretary shall issue the rule establishing the amended standard.

"(C) AMENDMENT OF STANDARD.—

"(i) " IN GENERAL.—Not later than 6 years after issuance of any final rule establishing or amending a standard, as required for a product under this part, the Secretary shall publish—

"(I)" a notice of the determination of the Secretary that standards for the product do not need to be amended, based on the criteria established under subparagraph (A); or

"(II) a notice of proposed rulemaking including new proposed standards based on the criteria and procedures established under subparagraph (B).

"(ii) NOTICE.—If the Secretary publishes a notice under clause (i), the Secretary shall—

"(I) publish a notice stating that the analysis of the Department is publicly available; and

"(II) provide an opportunity for written comment.

"(iii) AMENDMENT OF STANDARD; NEW DETERMINATION.—

"(I)" AMENDMENT OF STANDARD.—Not later than 2 years after a notice is issued under clause (i)(II), the Secretary shall publish a final rule amending the standard for the product.

"(II) NEW DETERMINATION.—Not later than 3 years after a determination under clause (i)(I), the Secretary shall make a new determination and publication under subclause (I) or (II) of clause (i).

"(iv) APPLICATION TO PRODUCTS.—An amendment prescribed under this subsection shall apply to products manufactured after a date that is the later of—

"(I)" the date that is 3 years after publication of the final rule establishing a new standard; or

"(II) the date that is 6 years after the effective date of the current standard for a covered product.

"(v) REPORTS.—The Secretary shall promptly submit to the Committee on Energy and Commerce of the House of Representatives and the Committee on Energy and Natural Resources of the Senate a progress report every 180 days on compliance with this subparagraph, including a specific plan to remedy any failures to comply with deadlines for action established under this subparagraph.".

Sec. 306. Regional Standards For Furnaces, Central Air Conditioners, And Heat Pumps.

(a) IN GENERAL.—Section 325(o) of the Energy Policy and Conservation Act (42 U.S.C. 6295(o)) is amended by adding at the end the following:

"(6) REGIONAL STANDARDS FOR FURNACES, CENTRAL AIR CONDITIONERS, AND HEAT PUMPS.—

"(A) IN GENERAL.—In any rulemaking to establish a new or amended standard, the Secretary may consider the establishment of separate standards by geographic region for furnaces (except boilers), central air conditioners, and heat pumps.

"(B) NATIONAL AND REGIONAL STANDARDS.—

"(i) NATIONAL STANDARD.—If the Secretary establishes a regional standard for a product, the Secretary shall establish a base national standard for the product.

"(ii) REGIONAL STANDARDS.—If the Secretary establishes a regional standard for a product, the Secretary may establish more restrictive standards for the product by geographic region as follows:

"(I) For furnaces, the Secretary may establish 1 additional standard that is applicable in a geographic region defined by the Secretary.

"(II) For any cooling product, the Secretary may establish 1 or 2 additional standards that are applicable in 1 or 2 geographic regions as may be defined by the Secretary.

"(C) BOUNDARIES OF GEOGRAPHIC REGIONS.—

"(i) IN GENERAL.—Subject to clause (ii), the boundaries of additional geographic regions established by the Secretary under this paragraph shall include only contiguous States.

"(ii) ALASKA AND HAWAII.—The States of Alaska and Hawaii may be included under this paragraph in a geographic region that the States are not contiguous to.

"(iii) INDIVIDUAL STATES.—Individual States shall be placed only into a single region under this paragraph.

"(D) PREREQUISITES.—In establishing additional regional standards under this paragraph, the Secretary shall—

"(i) establish additional regional standards only if the Secretary determines that—

"(I) the establishment of additional regional standards will produce significant energy savings in comparison to establishing only a single national standard; and

"(II) the additional regional standards are economically justified under this paragraph; and

"(ii) consider the impact of the additional regional standards on consumers, manufacturers, and other market participants, including product distributors, dealers, contractors, and installers.

"(E) APPLICATION; EFFECTIVE DATE.—

"(i) BASE NATIONAL STANDARD.—Any base national standard established for a product under this paragraph shall—

"(I) be the minimum standard for the product; and

"(II) apply to all products manufactured or imported into the United States on and after the effective date for the standard.

"(ii) REGIONAL STANDARDS.—Any additional and more restrictive regional standard established for a product under this paragraph shall apply to any such product installed on or after the effective date of the standard in States in which the Secretary has designated the standard to apply.

"(F) CONTINUATION OF REGIONAL STANDARDS.—

"(i) IN GENERAL.—In any subsequent rulemaking for any product for which a regional standard has been previously established, the Secretary shall determine whether to continue the establishment of separate regional standards for the product.

"(ii) REGIONAL STANDARD NO LONGER APPROPRIATE.—Except as provided in clause (iii), if the Secretary determines that regional standards are no longer appropriate for a product, beginning on the effective date of the amended standard for the product—

"(I) there shall be 1 base national standard for the product with Federal enforcement; and

"(II) State authority for enforcing a regional standard for the product shall terminate.

"(iii) REGIONAL STANDARD APPROPRIATE BUT STANDARD OR REGION CHANGED.—

"(I) " STATE NO LONGER CONTAINED IN REGION.—Subject to subclause (III), if a State is no longer contained in a region in which a regional standard that is more stringent than the base national standard applies, the authority of the State to enforce the regional standard shall terminate.

"(II) STANDARD OR REGION REVISED SO THAT EXISTING REGIONAL STANDARD EQUALS BASE NATIONAL STANDARD.—If the Secretary revises a base national standard for a product or the geographic definition of a region so that an existing regional standard for a State is equal to the revised base national standard—

"(aa) the authority of the State to enforce the regional standard shall terminate on the effective date of the revised base national standard; and

"(bb) the State shall be subject to the revised base national standard.

"(I)" STANDARD OR REGION REVISED SO THAT EXISTING REGIONAL STANDARD EQUALS BASE NATIONAL STANDARD.—If the Secretary revises a base national standard for a product or the geographic definition of a region so that the standard for a State is lower than the previously approved regional standard, the State may continue to enforce the previously approved standard level.

"(iv) WAIVER OF FEDERAL PREEMPTION.—Nothing in this paragraph diminishes the authority of a State to enforce a State regulation for which a waiver of Federal preemption has been granted under section 327(d).

"(II) ENFORCEMENT.—

"(i) BASE NATIONAL STANDARD.—

"(I)" IN GENERAL.—The Secretary shall enforce any base national standard.

"(II) TRADE ASSOCIATION CERTIFICATION PROGRAMS.—In enforcing the base national standard, the Secretary shall use, to the maximum extent practicable, national standard nationally recognized certification programs of trade associations.

"(ii) REGIONAL STANDARDS.—

"(I)" ENFORCEMENT PLAN.—Not later than 90 days after the date of the issuance of a final rule that establishes a regional standard, the Secretary shall initiate a rulemaking to develop and implement an effective enforcement plan for regional standards for the products that are covered by the final rule.

"(II) RESPONSIBLE ENTITIES.—Any rules regarding enforcement of a regional standard shall clearly specify which entities are legally responsible for compliance with the standards and for making any required information or labeling disclosures.

"(III) FINAL RULE.—Not later than 15 months after the date of the issuance of a final rule that establishes a regional standard for a product, the Secretary shall promulgate a final rule covering enforcement of regional standards for the product.

"(IV) INCORPORATION BY STATES AND LOCALITIES.—A State or locality may incorporate any Federal regional standard into State or local building codes or State appliance standards.

"(V) STATE ENFORCEMENT.—A State agency may seek enforcement of a Federal regional standard in a Federal court of competent jurisdiction.

"(III) INFORMATION DISCLOSURE.—

"(i) IN GENERAL.—Not later than 90 days after the date of the publication of a final rule that establishes a regional standard for a product, the Federal Trade Commission shall undertake a rulemaking to determine the appropriate 1 or more methods for disclosing information so that consumers, distributors, contractors, and installers can easily determine whether a specific piece of equipment that is installed in a specific building is in conformance with the regional standard that applies to the building.

"(ii) METHODS.—A method of disclosing information under clause (i) may include—

"(I)" modifications to the Energy Guide label; or

"(II) other methods that make it easy for consumers and installers to use and understand at the point of installation.

"(iii) COMPLETION OF RULEMAKING.—The rule-making shall be completed not later 15 months after the date of the publication of a final rule that establishes a regional standard for a product.".

(b) PROHIBITED ACTS.—Section 332(a) of the Energy Policy and Conservation Act (42 U.S.C. 6302(a)) is amended—

(1) in paragraph (4), by striking "or" after the semicolon at the end;

(2) in paragraph (5), by striking "part." and inserting "part, except to the extent that the new covered product is covered by a regional standard that is more stringent than the base national standard; or"; and

(3) by adding at the end the following:

"(6) for any manufacturer or private labeler to knowingly sell a product to a distributor, contractor, or dealer with knowledge that the entity routinely violates any regional standard applicable to the product.".

(c) CONSIDERATION OF PRICES AND OPERATING PATTERNS.—Section 342(a)(6)(B) of the Energy Policy and Conservation Act (42 U.S.C. 6313(a)(6)(B)) is amended by adding at the end the following:

"(iii) CONSIDERATION OF PRICES AND OPERATING PATTERNS.—If the Secretary is considering revised standards for air-cooled 3-phase central air conditioners and central air conditioning heat pumps with less 65,000 Btu per hour (cooling capacity), the Secretary shall use commercial energy prices and operating patterns in all analyses conducted by the Secretary.".

(3) by redesignating paragraphs (2) through (4) as paragraphs (1) through (3), respectively.

Sec. 308. Expedited Rulemakings.

(a) PROCEDURE FOR PRESCRIBING NEW OR AMENDED STANDARDS.—Section 325(p) of the Energy Policy and Conservation Act (42 U.S.C. 6295(p)) (as amended by section 307) is amended by adding at the end the following:

"(4) DIRECT FINAL RULES.—

"(A) IN GENERAL.—On receipt of a statement that is submitted jointly by interested persons that are fairly representative of relevant points of view (including representatives of manufacturers of covered products, States, and efficiency advocates), as determined by the Secretary, and contains recommendations with respect to an energy or water conservation standard—

"(i) if the Secretary determines that the recommended standard contained in the statement is in accordance with subsection (o) or section 342(a)(6)(B), as applicable, the Secretary may issue a final rule that establishes an energy or water conservation standard and is published simultaneously with a notice of proposed rulemaking that proposes a new or amended energy or water conservation standard that is identical to the standard established in the final rule to establish the recommended standard (referred to in this paragraph as a 'direct final rule'); or

"(ii) if the Secretary determines that a direct final rule cannot be issued based on the statement, the Secretary shall publish a notice of the determination, together with an explanation of the reasons for the determination.

"(B) PUBLIC COMMENT.—The Secretary shall solicit public comment for a period of at least 110 days with respect to each direct final rule issued by the Secretary under subparagraph (A)(i).

"(C) WITHDRAWAL OF DIRECT FINAL RULES.—

"(i) IN GENERAL.—Not later than 120 days after the date on which a direct final rule issued under subparagraph (A)(i) is published in the Federal Register, the Secretary shall withdraw the direct final rule if—

"(I) the Secretary receives 1 or more adverse public comments relating to the direct final rule under subparagraph (B)(i) or any alternative joint recommendation; and

"(II) based on the rulemaking record relating to the direct final rule, the Secretary determines that such adverse public comments or alternative joint recommendation may provide a reasonable basis for withdrawing the direct final rule under subsection (o), section 342(a)(6)(B), or any other applicable law.

"(ii) ACTION ON WITHDRAWAL.—On withdrawal of a direct final rule under clause (i), the Secretary shall—

"(I) proceed with the notice of proposed rule- making published simultaneously with the direct final rule as described in subparagraph (A)(i); and

"(II) publish in the Federal Register the reasons why the direct final rule was withdrawn.

"(iii) TREATMENT OF WITHDRAWN DIRECT FINAL RULES.—A direct final rule that is withdrawn under clause (i) shall not be considered to be a final rule for purposes of subsection (o).

"(D) EFFECT OF PARAGRAPH.—Nothing in this paragraph authorizes the Secretary to issue a direct final rule based solely on receipt of more than 1 statement containing recommended standards relating to the direct final rule.".

(b) CONFORMING AMENDMENT.—Section 345(b)(1) of the Energy Policy and Conservation Act (42 U.S.C. 6316(b)(1)) is amended in the first sentence by inserting "section 325(p)(5)," after "The provisions of''.

Sec. 309. Battery Chargers.

Section 325(u)(1)(E) of the Energy Policy and Conservation Act (42 U.S.C. 6295(u)(1)(E)) is amended—

(1) by striking "(E)(i) Not" and inserting the following:

"(E) EXTERNAL POWER SUPPLIES AND BATTERY CHARGERS.—

"(i) ENERGY CONSERVATION STANDARDS.—

"(I) EXTERNAL POWER SUPPLIES.—Not";

(2) by striking "3 years" and inserting "2 years";

(3) by striking "battery chargers and" each place it appears; and

(4) by adding at the end the following:

"(II) BATTERY CHARGERS.—Not later than July 1, 2011, the Secretary shall issue a final rule that prescribes energy conservation standards for battery chargers or classes of battery chargers or determine that no energy conservation standard is technically feasible and economically justified.".

Sec. 310. Standby Mode.

Section 325 of the Energy Policy and Conservation Act (42 U.S.C. 6295) is amended—

(1) in subsection (u)—

(A) by striking paragraphs (2), (3), and (4); and

(B) by redesignating paragraphs (5) and (6) as paragraphs (2) and (3), respectively;

(2) by redesignating subsection (gg) as subsection (hh);

(3) by inserting after subsection (ff) the following:

"(gg) STANDBY MODE ENERGY USE.—

"(1) DEFINITIONS.—

"(A) IN GENERAL.—Unless the Secretary determines otherwise pursuant to subparagraph (B), in this subsection:

"(i) ACTIVE MODE.—The term 'active mode' means the condition in which an energy-using product—

"(I) is connected to a main power source;

"(II) has been activated; and

"(III) provides 1 or more main functions.

"(ii) "(ii) OFF MODE.—The term 'off mode' means the condition in which an energy-using product—

"(I) is connected to a main power source; and

"(II) "(II) is not providing any standby or active mode function.

"(iii) "(iii) STANDBY MODE.—The term 'standby mode' means the condition in which an energy-using product—

"(I)" is connected to a main power source; and

"(II) offers 1 or more of the following user- oriented or protective functions:

"(aa) To facilitate the activation or deactivation of other functions (including active mode) by remote switch (including remote control), internal sensor, or timer.

"(bb) Continuous functions, including information or status displays (including clocks) or sensor-based functions.

"(B) AMENDED DEFINITIONS.—The Secretary may, by rule, amend the definitions under subparagraph (A), taking into consideration the most current versions of Standards 62301 and 62087 of the International Electrotechnical Commission.

"(2) TEST PROCEDURES.—

"(A) IN GENERAL.—Test procedures for all covered products shall be amended pursuant to section 323 to include standby mode and off mode energy consumption, taking into consideration the most current versions of

Standards 62301 and 62087 of the International Electrotechnical Commission, with such energy consumption integrated into the overall energy efficiency, energy consumption, or other energy descriptor for each covered product, unless the Secretary determines that—

"(i) the current test procedures for a covered product already fully account for and incorporate the standby mode and off mode energy consumption of the covered product; or

"(ii) such an integrated test procedure is technically infeasible for a particular covered product, in which case the Secretary shall prescribe a separate standby mode and off mode energy use test procedure for the covered product, if technically feasible.

"(B) DEADLINES.—The test procedure amendments required by subparagraph (A) shall be prescribed in a final rule no later than the following dates:

"(i) December 31, 2008, for battery chargers and external power supplies.

"(ii) March 31, 2009, for clothes dryers, room air conditioners, and fluorescent lamp ballasts.

"(iii) June 30, 2009, for residential clothes washers.

"(iv) September 30, 2009, for residential furnaces and boilers.

"(v) March 31, 2010, for residential water heaters, direct heating equipment, and pool heaters.

"(vi) March 31, 2011, for residential dishwashers, ranges and ovens, microwave ovens, and dehumidifiers.

"(C) PRIOR PRODUCT STANDARDS.—The test procedure amendments adopted pursuant to subparagraph (B) shall not be used to determine compliance with product standards established prior to the adoption of the amended test procedures.

"(3) INCORPORATION INTO STANDARD.—

"(A) IN GENERAL.—Subject to subparagraph (B), based on the test procedures required under paragraph (2), any final rule establishing or revising a standard for a covered product, adopted after July 1, 2010, shall incorporate standby mode and off mode energy use into a single amended or new standard, pursuant to subsection (o), if feasible.

"(B) SEPARATE STANDARDS.—If not feasible, the Secretary shall prescribe within the final rule a separate standard for standby mode and off mode energy consumption, if justified under subsection (o)."; and

(2) in paragraph (2) of subsection (hh) (as redesignated by paragraph (2)), by striking "(ff)" each place it appears and inserting "(gg)".

Sec. 311. Energy Standards For Home Appliances.

(a) APPLIANCES.—

(1) DEHUMIDIFIERS.—Section 325(cc) of the Energy Policy and Conservation Act (42 U.S.C. 6295(cc)) is amended by striking paragraph (2) and inserting the following:

"(9) DEHUMIDIFIERS MANUFACTURED ON OR AFTER OCTOBER 1, 2012.—Dehumidifiers manufactured on or after October 1, 2012, shall have an Energy Factor that meets or exceeds the following values:

"Product Capacity (pints/day):	Minimum Energy Factor (liters/ kWh)
Up to 35.00	1.35
35.01–45.00	1.50
45.01–54.00	1.60
54.01–75.00	1.70
Greater than 75.00	2.5.".

(2) RESIDENTIAL CLOTHES WASHERS AND RESIDENTIAL DISHWASHERS.—Section 325(g) of the Energy Policy and Conservation Act (42 U.S.C. 6295(g)) is amended by adding at the end the following:

"(9) RESIDENTIAL CLOTHES WASHERS MANUFACTURED ON OR AFTER JANUARY 1, 2011.—

"(A) IN GENERAL.—A top-loading or front-loading standard-size residential clothes washer manufactured on or after January 1, 2011, shall have—

"(i) a Modified Energy Factor of at least 1.26; and

"(ii) a water factor of not more than 9.5.

"(B) AMENDMENT OF STANDARDS.—

"(i) IN GENERAL.—Not later than December 31, 2011, the Secretary shall publish a final rule determining whether to amend the standards in effect for clothes washers manufactured on or after January 1, 2015.

"(ii) AMENDED STANDARDS.—The final rule shall contain any amended standards.

"(10) RESIDENTIAL DISHWASHERS MANUFACTURED ON OR AFTER JANUARY 1, 2010.—

"(A) IN GENERAL.—A dishwasher manufactured on or after January 1, 2010, shall—

"(i) for a standard size dishwasher not exceed 355 kWh/year and 6.5 gallons per cycle; and

"(ii) for a compact size dishwasher not exceed 260 kWh/year and 4.5 gallons per cycle.

"(B) AMENDMENT OF STANDARDS.—

"(i) IN GENERAL.—Not later than January 1, 2015, the Secretary shall publish a final rule determining whether to amend the standards for dishwashers manufactured on or after January 1, 2018.

"(ii) AMENDED STANDARDS.—The final rule shall contain any amended standards.".

(3) REFRIGERATORS AND FREEZERS.—Section 325(b) of the Energy Policy and Conservation Act (42 U.S.C. 6295(b)) is amended by adding at the end the following:

"(4) REFRIGERATORS AND FREEZERS MANUFACTURED ON OR AFTER JANUARY 1, 2014.—

"(A) IN GENERAL.—Not later than December 31, 2010, the Secretary shall publish a final rule determining whether to amend the standards in effect for refrigerators, refrigerator-freezers, and freezers manufactured on or after January 1, 2014.

"(B) AMENDED STANDARDS.—The final rule shall contain any amended standards.".

(b) ENERGY STAR.—Section 324A(d)(2) of the Energy Policy and Conservation Act (42 U.S.C. 6294a(d)(2)) is amended by striking "January 1, 2010" and inserting "July 1, 2009".

Sec. 312. Walk-In Coolers And Walk-In Freezers.

(a) DEFINITIONS.—Section 340 of the Energy Policy and Conservation Act (42 U.S.C. 6311) is amended—

(1) in paragraph (1)—

by redesignating subparagraphs (G) through (K) as subparagraphs (H) through (L), respectively; and

by inserting after subparagraph (F) the following: "(G) Walk-in coolers and walk-in freezers.";

(2) by redesignating paragraphs (20) and (21) as paragraphs (21) and (22), respectively; and

(3) by inserting after paragraph (19) the following: "(20) WALK-IN COOLER; WALK-IN FREEZER.—

(A) IN GENERAL.—The terms 'walk-in cooler' and 'walk- in freezer' mean an enclosed storage space refrigerated to temperatures, respectively, above, and at or below 32 degrees Fahrenheit that can be walked into, and has a total chilled storage area of less than 3,000 square feet.

(B) EXCLUSION.—The terms 'walk-in cooler' and 'walk- in freezer' do not include products designed and marketed exclusively for medical, scientific, or research purposes.".

(b) STANDARDS.—Section 342 of the Energy Policy and Conservation Act (42 U.S.C. 6313) is amended by adding at the end the following:

"(f) WALK-IN COOLERS AND WALK-IN FREEZERS.—

"(1) IN GENERAL.—Subject to paragraphs (2) through (5), each walk-in cooler or walk-in freezer manufactured on or after January 1, 2009, shall—

"(A) have automatic door closers that firmly close all walk-in doors that have been closed to within 1 inch of full closure, except that this subparagraph shall not apply to doors wider than 3 feet 9 inches or taller than 7 feet;

"(B) have strip doors, spring hinged doors, or other method of minimizing infiltration when doors are open;

"(C) contain wall, ceiling, and door insulation of at least R–25 for coolers and R–32 for freezers, except that this subparagraph shall not apply to glazed portions of doors nor to structural members;

"(D) contain floor insulation of at least R–28 for freezers;

"(E) for evaporator fan motors of under 1 horsepower and less than 460 volts, use—

"(i) electronically commutated motors (brushless direct current motors); or

"(ii) 3-phase motors;

"(F) for condenser fan motors of under 1 horsepower, use—

"(i) electronically commutated motors;

"(ii) permanent split capacitor-type motors; or

"(iii) 3-phase motors; and

"(G) for all interior lights, use light sources with an efficacy of 40 lumens per watt or more, including ballast losses (if any), except that light sources with an efficacy of 40 lumens per watt or less, including ballast losses (if any), may be used in conjunction with a timer or device that turns off the lights within 15 minutes of when the walk-in cooler or walk-in freezer is not occupied by people.

"(2) ELECTRONICALLY COMMUTATED MOTORS.—

"(A) IN GENERAL.—The requirements of paragraph Effective date. (1)(E)(i) for electronically commutated motors shall take effect January 1, 2009, unless, prior to that date, the Secretary determines that such motors are only available from 1 manufacturer.

"(B) OTHER TYPES OF MOTORS.—In carrying out paragraph (1)(E)(i) and subparagraph (A), the Secretary may allow other types of motors if the Secretary determines that, on average, those other motors use no more energy in evaporator fan applications than electronically commutated motors.

"(C)MAXIMUM ENERGY CONSUMPTION LEVEL.—The Sec Deadline. retary shall establish the maximum energy consumption level under subparagraph (B) not later than January 1, 2010.

"(3) ADDITIONAL SPECIFICATIONS.—Each walk-in cooler or walk-in freezer with transparent reach-in doors manufactured on or after January 1, 2009, shall also meet the following specifications:

"(A) Transparent reach-in doors for walk-in freezers and windows in walk-in freezer doors shall be of triple- pane glass with either heat-reflective treated glass or gas fill.

"(B) Transparent reach-in doors for walk-in coolers and windows in walk-in cooler doors shall be—

"(i) double-pane glass with heat-reflective treated glass and gas fill; or

"(ii) triple-pane glass with either heat-reflective treated glass or gas fill.

"(C) If the appliance has an antisweat heater without antisweat heat controls, the appliance shall have a total door rail, glass, and frame heater power draw of not more than 7.1 watts per square foot of door opening (for freezers) and 3.0 watts per square foot of door opening (for coolers).

"(D) If the appliance has an antisweat heater with antisweat heat controls, and the total door rail, glass, and frame heater power draw is more than 7.1 watts per square foot of door opening (for freezers) and 3.0 watts per square foot of door opening (for coolers), the antisweat heat controls shall reduce the energy use of the antisweat heater in a quantity corresponding to the relative humidity in the air outside the door or to the condensation on the inner glass pane.

"(4) PERFORMANCE-BASED STANDARDS.—

"(A) IN GENERAL.—Not later than January 1, 2012, the Secretary shall publish performance-based standards for walk-in coolers and walk-in freezers that achieve the maximum improvement in energy that the Secretary determines is technologically feasible and economically justified.

"(B)" APPLICATION.—

"(i) IN GENERAL.—Except as provided in clause (ii), the standards shall apply to products described in subparagraph (A) that are manufactured beginning on the date that is 3 years after the final rule is published.

"(ii) DELAYED EFFECTIVE DATE.—If the Secretary determines, by rule, that a 3-year period is inadequate, the Secretary may establish an effective date for products manufactured beginning on the date that is not more than 5 years after the date of publication of a final rule for the products.

"(5) AMENDMENT OF STANDARDS.—

"(A) IN GENERAL.—Not later than January 1, 2020, the Secretary shall publish a final rule to determine if the standards established under paragraph (4) should be amended.

"(B) APPLICATION.—

"(i) IN GENERAL.—Except as provided in clause (ii), the rule shall provide that the standards shall apply to products manufactured beginning on the date that is 3 years after the final rule is published.

"(ii) DELAYED EFFECTIVE DATE.—If the Secretary determines, by rule, that a 3-year period is inadequate, the Secretary may establish an effective date for products manufactured beginning on the date that is not more than 5 years after the date of publication of a final rule for the products.".

(c) TEST PROCEDURES.—Section 343(a) of the Energy Policy and Conservation Act (42 U.S.C. 6314(a)) is amended by adding at the end the following:

"(9) WALK-IN COOLERS AND WALK-IN FREEZERS.—

"(A) IN GENERAL.—For the purpose of test procedures for walk-in coolers and walk-in freezers:

"(i) The R value shall be the 1/K factor multiplied by the thickness of the panel.

"(ii) The K factor shall be based on ASTM test procedure C5 18–2004.

"(iii) For calculating the R value for freezers, the K factor of the foam at 20°F (average foam temperature) shall be used.

"(iv) For calculating the R value for coolers, the K factor of the foam at 55°F (average foam temperature) shall be used.

"(B) TEST PROCEDURE.—

"(i) IN GENERAL.—Not later than January 1, 2010, Deadline. the Secretary shall establish a test procedure to measure the energy-use of walk-in coolers and walk- in freezers.

"(ii) COMPUTER MODELING.—The test procedure may be based on computer modeling, if the computer model or models have been verified using the results of laboratory tests on a significant sample of walk- in coolers and walk-in freezers.".

(d) LABELING.—Section 344(e) of the Energy Policy and Conservation Act (42 U.S.C. 6315(e)) is amended by inserting "walk- in coolers and walk-in freezers," after "commercial clothes washers," each place it appears.

(e) ADMINISTRATION, PENALTIES, ENFORCEMENT, AND PREEMPTION.—Section 345 of the Energy Policy and Conservation Act (42 U.S.C. 6316) is amended—

(1) by striking "subparagraphs (B), (C), (D), (E), and (F)" each place it appears and inserting "subparagraphs (B) through (G)"; and

(2) by adding at the end the following:

"(h) WALK-IN COOLERS AND WALK-IN FREEZERS.—

"(1) COVERED TYPES.—

"(A) RELATIONSHIP TO OTHER LAW.—

"(i) IN GENERAL.—Except as otherwise provided Applicability.

in this subsection, section 327 shall apply to walk- in coolers and walk-in freezers for which standards have been established under paragraphs (1), (2), and (3) of section 342(f) to the same extent and in the same manner as the section applies under part A on the date of enactment of this subsection.

"(ii) STATE STANDARDS.—Any State standard prescribed before the date of enactment of this subsection shall not be preempted until the standards established under paragraphs (1) and (2) of section 342(f) take effect.

"(B) ADMINISTRATION.—In applying section 327 to Applicability. equipment under subparagraph (A), paragraphs (1), (2), and (3) of subsection (a) shall apply.

"(2) FINAL RULE NOT TIMELY.—

"(A) IN GENERAL.—If the Secretary does not issue a final rule for a specific type of walk-in cooler or walk- in freezer within the timeframe established under paragraph (4) or (5) of section 342(f), subsections (b) and (c) of section 327 shall no longer apply to the specific type of walk-in cooler or walk-in freezer during the period—

"(i) beginning on the day after the scheduled date for a final rule; and

"(ii) ending on the date on which the Secretary publishes a final rule covering the specific type of walk-in cooler or walk-in freezer.

"(B) STATE STANDARDS.—Any State standard issued before the publication of the final rule shall not be preempted until the standards established in the final rule take effect.

"(3) CALIFORNIA.—Any standard issued in the State of California before January 1, 2011, under title 20 of the California Code of Regulations, that refers to walk-in coolers and walk- in freezers, for which standards have been established under paragraphs (1), (2), and (3) of section 342(f), shall not be preempted until the standards established under section 342(f)(3) take effect.".

Sec. 313. Electric Motor Efficiency Standards.

(a) DEFINITIONS.—Section 340(13) of the Energy Policy and Conservation Act (42 U.S.C. 63 11(13)) is amended—

(1) by redesignating subparagraphs (B) through (H) as subparagraphs (C) through (I), respectively; and

(2)by striking "(13)(A)" and all that follows through the end of subparagraph (A) and inserting the following:

"(13) ELECTRIC MOTOR.—

(A) GENERAL PURPOSE ELECTRIC MOTOR (SUBTYPE I).—

The term 'general purpose electric motor (subtype I)' means any motor that meets the definition of 'General Purpose' as established in the final rule issued by the Department of Energy entitled 'Energy Efficiency Program for Certain Commercial and Industrial Equipment: Test Procedures, Labeling, and Certification Requirements for Electric Motors' (10 CFR 431), as in effect on the date of enactment of the Energy Independence and Security Act of 2007.

(B) GENERAL PURPOSE ELECTRIC MOTOR (SUBTYPE II).—
The term 'general purpose electric motor (subtype II)' means motors incorporating the design elements of a general purpose electric motor (subtype I) that are configured as 1 of the following:
"(i) A U-Frame Motor.
"(ii) A Design C Motor.
"(iii) A close-coupled pump motor.
"(iv) A Footless motor.
"(v) A vertical solid shaft normal thrust motor (as tested in a horizontal configuration).
"(vi) An 8-pole motor (900 rpm).
"(vii) A poly-phase motor with voltage of not more than 600 volts (other than 230 or 460 volts.".

(b) STANDARDS.—
(1) AMENDMENTS.—Section 342(b) of the Energy Policy and Conservation Act (42 U.S.C. 63 13(b)) is amended—
(A) by redesignating paragraphs (2) and (3) as paragraphs (3) and (4), respectively; and
(B) by inserting after paragraph (1) the following: "(2) ELECTRIC MOTORS.—
"(A) GENERAL PURPOSE ELECTRIC MOTORS (SUBTYPE I).—
Except as provided in subparagraph (B), each general purpose electric motor (subtype I) with a power rating of 1 horsepower or greater, but not greater than 200 horsepower, manufactured (alone or as a component of another piece of equipment) after the 3-year period beginning on the date of enactment of the Energy Independence and Security Act of 2007, shall have a nominal full load efficiency that is not less than as defined in NEMA MG– 1 (2006) Table 12–12.
"(B) FIRE PUMP MOTORS.—Each fire pump motor manufactured (alone or as a component of another piece of equipment) after the 3-year period beginning on the date of enactment of the Energy Independence and Security Act of 2007 shall have nominal full load efficiency that is not less than as defined in NEMA MG–1 (2006) Table 12–11.
"(C) GENERAL PURPOSE ELECTRIC MOTORS (SUBTYPE II).—Each general purpose electric motor (subtype II) with a power rating of 1 horsepower or greater, but not greater than 200 horsepower, manufactured (alone or as a component of another piece of equipment) after the 3-year period beginning on the date of enactment of the Energy Independence and Security Act of 2007, shall have a nominal full load efficiency that is not less than as defined in NEMA MG–1 (2006) Table 12–11.
"(D) NEMA DESIGN B, GENERAL PURPOSE ELECTRIC MOTORS.—Each NEMA Design B, general purpose electric motor with a power rating of more than 200 horsepower, but not greater than 500 horsepower, manufactured (alone or as a component of another piece of equipment) after the 3-year period beginning on the date of enactment of the Energy Independence and Security Act of 2007, shall have a nominal full load efficiency that is not less than as defined in NEMA MG–1 (2006) Table 12–11.".
(1) EFFECTIVE DATE.—The amendments made by paragraph (1) take effect on the date that is 3 years after the date of enactment of this Act.

Sec. 314. Standards For Single Package Vertical Air Conditioners And Heat Pumps.

(a) DEFINITIONS.—Section 340 of the Energy Policy and Conservation Act (42 U.S.C. 6311) is amended by adding at the end the following:

"(22) SINGLE PACKAGE VERTICAL AIR CONDITIONER.—The term 'single package vertical air conditioner' means air-cooled commercial package air conditioning and heating equipment that—

"(A) is factory-assembled as a single package that—

"(i) has major components that are arranged vertically;

"(ii) is an encased combination of cooling and optional heating components; and

"(iii) is intended for exterior mounting on, adjacent interior to, or through an outside wall;

"(B) is powered by a single- or 3-phase current;

"(C) may contain 1 or more separate indoor grilles, outdoor louvers, various ventilation options, indoor free air discharges, ductwork, well plenum, or sleeves; and

"(D) has heating components that may include electrical resistance, steam, hot water, or gas, but may not include reverse cycle refrigeration as a heating means.

"(23) SINGLE PACKAGE VERTICAL HEAT PUMP.—The term 'single package vertical heat pump' means a single package vertical air conditioner that—

"(A) uses reverse cycle refrigeration as its primary heat source; and

"(B) may include secondary supplemental heating by means of electrical resistance, steam, hot water, or gas.".

(b) STANDARDS.—Section 342(a) of the Energy Policy and Conservation Act (42 U.S.C. 63 13(a)) is amended—

(1) in the first sentence of each of paragraphs (1) and (2), by inserting "(including single package vertical air conditioners and single package vertical heat pumps)" after "heating equipment" each place it appears;

(2) in paragraph (1), by striking "but before January 1, 2010,";

(3) in the first sentence of each of paragraphs (7), (8), and (9), by inserting "(other than single package vertical air conditioners and single package vertical heat pumps)" after "heating equipment" each place it appears;

(4) in paragraph (7)—

(A) by striking "manufactured on or after January 1, 2010,";

(B) in each of subparagraphs (A), (B), and (C), by striking "The" and inserting (C) For equipment manufactured on or after January 1, 2010, the"; and by adding at the end the following:

"(D) For equipment manufactured on or after the later of January 1, 2008, or the date that is 180 days after the date of enactment of the Energy Independence and Security Act of 2007—

(i) the minimum seasonal energy efficiency ratio of air-cooled 3-phase electric central air conditioners and central air conditioning heat pumps less than 65,000 Btu per hour (cooling capacity), split systems, shall be 13.0;

(ii) the minimum seasonal energy efficiency ratio of air-cooled 3-phase electric central air conditioners and central air conditioning heat pumps less than 65,000 Btu per hour (cooling capacity), single package, shall be 13.0;

(iii) the minimum heating seasonal performance factor of air-cooled 3-phase electric central air conditioning heat pumps less than 65,000 Btu per hour (cooling capacity), split systems, shall be 7.7; and

(iv) the minimum heating seasonal performance factor of air-cooled 3-phase electric central air conditioning heat pumps less than 65,000 Btu per hour (cooling capacity), single package, shall be 7.7."; and

(5) by adding at the end the following:

"(10) SINGLE PACKAGE VERTICAL AIR CONDITIONERS AND SINGLE PACKAGE VERTICAL HEAT PUMPS.—

"(A) IN GENERAL.—Single package vertical air conditioners and single package vertical heat pumps manufactured on or after January 1, 2010, shall meet the following standards:

(i) The minimum energy efficiency ratio of single package vertical air conditioners less than 65,000 Btu per hour (cooling capacity), single-phase, shall be 9.0.

(ii) The minimum energy efficiency ratio of single package vertical air conditioners less than 65,000 Btu per hour (cooling capacity), 3-phase, shall be 9.0.

(iii) The minimum energy efficiency ratio of single package vertical air conditioners at or above 65,000 Btu per hour (cooling capacity) but less than 135,000 Btu per hour (cooling capacity), shall be 8.9.

(iv) The minimum energy efficiency ratio of single package vertical air conditioners at or above 135,000 Btu per hour (cooling capacity) but less than 240,000 Btu per hour (cooling capacity), shall be 8.6.

(v) The minimum energy efficiency ratio of single package vertical heat pumps less than 65,000 Btu per hour (cooling capacity), single-phase, shall be 9.0 and the minimum coefficient of performance in the heating mode shall be 3.0.

(vi) The minimum energy efficiency ratio of single package vertical heat pumps less than 65,000 Btu per hour (cooling capacity), 3-phase, shall be 9.0 and the minimum coefficient of performance in the heating mode shall be 3.0.

(vii) The minimum energy efficiency ratio of single package vertical heat pumps at or above 65,000 Btu per hour (cooling capacity) but less than 135,000 Btu per hour (cooling capacity), shall be 8.9 and the minimum coefficient of performance in the heating mode shall be 3.0.

(viii) The minimum energy efficiency ratio of single package vertical heat pumps at or above 135,000 Btu per hour (cooling capacity) but less than 240,000 Btu per hour (cooling capacity), shall be 8.6 and the minimum coefficient of performance in the heating mode shall be 2.9.

"(B) REVIEW.—Not later than 3 years after the date of enactment of this paragraph, the Secretary shall review the most recently published ASHRAE/IES Standard 90.1 with respect to single package vertical air conditioners and single package vertical heat pumps in accordance with the procedures established under paragraph (6).".

Sec. 315. Improved Energy Efficiency For Appliances And Buildings In Cold Climates.

(a) RESEARCH.—Section 911(a)(2) of the Energy Policy Act of 2005 (42 U.S.C. 16191(a)(2)) is amended—

(1) in subparagraph (C), by striking "and" at the end;

(2) in subparagraph (D), by striking the period at the end and inserting "; and"; and

(3) by adding at the end the following:

"(E) technologies to improve the energy efficiency of appliances and mechanical systems for buildings in cold climates, including combined heat and power units and increased use of renewable resources, including fuel.".

(b) REBATES.—Section 124 of the Energy Policy Act of 2005

(42 U.S.C. 15821) is amended—

(1) in subsection (b)(1), by inserting ", or products with improved energy efficiency in cold climates," after "residential Energy Star products"; and

(2) in subsection (e), by inserting "or product with improved energy efficiency in a cold climate" after "residential Energy Star product" each place it appears.

Sec. 316. Technical Corrections.

(a) DEFINITION OF F96T12 LAMP.—

(1) IN GENERAL.—Section 135(a)(1)(A)(ii) of the Energy Policy Act of 2005 (Public Law 109–58; 119 Stat. 624) is amended by striking "C78.1–1978 (R1984)" and inserting "C78.3–1978 (R1984)".

(2) EFFECTIVE DATE.—The amendment made by paragraph (1) takes effect on August 8, 2005.

(b) DEFINITION OF FLUORESCENT LAMP.—Section 321(30)(B)(viii) of the Energy Policy and Conservation Act (42 U.S.C. 6291(30)(B)(viii)) is amended by striking "82" and inserting "87".

(c) MERCURY VAPOR LAMP BALLASTS.—

(1) DEFINITIONS.—Section 321 of the Energy Policy and Conservation Act (42 U.S.C. 6291) (as amended by section 301(a)(2)) is amended—

(A) by striking paragraphs (46) through (48) and inserting the following:

"(46) HIGH INTENSITY DISCHARGE LAMP.—

"(A) IN GENERAL.—The term 'high intensity discharge lamp' means an electric-discharge lamp in which—

(i) the light-producing arc is stabilized by the arc tube wall temperature; and

(ii) the arc tube wall loading is in excess of 3 $Watts/cm^2$.

"(B) INCLUSIONS.—The term 'high intensity discharge lamp' includes mercury vapor, metal halide, and high-pressure sodium lamps described in subparagraph (A).

"(47) MERCURY VAPOR LAMP.—

"(A) IN GENERAL.—The term 'mercury vapor lamp' means a high intensity discharge lamp in which the major portion of the light is produced by radiation from mercury typically operating at a partial vapor pressure in excess of 100,000 Pa (approximately 1 atm).

"(B) INCLUSIONS.—The term 'mercury vapor lamp' includes clear, phosphor-coated, and self-ballasted screw base lamps described in subparagraph (A).

"(48) MERCURY VAPOR LAMP BALLAST.—The term 'mercury vapor lamp ballast' means a device that is designed and marketed to start and operate mercury vapor lamps intended for general illumination by providing the necessary voltage and current."; and

(B) by adding at the end the following:

"(53) SPECIALTY APPLICATION MERCURY VAPOR LAMP BALLAST.—The term 'specialty application mercury vapor lamp ballast' means a mercury vapor lamp ballast that—

"(A) is designed and marketed for operation of mercury vapor lamps used in quality inspection, industrial processing, or scientific use, including fluorescent microscopy and ultraviolet curing; and

"(B) in the case of a specialty application mercury vapor lamp ballast, the label of which—

(i) provides that the specialty application mercury vapor lamp ballast is 'For specialty applications only, not for general illumination'; and

(ii) specifies the specific applications for which the ballast is designed.".

(2) STANDARD SETTING AUTHORITY.—Section 325(ee) of the Energy Policy and Conservation Act (42 U.S.C. 6295(ee)) is amended by inserting "(other than specialty application mercury vapor lamp ballasts)" after "ballasts".

(d) ENERGY CONSERVATION STANDARDS.—Section 325 of the Energy Policy and Conservation Act (42 U.S.C. 6295) is amended—

(1) in subsection (v)—

(A) in the subsection heading, by striking "CEILING FANS AND";

(B) by striking paragraph (1); and

(C) by redesignating paragraphs (2) through (4) as paragraphs (1) through (3), respectively; and

(2) in subsection (ff)—

(A) in paragraph (1)(A)—

(i) by striking clause (iii);

(ii) by redesignating clause (iv) as clause (iii); and

(iii) in clause (iii)(II) (as so redesignated), by inserting "fans sold for" before "outdoor"; and

(B) in paragraph (4)(C)—

(i) in the matter preceding clause (i), by striking "subparagraph (B)" and inserting "subparagraph (A)"; and

(ii) by striking clause (ii) and inserting the following:

"(ii) shall be packaged with lamps to fill all sockets.";

(C) in paragraph (6), by redesignating subparagraphs (C) and (D) as clauses (i) and (ii), respectively, of subparagraph (B); and

(D) in paragraph (7), by striking "327" the second place it appears and inserting "324".

Subtitle B—Lighting Energy Efficiency

Sec. 321. Efficient Light Bulbs.

(a) ENERGY EFFICIENCY STANDARDS FOR GENERAL SERVICE INCANDESCENT LAMPS.—

(1) DEFINITION OF GENERAL SERVICE INCANDESCENT LAMP.—

Section 321(30) of the Energy Policy and Conservation Act (42 U.S.C. 6291(30)) is amended—

(A) by striking subparagraph (D) and inserting the following:

"(D) GENERAL SERVICE INCANDESCENT LAMP.—

"(i) IN GENERAL.—The term 'general service incandescent lamp' means a standard incandescent or halogen type lamp that—

(I) is intended for general service applications;

(II) has a medium screw base;

(III) has a lumen range of not less than 310 lumens and not more than 2,600 lumens; and

(IV) is capable of being operated at a voltage range at least partially within 110 and 130 volts.

"(ii) EXCLUSIONS.—The term 'general service incandescent lamp' does not include the following incandescent lamps:

"(I) An appliance lamp.

"(II) A black light lamp.

"(III) A bug lamp.

"(IV) A colored lamp.

"(V) An infrared lamp.

"(VI) A left-hand thread lamp.

"(VII) A marine lamp.

"(VIII) A marine signal service lamp.

"(IX) A mine service lamp.

"(X) A plant light lamp.

"(XI) A reflector lamp.

"(XII) A rough service lamp.

"(XIII) A shatter-resistant lamp (including a shatter-proof lamp and a shatter-protected lamp).

"(XIV) A sign service lamp.

"(XV) A silver bowl lamp.

"(XVI) A showcase lamp.

"(XVII) A 3-way incandescent lamp.

"(XVIII) A traffic signal lamp.

"(XIX) A vibration service lamp.

"(XX) A G shape lamp (as defined in ANSI C78.20–2003 and C79.1–2002 with a diameter of 5 inches or more.

"(XXI) A T shape lamp (as defined in ANSI C78.20–2003 and C79.1–2002) and that uses not more than 40 watts or has a length of more than 10 inches.

"(XXII) A B, BA, CA, F, G16–1/2, G–25, G30, S, or M–14 lamp (as defined in ANSI C79.1–2002 and ANSI C78.20–2003) of 40 watts or less."; and

(B) by adding at the end the following:

"(T) APPLIANCE LAMP.—The term 'appliance lamp' means any lamp that—

is specifically designed to operate in a household appliance, has a maximum wattage of 40 watts, and is sold at retail, including an oven lamp, refrigerator lamp, and vacuum cleaner lamp; and

"(i) is designated and marketed for the intended application, with—

"(I) the designation on the lamp packaging; and

"(II) marketing materials that identify the lamp as being for appliance use.

"(U) CANDELABRA BASE INCANDESCENT LAMP.—The term 'candelabra base incandescent lamp' means a lamp that uses candelabra screw base as described in ANSI C81.61–2006, Specifications for Electric Bases, common designations E11 and E12.

"(V) INTERMEDIATE BASE INCANDESCENT LAMP.—The term 'intermediate base incandescent lamp' means a lamp that uses an intermediate screw base as described in ANSI C81.61–2006, Specifications for Electric Bases, common designation E17.

"(W) MODIFIED SPECTRUM.—The term 'modified spectrum' means, with respect to an incandescent lamp, an incandescent lamp that—

"(i) is not a colored incandescent lamp; and

"(ii) when operated at the rated voltage and wattage of the incandescent lamp—

"(I) has a color point with (x,y) chromaticity coordinates on the Commission Internationale de l'Eclairage (C.I.E.) 1931 chromaticity diagram that lies below the black-body locus; and

"(II) has a color point with (x,y) chromaticity coordinates on the C.I.E. 1931 chromaticity diagram that lies at least 4 MacAdam steps (as referenced in IESNA LM16) distant from the color point of a clear lamp with the same filament and bulb shape, operated at the same rated voltage and wattage.

"(X) ROUGH SERVICE LAMP.—The term 'rough service lamp' means a lamp that—

"(i) has a minimum of 5 supports with filament configurations that are C–7A, C–11, C–17, and C– 22 as listed in Figure 6–12 of the 9th edition of the IESNA Lighting handbook, or similar configurations where lead wires are not counted as supports; and

"(ii) is designated and marketed specifically for 'rough service' applications, with—

"(I) the designation appearing on the lamp packaging; and

"(II) marketing materials that identify the lamp as being for rough service.

"(Y) 3-WAY INCANDESCENT LAMP.—The term '3-way incandescent lamp' includes an incandescent lamp that—

"(i) employs 2 filaments, operated separately and in combination, to provide 3 light levels; and

"(ii) is designated on the lamp packaging and marketing materials as being a 3-way incandescent lamp.

"(Z) SHATTER-RESISTANT LAMP, SHATTER-PROOF LAMP, OR SHATTER-PROTECTED LAMP.—The terms 'shatter-resistant lamp', 'shatter-proof lamp', and 'shatter-protected lamp' mean a lamp that—

"(i) has a coating or equivalent technology that is compliant with NSF/ANSI 51 and is designed to contain the glass if the glass envelope of the lamp is broken; and

"(ii) is designated and marketed for the intended application, with—

"(I) the designation on the lamp packaging;

"(II) marketing materials that identify the lamp as being shatter-resistant, shatter-proof, or shatter-protected.

"(AA) VIBRATION SERVICE LAMP.—The term 'vibration service lamp' means a lamp that—

"(i) has filament configurations that are C–5, C– 7A, or C–9, as listed in Figure 6–12 of the 9th Edition of the IESNA Lighting Handbook or similar configurations;

"(ii) has a maximum wattage of 60 watts;

"(iii) is sold at retail in packages of 2 lamps or less; and

"(iv) is designated and marketed specifically for vibration service or vibration-resistant applications, with—

"(I) the designation appearing on the lamp packaging; and

"(II) marketing materials that identify the lamp as being vibration service only.

"(BB) GENERAL SERVICE LAMP.—

"(i) IN GENERAL.—The term 'general service lamp' includes—

"(I) general service incandescent lamps;

"(II) compact fluorescent lamps;

"(III) general service light-emitting diode (LED or OLED) lamps; and

"(IV) any other lamps that the Secretary determines are used to satisfy lighting applications traditionally served by general service incandescent lamps.

"(ii) EXCLUSIONS.—The term 'general service lamp' does not include—

"(I) any lighting application or bulb shape described in any of subclauses (I) through (XXII) of subparagraph (D)(ii); or

"(II) any general service fluorescent lamp or incandescent reflector lamp.

"(CC) LIGHT-EMITTING DIODE; LED.—

"(i) IN GENERAL.—The terms 'light-emitting diode' and 'LED' means a p-n junction solid state device the radiated output of which is a function of the physical construction, material used, and exciting current of the device.

"(ii) OUTPUT.—The output of a light-emitting diode may be in—

"(I) the infrared region;

"(II) the visible region; or

"(III) the ultraviolet region.

"(DD) ORGANIC LIGHT-EMITTING DIODE; OLED.—The terms 'organic light-emitting diode' and 'OLED' mean a thin-film light-emitting device that typically consists of a series of organic layers between 2 electrical contacts (electrodes).

"(EE) COLORED INCANDESCENT LAMP.—The term 'colored incandescent lamp' means an incandescent lamp designated and marketed as a colored lamp that has—

"(i) a color rendering index of less than 50, as determined according to the test method given in C.I.E. publication 13.3–1995; or

"(ii) a correlated color temperature of less than 2,500K, or greater than 4,600K, where correlated temperature is computed according to the Journal of Optical Society of America, Vol. 58, pages 1528–1595 (1986).".

(2) COVERAGE.—Section 322(a)(14) of the Energy Policy and Conservation Act (42 U.S.C. 6292(a)(14)) is amended by inserting ", general service incandescent lamps," after "fluorescent lamps".

(3) ENERGY CONSERVATION STANDARDS.—Section 325 of the Energy Policy and Conservation Act (42 U.S.C. 6295) is amended—

(A) in subsection (i)—

(i) in the section heading, by inserting ", GENERAL SERVICE INCANDESCENT LAMPS, INTERMEDIATE BASE INCANDESCENT LAMPS, CANDELABRA BASE INCANDESCENT LAMPS," after "FLUORESCENT LAMPS";

(ii) in paragraph (1)—

(I) in subparagraph (A)—

(aa) by inserting ", general service incandescent lamps, intermediate base incandescent lamps, candelabra base incandescent lamps," after "fluorescent lamps";

(bb) by inserting ", new maximum wattage," after "lamp efficacy"; and

(cc) by inserting after the table entitled "INCANDESCENT REFLECTOR LAMPS" the following:

"General Service Incandescent Lamps

Rated Lumen Ranges	Maximum Rate Wattage	Minimum Rate Life-time	Effective Date
1490–2600	72	1,000 hrs	1/1/2012
1050–1489	53	1,000 hrs	1/1/2013
750–1049	43	1,000 hrs	1/1/2014
310–749	29	1,000 hrs	1/1/2014

Modified Spectrum General Service Incandescent Lamps

Rated Lumen Ranges	Maximum Rate Wattage	Minimum Rate Life-time	Effective Date
1118–	72	1,000 hrs	1/1/2012
1950 788–	53	1,000 hrs	1/1/2013
1117 563–	43	1,000 hrs	1/1/2014
787 232–	29	1,000 hrs	1/1/2014";

and

(II) by striking subparagraph (B) and inserting the following:

"(B) APPLICATION.—

"(i) APPLICATION CRITERIA.—This subparagraph applies to each lamp that—

"(I) is intended for a general service or general illumination application (whether incandescent or not);

"(II) has a medium screw base or any other screw base not defined in ANSI C81.61–2006;

"(III) is capable of being operated at a voltage at least partially within the range of 110 to 130 volts; and

"(IV) is manufactured or imported after December 31, 2011.

"(ii) REQUIREMENT.—For purposes of this paragraph, each lamp described in clause (i) shall have a color rendering index that is greater than or equal to—

"(I) 80 for nonmodified spectrum lamps; or

"(II) 75 for modified spectrum lamps.

"(C) CANDELABRA INCANDESCENT LAMPS AND INTERMEDIATE BASE INCANDESCENT LAMPS.—

"(i) CANDELABRA BASE INCANDESCENT LAMPS.—A candelabra base incandescent lamp shall not exceed 60 rated watts.

"(ii) INTERMEDIATE BASE INCANDESCENT LAMPS.— An intermediate base incandescent lamp shall not exceed 40 rated watts.

"(D) EXEMPTIONS.—

"(i) PETITION.—Any person may petition the Secretary for an exemption for a type of general service lamp from the requirements of this subsection.

"(ii) CRITERIA.—The Secretary may grant an exemption under clause (i) only to the extent that the Secretary finds, after a hearing and opportunity for public comment, that it is not technically feasible to serve a specialized lighting application (such as a military, medical, public safety, or certified historic lighting application) using a lamp that meets the requirements of this subsection.

"(iii) ADDITIONAL CRITERION.—To grant an exemption for a product under this subparagraph, the Secretary shall include, as an additional criterion, that the exempted product is unlikely to be used in a general service lighting application.

"(E) EXTENSION OF COVERAGE.—

"(i) PETITION.—Any person may petition the Secretary to establish standards for lamp shapes or bases that are excluded from the definition of general service lamps.

"(ii) INCREASED SALES OF EXEMPTED LAMPS.—The petition shall include evidence that the availability or sales of exempted incandescent lamps have increased significantly since the date on which the standards on general service incandescent lamps were established.

"(iii) CRITERIA.—The Secretary shall grant a petition under clause (i) if the Secretary finds that—

"(I) the petition presents evidence that demonstrates that commercial availability or sales of exempted incandescent lamp types have increased significantly since the standards on general service lamps were established and likely are being widely used in general lighting applications; and

"(II) significant energy savings could be achieved by covering exempted products, as determined by the Secretary based on sales data provided to the Secretary from manufacturers and importers.

"(iv) NO PRESUMPTION.—The grant of a petition under this subparagraph shall create no presumption with respect to the determination of the Secretary with respect to any criteria under a rulemaking conducted under this section.

"(v) EXPEDITED PROCEEDING.—If the Secretary grants a petition for a lamp shape or base under this subparagraph, the Secretary shall—

"(I) conduct a rulemaking to determine standards for the exempted lamp shape or base; and

"(II) complete the rulemaking not later than 18 months after the date on which notice is provided granting the petition.

"(F) DEFINITION OF EFFECTIVE DATE.—In this paragraph, except as otherwise provided in a table contained in subparagraph (A), the term 'effective date' means the last day of the month specified in the table that follows October 24, 1992.";

(iii)in paragraph (5), in the first sentence, by striking "and general service incandescent lamps";

(iv) by redesignating paragraphs (6) and (7) as paragraphs (7) and (8), respectively; and

(iv) by inserting after paragraph (5) the following:

"(6) STANDARDS FOR GENERAL SERVICE LAMPS.—

"(A) RULEMAKING BEFORE JANUARY 1, 2014.—

"(i) IN GENERAL.—Not later than January 1, 2014, the Secretary shall initiate a rulemaking procedure to determine whether—

"(I)vstandards in effect for general service lamps should be amended to establish more stringent standards than the standards specified in paragraph (1)(A); and

"(II) the exemptions for certain incandescent lamps should be maintained or discontinued based, in part, on exempted lamp sales collected by the Secretary from manufacturers.

"(ii) SCOPE.—The rulemaking—

"(I) shall not be limited to incandescent lamp technologies; and

"(II) shall include consideration of a minimum standard of 45 lumens per watt for general service lamps.

"(iii) AMENDED STANDARDS.—If the Secretary determines that the standards in effect for general service incandescent lamps should be amended, the Secretary shall publish a final rule not later than January 1, 2017, with an effective date that is not earlier than 3 years after the date on which the final rule is published.

"(iv) PHASED-IN EFFECTIVE DATES.—The Secretary shall consider phased-in effective dates under this subparagraph after considering—

"(I) the impact of any amendment on manufacturers, retiring and repurposing existing equipment, stranded investments, labor contracts, workers, and raw materials; and

"(II) the time needed to work with retailers and lighting designers to revise sales and marketing strategies.

"(v) BACKSTOP REQUIREMENT.—If the Secretary fails to complete a rulemaking in accordance with clauses (i) through (iv) or if the final rule does not produce savings that are greater than or equal to the savings from a minimum efficacy standard of 45 lumens per watt, effective beginning January 1, 2020, the Secretary shall prohibit the sale of any general service lamp that does not meet a minimum efficacy standard of 45 lumens per watt.

"(vi) STATE PREEMPTION.—Neither section 327(b) nor any other provision of law shall preclude California or Nevada from adopting, effective beginning on or after January 1, 2018—

"(I) a final rule adopted by the Secretary in accordance with clauses (i) through (iv);

"(II) if a final rule described in subclause (I) has not been adopted, the backstop requirement under clause (v); or

"(III) in the case of California, if a final rule described in subclause (I) has not been adopted, any California regulations relating to these covered products adopted pursuant to State statute in effect as of the date of enactment of the Energy Independence and Security Act of 2007.

"(B) RULEMAKING BEFORE JANUARY 1, 2020.—

"(i) IN GENERAL.—Not later than January 1, 2020, the Secretary shall initiate a rulemaking procedure to determine whether—

"(I) standards in effect for general service incandescent lamps should be amended to reflect lumen ranges with more stringent maximum wattage than the standards specified in paragraph (1)(A); and

"(II) the exemptions for certain incandescent lamps should be maintained or discontinued based, in part, on exempted lamp sales data collected by the Secretary from manufacturers.

"(ii) SCOPE.—The rulemaking shall not be limited to incandescent lamp technologies.

"(iii) AMENDED STANDARDS.—If the Secretary determines that the standards in effect for general service incandescent lamps should be amended, the Secretary shall publish a final rule not later than January 1, 2022, with an effective date that is not earlier than 3 years after the date on which the final rule is published.

"(iv) PHASED-IN EFFECTIVE DATES.—The Secretary shall consider phased-in effective dates under this subparagraph after considering—

"(I) the impact of any amendment on manufacturers, retiring and repurposing existing equipment, stranded investments, labor contracts, workers, and raw materials; and

"(II) the time needed to work with retailers and lighting designers to revise sales and marketing strategies."; and

(B) in subsection (l), by adding at the end the following:

"(4) ENERGY EFFICIENCY STANDARDS FOR CERTAIN LAMPS.—

"(A) IN GENERAL.—The Secretary shall prescribe an energy efficiency standard for rough service lamps, vibration service lamps, 3-way incandescent lamps, 2,601–3,300 lumen general service incandescent lamps, and shatter- resistant lamps only in accordance with this paragraph.

"(B) BENCHMARKS.—Not later than 1 year after the date of enactment of this paragraph, the Secretary, in consultation with the National Electrical Manufacturers Association, shall—

"(i) collect actual data for United States unit sales for each of calendar years 1990 through 2006 for each of the 5 types of lamps described in subparagraph (A) to determine the historical growth rate of the type of lamp; and

"(ii) construct a model for each type of lamp based on coincident economic indicators that closely match the historical annual growth rate of the type of lamp to provide a neutral comparison benchmark to model future unit sales after calendar year 2006.

"(C) ACTUAL SALES DATA.—

"(i) IN GENERAL.—Effective for each of calendar years 2010 through 2025, the Secretary, in consultation with the National Electrical Manufacturers Association, shall—

"(I) collect actual United States unit sales data for each of 5 types of lamps described in subparagraph (A); and

"(II) not later than 90 days after the end of each calendar year, compare the lamp sales in that year with the sales predicted by the comparison benchmark for each of the 5 types of lamps described in subparagraph (A).

"(ii) CONTINUATION OF TRACKING.—

"(I) DETERMINATION.—Not later than January 1, 2023, the Secretary shall determine if actual sales data should be tracked for the lamp types described in subparagraph (A) after calendar year 2025.

"(II) CONTINUATION.—If the Secretary finds that the market share of a lamp type described in subparagraph (A) could significantly erode the market share for general service lamps, the Secretary shall continue to track the actual sales data for the lamp type.

"(D) ROUGH SERVICE LAMPS.—

"(i) IN GENERAL.—Effective beginning with the first year that the reported annual sales rate for rough service lamps demonstrates actual unit sales of rough service lamps that achieve levels that are at least 100 percent higher than modeled unit sales for that same year, the Secretary shall—

"(I) not later than 90 days after the end of the previous calendar year, issue a finding that the index has been exceeded; and

"(II) not later than the date that is 1 year after the end of the previous calendar year, complete an accelerated rulemaking to establish an energy conservation standard for rough service lamps.

"(ii) BACKSTOP REQUIREMENT.—If the Secretary fails to complete an accelerated rulemaking in accordance with clause (i)(II), effective beginning 1 year after the date of the issuance of the finding under clause (i)(I), the Secretary shall require rough service lamps to—

"(I) have a shatter-proof coating or equivalent technology that is compliant with NSF/ANSI 51 and is designed to contain the glass if the glass envelope of the lamp is broken and to provide effective containment over the life of the lamp;

"(II) have a maximum 40-watt limitation; and

"(III) be sold at retail only in a package containing 1 lamp.

"(E) VIBRATION SERVICE LAMPS.—

"(i) IN GENERAL.—Effective beginning with the first year that the reported annual sales rate for vibration service lamps demonstrates actual unit sales of vibration service lamps that achieve levels that are at least 100 percent higher than modeled unit sales for that same year, the Secretary shall—

"(I) not later than 90 days after the end of the previous calendar year, issue a finding that the index has been exceeded; and

"(II) not later than the date that is 1 year after the end of the previous calendar year, complete an accelerated rulemaking to establish an energy conservation standard for vibration service lamps.

"(ii) BACKSTOP REQUIREMENT.—If the Secretary fails to complete an accelerated rulemaking in accordance with clause (i)(II), effective beginning 1 year after the date of the issuance of the finding under clause (i)(I), the Secretary shall require vibration service lamps to—

"(I) have a maximum 40-watt limitation; and

"(II) be sold at retail only in a package containing 1 lamp.

"(F) 3-WAY INCANDESCENT LAMPS.—

"(i) IN GENERAL.—Effective beginning with the first year that the reported annual sales rate for 3-way incandescent lamps demonstrates actual unit sales of 3-way incandescent lamps that achieve levels that are at least 100 percent higher than modeled unit sales for that same year, the Secretary shall—

"(I) not later than 90 days after the end of the previous calendar year, issue a finding that the index has been exceeded; and

"(II) not later than the date that is 1 year after the end of the previous calendar year, complete an accelerated rulemaking to establish an energy conservation standard for 3-way incandescent lamps.

"(ii) BACKSTOP REQUIREMENT.—If the Secretary fails to complete an accelerated rulemaking in accordance with clause (i)(II), effective beginning 1 year after the date of issuance of the finding under clause (i)(I), the Secretary shall require that—

"(I) each filament in a 3-way incandescent lamp meet the new maximum wattage requirements for the respective lumen range established under subsection (i)(1)(A); and

"(II) 3-way lamps be sold at retail only in a package containing 1 lamp.

"(G) 2,601–3,300 LUMEN GENERAL SERVICE INCANDESCENT LAMPS.—Effective beginning with the first year that the reported annual sales rate demonstrates actual unit sales of 2,601–3,300 lumen general service incandescent lamps in the lumen range of 2,601 through 3,300 lumens (or, in the case of a modified spectrum, in the lumen range of 1,951 through 2,475 lumens) that achieve levels that are at least 100 percent higher than modeled unit sales for that same year, the Secretary shall impose—

"(i) a maximum 95-watt limitation on general service incandescent lamps in the lumen range of 2,601 through 3,300 lumens; and

"(ii) a requirement that those lamps be sold at retail only in a package containing 1 lamp.

"(H) SHATTER-RESISTANT LAMPS.—

"(i) IN GENERAL.—Effective beginning with the first year that the reported annual sales rate for shatter- resistant lamps demonstrates actual unit sales of shatter-resistant lamps that achieve levels that are at least 100 percent higher than modeled unit sales for that same year, the Secretary shall—

"(I) not later than 90 days after the end of the previous calendar year, issue a finding that the index has been exceeded; and

"(II) not later than the date that is 1 year after the end of the previous calendar year, complete an accelerated rulemaking to establish an energy conservation standard for shatter-resistant lamps.

"(ii) BACKSTOP REQUIREMENT.—If the Secretary fails to complete an accelerated rulemaking in accordance with clause (i)(II), effective beginning 1 year after the date of issuance of the finding under clause (i)(I), the Secretary shall impose—

"(I) a maximum wattage limitation of 40 watts on shatter resistant lamps; and

"(II) a requirement that those lamps be sold at retail only in a package containing 1 lamp.

"(I) RULEMAKINGS BEFORE JANUARY 1, 2025.—

"(i) IN GENERAL.—Except as provided in clause (ii), if the Secretary issues a final rule prior to January 1, 2025, establishing an energy conservation standard for any of the 5 types of lamps for which data collection is required under any of subparagraphs (D) through (G), the requirement to collect and model data for that type of lamp shall terminate unless, as part of the rulemaking, the Secretary determines that continued tracking is necessary.

"(ii) BACKSTOP REQUIREMENT.—If the Secretary imposes a backstop requirement as a result of a failure to complete an accelerated rulemaking in accordance with clause (i)(II) of any of subparagraphs (D) through (G), the requirement to collect and model data for the applicable type of lamp shall continue for an additional 2 years after the effective date of the backstop requirement.".

(b) CONSUMER EDUCATION AND LAMP LABELING.—Section 324(a)(2)(C) of the Energy Policy and Conservation Act (42 U.S.C. 6294(a)(2)(C)) is amended by adding at the end the following:

"(iii) RULEMAKING TO CONSIDER EFFECTIVENESS OF LAMP LABELING.—

"(I) IN GENERAL.—Not later than 1 year after the date of enactment of this clause, the Commission shall initiate a rulemaking to consider—

"(aa) the effectiveness of current lamp labeling for power levels or watts, light output or lumens, and lamp lifetime; and

"(bb) alternative labeling approaches that will help consumers to understand new high- efficiency lamp products and to base the purchase decisions of the consumers on the most appropriate source that meets the requirements of the consumers for lighting level, light quality, lamp lifetime, and total lifecycle cost.

"(I) COMPLETION.—The Commission shall—

"(aa) complete the rulemaking not later than the date that is 30 months after the date of enactment of this clause; and

"(bb) consider reopening the rulemaking not later than 180 days before the effective dates of the standards for general service incandescent lamps established under section 325(i)(1)(A), if the Commission determines that further labeling changes are needed to help consumers understand lamp alternatives.".

(c) MARKET ASSESSMENTS AND CONSUMER AWARENESS PROGRAM.—

(1) IN GENERAL.—In cooperation with the Administrator of the Environmental Protection Agency, the Secretary of Commerce, the Federal Trade Commission, lighting and retail industry associations, energy efficiency organizations, and any other entities that the Secretary of Energy determines to be appropriate, the Secretary of Energy shall—

(A) conduct an annual assessment of the market for general service lamps and compact fluorescent lamps—

(i) to identify trends in the market shares of lamp types, efficiencies, and light output levels purchased by residential and nonresidential consumers; and

(ii) to better understand the degree to which consumer decisionmaking is based on lamp power levels or watts, light output or lumens, lamp lifetime, and other factors, including information required on labels mandated by the Federal Trade Commission;

(B) provide the results of the market assessment to the Federal Trade Commission for consideration in the rulemaking described in section 324(a)(2)(C)(iii) of the Energy Policy and Conservation Act (42 U.S.C. 6294(a)(2)(C)(iii)); and

(C) in cooperation with industry trade associations, lighting industry members, utilities, and other interested parties, carry out a proactive national program of consumer awareness, information, and education that broadly uses the media and other effective communication techniques over an extended period of time to help consumers understand the lamp labels and make energy-efficient lighting choices that meet the needs of consumers.

(2) AUTHORIZATION OF APPROPRIATIONS.—There is authorized to be appropriated to carry out this subsection $10,000,000 for each of fiscal years 2009 through 2012.

(d) GENERAL RULE OF PREEMPTION FOR ENERGY CONSERVATION STANDARDS BEFORE FEDERAL STANDARD BECOMES EFFECTIVE FOR A PRODUCT.—Section 327(b)(1) of the Energy Policy and Conservation Act (42 U.S.C. 6297(b)(1)) is amended—

(1) by inserting "(A)" after "(1)";

(2) by inserting "or" after the semicolon at the end; and

(3) by adding at the end the following:

"(B) in the case of any portion of any regulation that establishes requirements for general service incandescent lamps, intermediate base incandescent lamps, or candelabra base lamps, was enacted or adopted by the State of California or Nevada before December 4, 2007, except that—

"(i) the regulation adopted by the California Energy Commission with an effective date of January 1, 2008, shall only be effective until the effective date of the Federal standard for the applicable lamp category under subparagraphs (A), (B), and (C) of section 325(i)(1);

"(ii) the States of California and Nevada may, at any time, modify or adopt a State standard for general service lamps to conform with Federal standards with effective dates no earlier than 12 months prior to the Federal effective dates prescribed under subparagraphs (A), (B), and (C) of section 325(i)(1), at which time any prior regulations adopted by the State of California or Nevada shall no longer be effective; and

"(iii) all other States may, at any time, modify or adopt a State standard for general service lamps to conform with Federal standards and effective dates.".

(e) PROHIBITED ACTS.—Section 332(a) of the Energy Policy and Conservation Act (42 U.S.C. 6302(a)) is amended—

(1) in paragraph (4), by striking "or" at the end;

(2) in paragraph (5), by striking the period at the end and inserting "; or"; and

(3) by adding at the end the following:

"(6) for any manufacturer, distributor, retailer, or private labeler to distribute in commerce an adapter that—

"(A) is designed to allow an incandescent lamp that does not have a medium screw base to be installed into a fixture or lampholder with a medium screw base socket; and

"(B) is capable of being operated at a voltage range at least partially within 110 and 130 volts.".

(f) ENFORCEMENT.—Section 334 of the Energy Policy and Conservation Act (42 U.S.C. 6304) is amended by inserting after the second sentence the following: "Any such action to restrain any person from distributing in commerce a general service incandescent lamp that does not comply with the applicable standard established under section 325(i) or an adapter prohibited under section 332(a)(6) may also be brought by the attorney general of a State in the name of the State.".

(g) RESEARCH AND DEVELOPMENT PROGRAM.—

(1) IN GENERAL.—The Secretary may carry out a lighting technology research and development program—

(A) to support the research, development, demonstration, and commercial application of lamps and related technologies sold, offered for sale, or otherwise made available in the United States; and

(B) to assist manufacturers of general service lamps in the manufacturing of general service lamps that, at a minimum, achieve the wattage requirements imposed as a result of the amendments made by subsection (a).

(2) AUTHORIZATION OF APPROPRIATIONS.—There are authorized to be appropriated to carry out this subsection $10,000,000 for each of fiscal years 2008 through 2013.

(3) TERMINATION OF AUTHORITY.—The program under this subsection shall terminate on September 30, 2015.

(h) REPORTS TO CONGRESS.—

(1) REPORT ON MERCURY USE AND RELEASE.—Not later than 1 year after the date of enactment of this Act, the Secretary, in cooperation with the Administrator of the Environmental Protection Agency, shall submit to Congress a report describing recommendations relating to the means by which the Federal

Government may reduce or prevent the release of mercury during the manufacture, transportation, storage, or disposal of light bulbs.

(2) REPORT ON RULEMAKING SCHEDULE.—Beginning on July 1, 2013, and semiannually through July 1, 2016, the Secretary shall submit to the Committee on Energy and Commerce of the House of Representatives and the Committee on Energy and Natural Resources of the Senate a report on—

(A) whether the Secretary will meet the deadlines for the rulemakings required under this section;

(B) a description of any impediments to meeting the deadlines; and

(C) a specific plan to remedy any failures, including recommendations for additional legislation or resources.

(3) NATIONAL ACADEMY REVIEW.—

(A) IN GENERAL.—Not later than December 31, 2009, the Secretary shall enter into an arrangement with the National Academy of Sciences to provide a report by December 31, 2013, and an updated report by July 31, 2015. The report should include—

(i) the status of advanced solid state lighting research, development, demonstration and commercialization;

(ii) the impact on the types of lighting available to consumers of an energy conservation standard requiring a minimum of 45 lumens per watt for general service lighting effective in 2020; and

(iii) the time frame for the commercialization of lighting that could replace current incandescent and halogen incandescent lamp technology and any other new technologies developed to meet the minimum standards required under subsection (a)(3) of this section.

(A) REPORTS.—The reports shall be transmitted to the Committee on Energy and Commerce of the House of Representatives and the Committee on Energy and Natural Resources of the Senate.

Sec. 322. Incandescent Reflector Lamp Efficiency Standards.

(a) DEFINITIONS.—Section 321 of the Energy Policy and Conservation Act (42 U.S.C. 6291) (as amended by section 316(c)(1)(D)) is amended—

(1) in paragraph (30)(C)(ii)—

(A) in the matter preceding subclause (I)—

(i) by striking "or similar bulb shapes (excluding ER or BR)" and inserting "ER, BR, BPAR, or similar bulb shapes"; and

(ii) by striking "2.75" and inserting "2.25"; and

(B) by striking "is either—" and all that follows through subclause (II) and inserting "has a rated wattage that is 40 watts or higher"; and

(2) by adding at the end the following:

"(54) BPAR INCANDESCENT REFLECTOR LAMP.—The term 'BPAR incandescent reflector lamp' means a reflector lamp as shown in figure C78.21–278 on page 32 of ANSI C78.21– 2003.

"(55) BR INCANDESCENT REFLECTOR LAMP; BR30; BR40.—

"(A) BR INCANDESCENT REFLECTOR LAMP.—The term 'BR incandescent reflector lamp' means a reflector lamp that has—

"(i) a bulged section below the major diameter of the bulb and above the approximate baseline of the bulb, as shown in figure 1 (RB) on page 7 of ANSI C79.1–1994, incorporated by reference in section 430.22 of title 10, Code of Federal Regulations (as in effect on the date of enactment of this paragraph); and

"(ii) a finished size and shape shown in ANSI C78.21–1989, including the referenced reflective characteristics in part 7 of ANSI C78.21–1989, incorporated by reference in section 430.22 of title 10, Code of Federal Regulations (as in effect on the date of enactment of this paragraph).

"(B) BR30.—The term 'BR30' means a BR incandescent reflector lamp with a diameter of 30/8ths of an inch.

"(C) BR40.—The term 'BR40' means a BR incandescent reflector lamp with a diameter of 40/8ths of an inch.

"(56) ER INCANDESCENT REFLECTOR LAMP; ER30; ER40.—

"(A) ER INCANDESCENT REFLECTOR LAMP.—The term 'ER incandescent reflector lamp' means a reflector lamp that has—

"(i) an elliptical section below the major diameter of the bulb and above the approximate baseline of the bulb, as shown in figure 1 (RE) on page 7 of ANSI C79.1–1994, incorporated by reference in section 430.22 of title 10, Code of Federal Regulations (as in effect on the date of enactment of this paragraph); and

"(ii) a finished size and shape shown in ANSI C78.21–1989, incorporated by reference in section 430.22 of title 10, Code of Federal Regulations (as in effect on the date of enactment of this paragraph).

"(B) ER30.—The term 'ER30' means an ER incandescent reflector lamp with a diameter of 30/8ths of an inch.

"(C) ER40.—The term 'ER40' means an ER incandescent reflector lamp with a diameter of 40/8ths of an inch.

"(57) R20 INCANDESCENT REFLECTOR LAMP.—The term 'R20 incandescent reflector lamp' means a reflector lamp that has a face diameter of approximately 2.5 inches, as shown in figure 1(R) on page 7 of ANSI C79.1–1994.".

(b) STANDARDS FOR FLUORESCENT LAMPS AND INCANDESCENT REFLECTOR LAMPS.—Section 325(i) of the Energy Policy and Conservation Act (42 U.S.C. 6995(i)) is amended by striking paragraph (1) and inserting the following:

"(1) STANDARDS.—

"(A) DEFINITION OF EFFECTIVE DATE.—In this paragraph (other than subparagraph (D)), the term 'effective date' means, with respect to each type of lamp specified in a table contained in subparagraph (B), the last day of the period of months corresponding to that type of lamp (as specified in the table) that follows October 24, 1992.

"(B) MINIMUM STANDARDS.—Each of the following general service fluorescent lamps and incandescent reflector lamps manufactured after the effective date specified in the tables contained in this paragraph shall meet or exceed the following lamp efficacy and CRI standards:

"(C) EXEMPTIONS.—The standards specified in subparagraph (B) shall not apply to the following types of incandescent reflector lamps:

"(i) Lamps rated at 50 watts or less that are ER30, BR30, BR40, or ER40 lamps.

"(ii) Lamps rated at 65 watts that are BR30, BR40, or ER40 lamps.

"(iii) R20 incandescent reflector lamps rated 45 watts or less.

''(D) EFFECTIVE DATES.—

"(i) ER, BR, AND BPAR LAMPS.—The standards specified in subparagraph (B) shall apply with respect to ER incandescent reflector lamps, BR incandescent reflector lamps, BPAR incandescent reflector lamps, and similar bulb shapes on and after January 1, 2008.

"(ii) LAMPS BETWEEN 2.25–2.75 INCHES IN DIAMETER.—The standards specified in subparagraph (B) shall apply with respect to incandescent reflector lamps with a diameter of more than 2.25 inches, but not more than 2.75 inches, on and after the later of January 1, 2008, or the date that is 180 days after the date of enactment of the Energy Independence and Security Act of 2007.".

"Fluorescent Lamps

Lamp Type	Nominal Lamp Wattage	Minimum CRI	Minimum Average Lamp Efficacy (LPW)	Effective Date (Period of Months)
4-foot medium bi-pin	>35 W	69	75.0	36
	≤35 W	45	75.0	36
2-foot U-shaped	>35 W	69	68.0	36
	≤35 W	45	64.0	36
8-foot slimline	65 W	69	80.0	18
	≤65 W	45	80.0	18
8-foot high output	>100 W	69	80.0	18
	≤100 W	45	80.0	18

''Incandescent Reflector Lamps

Nominal Lamp Wattage	Minimum Average Lamp Efficacy (LPW)	Effective Date (Period of Months)
40–50	10.5	36
51–66	11.0	36
67–85	12.5	36
86–115	14.0	36
116–155	14.5	36
156–205	15.0	36

Sec. 323. Public Building Energy Efficient And Renewable Energy Systems.

(a) ESTIMATE OF ENERGY PERFORMANCE IN PROSPECTUS.—Section 3307(b) of title 40, United States Code, is amended—

(1) by striking ''and" at the end of paragraph (5);

(2) by striking the period at the end of paragraph (6) and inserting ''; and"; and

(3) by inserting after paragraph (6) the following:

"(7) with respect to any prospectus for the construction, alteration, or acquisition of any building or space to be leased, an estimate of the future energy performance of the building or space and a specific description of the use of energy efficient and renewable energy systems, including photovoltaic systems, in carrying out the project.".

(b) MINIMUM PERFORMANCE REQUIREMENTS FOR LEASED SPACE.—Section 3307 of such title is amended—

(1) by redesignating subsections (f) and (g) as subsections (g) and (h), respectively; and

(2) by inserting after subsection (e) the following:

"(f) MINIMUM PERFORMANCE REQUIREMENTS FOR LEASED SPACE.—With respect to space to be leased, the Administrator shall include, to the maximum extent practicable, minimum performance requirements requiring energy efficiency and the use of renewable energy.".

(c) USE OF ENERGY EFFICIENT LIGHTING FIXTURES AND BULBS.— (1) IN GENERAL.—Chapter 33 of such title is amended—

"(A) by redesignating sections 3313, 3314, and 3315 as sections 3314, 3315, and 3316, respectively; and

"(B) by inserting after section 3312 the following:

"§ 3313. Use of energy efficient lighting fixtures and bulbs

"(a) CONSTRUCTION, ALTERATION, AND ACQUISITION OF PUBLIC BUILDINGS.—Each public building constructed, altered, or acquired by the Administrator of General Services shall be equipped, to the maximum extent feasible as determined by the Administrator, with lighting fixtures and bulbs that are energy efficient.

"(b) MAINTENANCE OF PUBLIC BUILDINGS.—Each lighting fixture or bulb that is replaced by the Administrator in the normal course of maintenance of public buildings shall be replaced, to the maximum extent feasible, with a lighting fixture or bulb that is energy efficient.

"(c) CONSIDERATIONS.—In making a determination under this section concerning the feasibility of installing a lighting fixture or bulb that is energy efficient, the Administrator shall consider—

(1) the life-cycle cost effectiveness of the fixture or bulb;

(2) the compatibility of the fixture or bulb with existing equipment;

(3) whether use of the fixture or bulb could result in interference with productivity;

(4) the aesthetics relating to use of the fixture or bulb; and

(5) such other factors as the Administrator determines appropriate.

"(d) ENERGY STAR.—A lighting fixture or bulb shall be treated as being energy efficient for purposes of this section if—

"(1) the fixture or bulb is certified under the Energy Star program established by section 324A of the Energy Policy and Conservation Act (42 U.S.C. 6294a);

"(2) in the case of all light-emitting diode (LED) luminaires, lamps, and systems whose efficacy (lumens per watt) and Color Rendering Index (CRI) meet the Department of Energy requirements for minimum luminaire efficacy and CRI for the Energy Star certification, as verified by an independent third-party testing laboratory that the Administrator and the Secretary of Energy determine conducts its tests according to the procedures and recommendations of the Illuminating

Engineering Society of North America, even if the luminaires, lamps, and systems have not received such certification; or

"(3) the Administrator and the Secretary of Energy have otherwise determined that the fixture or bulb is energy efficient.

"(e) ADDITIONAL ENERGY EFFICIENT LIGHTING DESIGNATIONS.— The Administrator of the Environmental Protection Agency and the Secretary of Energy shall give priority to establishing Energy Star performance criteria or Federal Energy Management Program designations for additional lighting product categories that are appropriate for use in public buildings.

"(f) GUIDELINES.—The Administrator shall develop guidelines for the use of energy efficient lighting technologies that contain mercury in child care centers in public buildings.

"(g) APPLICABILITY OF BUY AMERICAN ACT.—Acquisitions carried out pursuant to this section shall be subject to the requirements of the Buy American Act (41 U.S.C. 10c et seq.).

"(h) EFFECTIVE DATE.—The requirements of subsections (a) and (b) shall take effect 1 year after the date of enactment of this subsection.".

(2) CLERICAL AMENDMENT.—The analysis for such chapter is amended by striking the items relating to sections 3313, 3314, and 3315 and inserting the following:

"3313. Use of energy efficient lighting fixtures and bulbs. "3314. Delegation.

"3315. Report to Congress.

"3316. Certain authority not affected.".

(d) EVALUATION FACTOR.—Section 3310 of such title is amended—

"(1) by redesignating paragraphs (3), (4), and (5) as paragraphs (4), (5), and (6), respectively; and

"(2) by inserting after paragraph (2) the following:

"(3) shall include in the solicitation for any lease requiring a prospectus under section 3307 an evaluation factor considering the extent to which the offeror will promote energy efficiency and the use of renewable energy;".

Sec. 324. Metal Halide Lamp Fixtures.

(a) DEFINITIONS.—Section 321 of the Energy Policy and Conservation Act (42 U.S.C. 6291) (as amended by section 322(a)(2)) is amended by adding at the end the following:

"(58) BALLAST.—The term 'ballast' means a device used with an electric discharge lamp to obtain necessary circuit conditions (voltage, current, and waveform) for starting and operating.

"(59) BALLAST EFFICIENCY.—

"(A) IN GENERAL.—The term 'ballast efficiency' means, in the case of a high intensity discharge fixture, the efficiency of a lamp and ballast combination, expressed as a percentage, and calculated in accordance with the following formula: Efficiency = Pout/Pin.

"(B) EFFICIENCY FORMULA.—For the purpose of subparagraph (A)—

"(i) Pout shall equal the measured operating lamp wattage;

"(ii) $_{\text{Pin}}$ shall equal the measured operating input wattage;

"(iii) the lamp, and the capacitor when the capacitor is provided, shall constitute a nominal system in accordance with the ANSI Standard C78.43–2004;

"(iv) for ballasts with a frequency of 60 Hz, Pin and $_{Pout}$ shall be measured after lamps have been stabilized according to section 4.4 of ANSI Standard C82.6–2005 using a wattmeter with accuracy specified in section 4.5 of ANSI Standard C82.6–2005; and

"(v) for ballasts with a frequency greater than 60 Hz, $_{Pin}$ ^{and} $_{Pout}$ shall have a basic accuracy of ± 0.5 percent at the higher of—

"(I) 3 times the output operating frequency of the ballast; or

"(II) 2 kHz for ballast with a frequency greater than 60 Hz.

"(C) MODIFICATION.—The Secretary may, by rule, modify the definition of 'ballast efficiency' if the Secretary determines that the modification is necessary or appropriate to carry out the purposes of this Act.

"(60) ELECTRONIC BALLAST.—The term 'electronic ballast' means a device that uses semiconductors as the primary means to control lamp starting and operation.

"(61) GENERAL LIGHTING APPLICATION.—The term 'general lighting application' means lighting that provides an interior or exterior area with overall illumination.

"(62) METAL HALIDE BALLAST.—The term 'metal halide ballast' means a ballast used to start and operate metal halide lamps.

"(63) METAL HALIDE LAMP.—The term 'metal halide lamp' means a high intensity discharge lamp in which the major portion of the light is produced by radiation of metal halides and their products of dissociation, possibly in combination with metallic vapors.

"(64) METAL HALIDE LAMP FIXTURE.—The term 'metal halide lamp fixture' means a light fixture for general lighting application designed to be operated with a metal halide lamp and a ballast for a metal halide lamp.

"(65) PROBE-START METAL HALIDE BALLAST.—The term 'probe-start metal halide ballast' means a ballast that—

"(A) starts a probe-start metal halide lamp that contains a third starting electrode (probe) in the arc tube; and

"(B) does not generally contain an igniter but instead starts lamps with high ballast open circuit voltage.

"(66) PULSE-START METAL HALIDE BALLAST.—

"(A) IN GENERAL.—The term 'pulse-start metal halide ballast' means an electronic or electromagnetic ballast that starts a pulse-start metal halide lamp with high voltage pulses.

"(B) STARTING PROCESS.—For the purpose of subparagraph (A)—

"(i) lamps shall be started by first providing a high voltage pulse for ionization of the gas to produce a glow discharge; and

"(ii) to complete the starting process, power shall be provided by the ballast to sustain the discharge through the glow-to-arc transition.".

(b) COVERAGE.—Section 322(a) of the Energy Policy and Conservation Act (42 U.S.C. 6292(a)) is amended—

(1) by redesignating paragraph (19) as paragraph (20); and

(2) by inserting after paragraph (18) the following: "(19) Metal halide lamp fixtures.".

(c) TEST PROCEDURES.—Section 323(b) of the Energy Policy and Conservation Act (42 U.S.C. 6293(b)) (as amended by section 30 1(b)) is amended by adding at the end the following:

"(18) METAL HALIDE LAMP BALLASTS.—Test procedures for metal halide lamp ballasts shall be based on ANSI Standard C82.6–2005, entitled 'Ballasts for High Intensity Discharge Lamps—Method of Measurement'.".

(d) LABELING.—Section 324(a)(2) of the Energy Policy and Conservation Act (42 U.S.C. 6294(a)(2)) is amended—

(1) by redesignating subparagraphs (C) through (G) as subparagraphs (D) through (H), respectively; and

(2) by inserting after subparagraph (B) the following:

"(C) METAL HALIDE LAMP FIXTURES.—

"(i) IN GENERAL.—The Commission shall issue Regulations. labeling rules under this section applicable to the covered product specified in section 322(a)(19) and to which standards are applicable under section 325.

"(ii) LABELING.—The rules shall provide that the Deadlines. labeling of any metal halide lamp fixture manufactured on or after the later of January 1, 2009, or the date that is 270 days after the date of enactment of this subparagraph, shall indicate conspicuously, in a manner prescribed by the Commission under subsection (b) by July 1, 2008, a capital letter 'E' printed within a circle on the packaging of the fixture, and on the ballast contained in the fixture.".

(e) STANDARDS.—Section 325 of the Energy Policy and Conservation Act (42 U.S.C. 6295) (as amended by section 310) is amended—

(1) by redesignating subsection (hh) as subsection (ii);

(2) by inserting after subsection (gg) the following: "(hh) METAL HALIDE LAMP FIXTURES.—

"(1) STANDARDS.—

"(A) IN GENERAL.—Subject to subparagraphs (B) and (C), metal halide lamp fixtures designed to be operated with lamps rated greater than or equal to 150 watts but less than or equal to 500 watts shall contain—

"(i) a pulse-start metal halide ballast with a minimum ballast efficiency of 88 percent;

"(ii) a magnetic probe-start ballast with a minimum ballast efficiency of 94 percent; or

"(iii) a nonpulse-start electronic ballast with—

"(I) a minimum ballast efficiency of 92 percent for wattages greater than 250 watts; and

"(II) a minimum ballast efficiency of 90 percent for wattages less than or equal to 250 watts.

"(B) EXCLUSIONS.—The standards established under subparagraph (A) shall not apply to—

"(i) fixtures with regulated lag ballasts;

"(ii) fixtures that use electronic ballasts that operate at 480 volts; or

"(iii) fixtures that—

"(I) are rated only for 150 watt lamps;

"(II) are rated for use in wet locations, as specified by the National Electrical Code 2002, section 4 10.4(A); and

"(III) contain a ballast that is rated to operate at ambient air temperatures above 50°C, as specified by UL 1029–2001.

"(C) APPLICATION.—The standards established under subparagraph (A) shall apply to metal halide lamp fixtures manufactured on or after the later of—

"(i) January 1, 2009; or

"(ii) the date that is 270 days after the date of enactment of this subsection.

"(2) FINAL RULE BY JANUARY 1, 2012.—

"(A) IN GENERAL.—Not later than January 1, 2012, the Secretary shall publish a final rule to determine whether the standards established under paragraph (1) should be amended.

"(B) ADMINISTRATION.—The final rule shall—

"(i) contain any amended standard; and

"(ii) apply to products manufactured on or after January 1, 2015.

"(3) FINAL RULE BY JANUARY 1, 2019.—

"(A) IN GENERAL.—Not later than January 1, 2019, the Secretary shall publish a final rule to determine whether the standards then in effect should be amended. Applicability.

"(B) ADMINISTRATION.—The final rule shall—

"(i) contain any amended standards; and

"(ii) apply to products manufactured after January 1, 2022.

"(4) DESIGN AND PERFORMANCE REQUIREMENTS.—Notwithstanding any other provision of law, any standard established pursuant to this subsection may contain both design and performance requirements."; and

(3) in paragraph (2) of subsection (ii) (as redesignated by paragraph (2)), by striking "(gg)" each place it appears and inserting "(hh)".

(f) EFFECT ON OTHER LAW.—Section 327(c) of the Energy Policy and Conservation Act (42 U.S.C. 6297(c)) is amended—

(1) in paragraph (8)(B), by striking the period at the end and inserting "; and"; and

(2) by adding at the end the following:

"(9) is a regulation concerning metal halide lamp fixtures adopted by the California Energy Commission on or before January 1, 2011, except that—

"(A) if the Secretary fails to issue a final rule within 180 days after the deadlines for rulemakings in section 325(hh), notwithstanding any other provision of this section, preemption shall not apply to a regulation concerning metal halide lamp fixtures adopted by the California Energy Commission—

"(i) on or before July 1, 2015, if the Secretary fails to meet the deadline specified in section 325(hh)(2); or

"(ii) on or before July 1, 2022, if the Secretary fails to meet the deadline specified in section 325(hh)(3).".

Sec. 325. Energy Efficiency Labeling For Consumer Electronic Products.

(a) IN GENERAL.—Section 324(a) of the Energy Policy and Conservation Act (42 U.S.C. 6294(a)) (as amended by section 324(d)) is amended—

(1) in paragraph (2), by adding at the end the following: "(I) LABELING REQUIREMENTS.—

"(i) IN GENERAL.—Subject to clauses (ii) through Deadline.

(iv), not later than 18 months after the date of issuance of applicable Department of Energy testing procedures, the Commission, in consultation with the Secretary and the Administrator of the Environmental Protection Agency (acting through the Energy Star program), shall, by regulation, prescribe labeling or other disclosure requirements for the energy use of—

"(I) televisions;

"(II) personal computers;

"(III) cable or satellite set-top boxes;

"(IV) stand-alone digital video recorder boxes; and

"(V) personal computer monitors.

"(ii) ALTERNATE TESTING PROCEDURES.—In the absence of applicable testing procedures described in clause (i) for products described in subclauses (I) through (V) of that clause, the Commission may, by regulation, prescribe labeling or other disclosure requirements for a consumer product category described in clause (i) if the Commission—

(I) identifies adequate non-Department of Energy testing procedures for those products; and

(II) determines that labeling of, or other disclosures relating to, those products is likely to assist consumers in making purchasing decisions.

"(iii) DEADLINE AND REQUIREMENTS FOR LABELING.—

"(I) DEADLINE.—Not later than 18 months after the date of promulgation of any requirements under clause (i) or (ii), the Commission shall require labeling of, or other disclosure requirements for, electronic products described in clause (i).

"(II) REQUIREMENTS.—The requirements prescribed under clause (i) or (ii) may include specific requirements for each electronic product to be labeled with respect to the placement, size, and content of Energy Guide labels.

"(iv) DETERMINATION OF FEASIBILITY.—Clause (i) or (ii) shall not apply in any case in which the Commission determines that labeling in accordance with this subsection—

"(I) is not technologically or economically feasible; or

"(II) is not likely to assist consumers in making purchasing decisions."; and

(2) by adding at the end the following:

"(6) AUTHORITY TO INCLUDE ADDITIONAL PRODUCT CATEGORIES.—The Commission may, by regulation, require labeling or other disclosures in accordance with this subsection for any consumer product not specified in this subsection or section 322 if the Commission determines that labeling for the product is likely to assist consumers in making purchasing decisions.". (b) CONTENT OF LABEL.—Section 324(c) of the Energy Policy and Conservation Act (42 U.S.C. 6924(c)) is amended by adding at the end the following:

"(9) DISCRETIONARY APPLICATION.—The Commission may apply paragraphs (1), (2), (3), (5), and (6) of this subsection to the labeling of any product covered by paragraph (2)(I) or (6) of subsection (a).".

TITLE IV—ENERGY SAVINGS IN BUILDINGS AND INDUSTRY

Sec. 401. Definitions.

In this title:

(1) ADMINISTRATOR.—The term "Administrator" means the Administrator of General Services.

(2) ADVISORY COMMITTEE.—The term "Advisory Committee" means the Green Building Advisory Committee established under section 484.

(3) COMMERCIAL DIRECTOR.—The term "Commercial Director" means the individual appointed to the position established under section 421.

(4) CONSORTIUM.—The term "Consortium" means the High- Performance Green Building Partnership Consortium created in response to section 436(c)(1) to represent the private sector in a public-private partnership to promote high-performance green buildings and zero-net-energy commercial buildings.

(5) COST-EFFECTIVE LIGHTING TECHNOLOGY.—

(A) IN GENERAL.—The term "cost-effective lighting technology" means a lighting technology that—

(i) will result in substantial operational cost savings by ensuring an installed consumption of not more than 1 watt per square foot; or

(ii) is contained in a list under—

(I) section 553 of Public Law 95–619 (42 U.S.C. 8259b);

(II) Federal acquisition regulation 23–203; and

(III) is at least as energy-conserving as required by other provisions of this Act, including the requirements of this title and title III which shall be applicable to the extent that they would achieve greater energy savings than provided under clause (i) or this clause.

(B) INCLUSIONS.—The term "cost-effective lighting technology" includes—

(i) lamps;

(ii) ballasts;

(iii) luminaires;

(iv) lighting controls;

(v) daylighting; and

(vi) early use of other highly cost-effective lighting technologies.

(6) COST-EFFECTIVE TECHNOLOGIES AND PRACTICES.—The term "cost-effective technologies and practices" means a technology or practice that—

(A) will result in substantial operational cost savings by reducing electricity or fossil fuel consumption, water, or other utility costs, including use of geothermal heat pumps;

(B) complies with the provisions of section 553 of Public Law 95–619 (42 U.S.C. 8259b) and Federal acquisition regulation 23–203; and

(C) is at least as energy and water conserving as required under this title, including sections 431 through 435, and title V, including sections 511 through 525, which shall be applicable to the extent that they are more stringent or require greater energy or water savings than required by this section.

(7) FEDERAL DIRECTOR.—The term "Federal Director" means the individual appointed to the position established under section 436(a).

(8) FEDERAL FACILITY.—The term "Federal facility" means any building that is constructed, renovated, leased, or purchased in part or in whole for use by the Federal Government.

(9) OPERATIONAL COST SAVINGS.—

(A) IN GENERAL.—The term "operational cost savings" means a reduction in end-use operational costs through the application of cost-effective technologies and practices or geothermal heat pumps, including a reduction in electricity consumption relative to consumption by the same customer or at the same facility in a given year, as defined in guidelines promulgated by the Administrator pursuant to section 329(b) of the Clean Air Act, that achieves cost savings sufficient to pay the incremental additional costs of using cost-effective technologies and practices including geothermal heat pumps by not later than the later of the date established under sections 431 through 434, or—

(i) for cost-effective technologies and practices, the date that is 5 years after the date of installation; and

(ii) for geothermal heat pumps, as soon as practical after the date of installation of the applicable geothermal heat pump.

(B) INCLUSIONS.—The term "operational cost savings" includes savings achieved at a facility as a result of—

(i) the installation or use of cost-effective technologies and practices; or

(ii) the planting of vegetation that shades the facility and reduces the heating, cooling, or lighting needs of the facility.

(C) EXCLUSION.—The term "operational cost savings" does not include savings from measures that would likely be adopted in the absence of cost-effective technology and practices programs, as determined by the Administrator.

(10) GEOTHERMAL HEAT PUMP.—The term "geothermal heat pump" means any heating or air conditioning technology that—

(A) uses the ground or ground water as a thermal energy source to heat, or as a thermal energy sink to cool, a building; and

(B) meets the requirements of the Energy Star program of the Environmental Protection Agency applicable to geothermal heat pumps on the date of purchase of the technology.

(11) GSA FACILITY.—

(A) IN GENERAL.—The term "GSA facility" means any building, structure, or facility, in whole or in part (including the associated support systems of the building, structure, or facility) that—

(i) is constructed (including facilities constructed for lease), renovated, or purchased, in whole or in part, by the Administrator for use by the Federal Government; or

(ii) is leased, in whole or in part, by the Administrator for use by the Federal Government—

(I) except as provided in subclause (II), for a term of not less than 5 years; or

(II) for a term of less than 5 years, if the Administrator determines that use of cost-effective technologies and practices would result in the payback of expenses.

(B) INCLUSION.—The term ''GSA facility'' includes any group of buildings, structures, or facilities described in subparagraph (A) (including the associated energy-consuming support systems of the buildings, structures, and facilities).

(C) EXEMPTION.—The Administrator may exempt from the definition of ''GSA facility'' under this paragraph a building, structure, or facility that meets the requirements of section 543(c) of Public Law 95–619 (42 U.S.C. 8253(c)).

(12) HIGH-PERFORMANCE BUILDING.—The term ''high- performance building'' means a building that integrates and optimizes on a life cycle basis all major high performance attributes, including energy conservation, environment, safety, security, durability, accessibility, cost-benefit, productivity, sustainability, functionality, and operational considerations.

(13) HIGH-PERFORMANCE GREEN BUILDING.—The term ''high-performance green building'' means a high-performance building that, during its life-cycle, as compared with similar buildings (as measured by Commercial Buildings Energy Consumption Survey or Residential Energy Consumption Survey data from the Energy Information Agency)—

(A) reduces energy, water, and material resource use;

(B) improves indoor environmental quality, including reducing indoor pollution, improving thermal comfort, and improving lighting and acoustic environments that affect occupant health and productivity;

(C) reduces negative impacts on the environment throughout the life-cycle of the building, including air and water pollution and waste generation;

(D) increases the use of environmentally preferable products, including biobased, recycled content, and nontoxic products with lower life-cycle impacts;

(E) increases reuse and recycling opportunities;

(F) integrates systems in the building;

(G) reduces the environmental and energy impacts of transportation through building location and site design that support a full range of transportation choices for users of the building; and

(H) considers indoor and outdoor effects of the building on human health and the environment, including—

(i) improvements in worker productivity;

(ii) the life-cycle impacts of building materials and operations; and

(iii) other factors that the Federal Director or the Commercial Director consider to be appropriate.

(14) LIFE-CYCLE.—The term "life-cycle", with respect to a high-performance green building, means all stages of the useful life of the building (including components, equipment, systems, and controls of the building) beginning at conception of a high- performance green building project and continuing through site selection, design, construction, landscaping, commissioning, operation, maintenance, renovation, deconstruction or demolition, removal, and recycling of the high-performance green building.

(15) LIFE-CYCLE ASSESSMENT.—The term "life-cycle assessment" means a comprehensive system approach for measuring the environmental performance of

a product or service over the life of the product or service, beginning at raw materials acquisition and continuing through manufacturing, transportation, installation, use, reuse, and end-of-life waste management.

(16) LIFE-CYCLE COSTING.—The term “life-cycle costing”, with respect to a high-performance green building, means a technique of economic evaluation that—

(A) sums, over a given study period, the costs of initial investment (less resale value), replacements, operations (including energy use), and maintenance and repair of an investment decision; and

(B) is expressed—

(i) in present value terms, in the case of a study period equivalent to the longest useful life of the building, determined by taking into consideration the typical life of such a building in the area in which the building is to be located; or

(ii) in annual value terms, in the case of any other study period.

(17) OFFICE OF COMMERCIAL HIGH-PERFORMANCE GREEN BUILDINGS.—The term “Office of Commercial High-Performance Green Buildings” means the Office of Commercial High- Performance Green Buildings established under section 421(a).

(18) OFFICE OF FEDERAL HIGH-PERFORMANCE GREEN BUILDINGS.—The term “Office of Federal High-Performance Green Buildings” means the Office of Federal High-Performance Green Buildings established under section 436(a).

(19) PRACTICES.—The term “practices” means design, financing, permitting, construction, commissioning, operation and maintenance, and other practices that contribute to achieving zero-net-energy buildings or facilities.

(20) ZERO-NET-ENERGY COMMERCIAL BUILDING.—The term “zero-net-energy commercial building” means a commercial building that is designed, constructed, and operated to—

(A) require a greatly reduced quantity of energy to operate;

(B) meet the balance of energy needs from sources of energy that do not produce greenhouse gases;

(C) therefore result in no net emissions of greenhouse gases; and

(D) be economically viable.

Subtitle A—Residential Building Efficiency

Sec. 411. Reauthorization Of Weatherization Assistance Program.

(a) IN GENERAL.—Section 422 of the Energy Conservation and Production Act (42 U.S.C. 6872) is amended by striking “appropriated $500,000,000 for fiscal year 2006, $600,000,000 for fiscal year 2007, and $700,000,000 for fiscal year 2008” and inserting “appropriated—

“(1) $750,000,000 for fiscal year 2008;

“(2) $900,000,000 for fiscal year 2009;

“(3) $1,050,000,000 for fiscal year 2010;

“(4) $1,200,000,000 for fiscal year 2011; and

“(5) $1,400,000,000 for fiscal year 2012.”.

(b) SUSTAINABLE ENERGY RESOURCES FOR CONSUMERS GRANTS.—

(1) IN GENERAL.—The Secretary may make funding available to local weatherization agencies from amounts authorized under the amendment made by subsection (a) to expand the weatherization assistance program for residential buildings to include materials, benefits, and renewable and domestic energy technologies not covered by the program (as of the date of enactment of this Act), if the State weatherization grantee certifies that the applicant has the capacity to carry out the proposed activities and that the grantee will include the project in the financial oversight of the grantee of the weatherization assistance program.

(2) PRIORITY.—In selecting grant recipients under this subsection, the Secretary shall give priority to—

(A) the expected effectiveness and benefits of the proposed project to low- and moderate-income energy consumers;

(B) the potential for replication of successful results;

(C) the impact on the health and safety and energy costs of consumers served; and

(D) the extent of partnerships with other public and private entities that contribute to the resources and implementation of the program, including financial partnerships.

(3) FUNDING.—

(A) IN GENERAL.—Except as provided in paragraph (2), the amount of funds used for projects described in paragraph (1) may equal up to 2 percent of the amount of funds made available for any fiscal year under section 422 of the Energy Conservation and Production Act (42 U.S.C. 6872).

(B) EXCEPTION.—No funds may be used for sustainable energy resources for consumers grants for a fiscal year under this subsection if the amount of funds made available for the fiscal year to carry out the Weatherization Assistance Program for Low-Income Persons established under part A of title IV of the Energy Conservation and Production Act (42 U.S.C. 6861 et seq.) is less than $275,000,000.

(c) DEFINITION OF STATE.—Section 412 of the Energy Conservation and Production Act (42 U.S.C. 6862) is amended by striking paragraph (8) and inserting the following:

"(8) STATE.—The term 'State' means—

"(A) a State;

"(B) the District of Columbia;

"(C) the Commonwealth of Puerto Rico; and

"(D) any other territory or possession of the United States.".

Sec. 412. Study Of Renewable Energy Rebate Programs.

(a) IN GENERAL.—Not later than 120 days after the date of enactment of this Act, the Secretary shall conduct, and submit to Congress a report on, a study regarding the rebate programs established under sections 124 and 206(c) of the Energy Policy Act of 2005 (42 U.S.C. 15821, 15853).

(b) COMPONENTS.—In conducting the study, the Secretary shall—

(1) develop a plan for how the rebate programs would be carried out if the programs were funded; and

(2) determine the minimum amount of funding the program would need to receive in order to accomplish the goals of the programs.

Sec. 413. Energy Code Improvements Applicable To Manufactured Housing.

(a) ESTABLISHMENT OF STANDARDS.—

(1) IN GENERAL.—Not later than 4 years after the date of enactment of this Act, the Secretary shall by regulation establish standards for energy efficiency in manufactured housing.

(2) NOTICE, COMMENT, AND CONSULTATION.—Standards described in paragraph (1) shall be established after—

(A) notice and an opportunity for comment by manufacturers of manufactured housing and other interested parties; and

(B) consultation with the Secretary of Housing and Urban Development, who may seek further counsel from the Manufactured Housing Consensus Committee.

(b) REQUIREMENTS.—

(1) INTERNATIONAL ENERGY CONSERVATION CODE.—The energy conservation standards established under this section shall be based on the most recent version of the International Energy Conservation Code (including supplements), except in cases in which the Secretary finds that the code is not cost- effective, or a more stringent standard would be more cost- effective, based on the impact of the code on the purchase price of manufactured housing and on total life-cycle construction and operating costs.

(2) CONSIDERATIONS.—The energy conservation standards established under this section may—

(A) take into consideration the design and factory construction techniques of manufactured homes;

(B) be based on the climate zones established by the Department of Housing and Urban Development rather than the climate zones under the International Energy Conservation Code; and provide for alternative practices that result in net estimated energy consumption equal to or less than the specified standards.

(3) UPDATING.—The energy conservation standards established under this section shall be updated not later than—

(A) 1 year after the date of enactment of this Act; and

(B) 1 year after any revision to the International Energy Conservation Code.

(a) ENFORCEMENT.—Any manufacturer of manufactured housing that violates a provision of the regulations under subsection (a) is liable to the United States for a civil penalty in an amount not exceeding 1 percent of the manufacturer's retail list price of the manufactured housing.

Subtitle B—High-Performance Commercial Buildings

Sec. 421. Commercial High-Performance Green Buildings.

(a) DIRECTOR OF COMMERCIAL HIGH-PERFORMANCE GREEN BUILDINGS.— Notwithstanding any other provision of law, the Secretary, acting through the Assistant Secretary of Energy Efficiency and Renewable Energy, shall appoint a Director of Commercial High-Performance Green Buildings to a position in the career- reserved Senior Executive service, with the principal responsibility to—

(1) establish and manage the Office of Commercial High- Performance Green Buildings; and

(2) carry out other duties as required under this subtitle.

(b) QUALIFICATIONS.—The Commercial Director shall be an individual, who by reason of professional background and experience, is specifically qualified to carry out the duties required under this subtitle.

(c) DUTIES.—The Commercial Director shall, with respect to development of high-performance green buildings and zero-energy commercial buildings nationwide—

(1) coordinate the activities of the Office of Commercial High-Performance Green Buildings with the activities of the Office of Federal High-Performance Green Buildings;

(2) develop the legal predicates and agreements for, negotiate, and establish one or more public-private partnerships with the Consortium, members of the Consortium, and other capable parties meeting the qualifications of the Consortium, to further such development;

(3) represent the public and the Department in negotiating and performing in accord with such public-private partnerships;

(4) use appropriated funds in an effective manner to encourage the maximum investment of private funds to achieve such development;

(5) promote research and development of high-performance green buildings, consistent with section 423; and

(6) jointly establish with the Federal Director a national high-performance green building clearinghouse in accordance with section 423(1), which shall provide high-performance green building information and disseminate research results through—

(A) outreach;

(B) education; and

(C) the provision of technical assistance.

(d) REPORTING.—The Commercial Director shall report directly to the Assistant Secretary for Energy Efficiency and Renewable Energy, or to other senior officials in a way that facilitates the integrated program of this subtitle for both energy efficiency and renewable energy and both technology development and technology deployment.

(e) COORDINATION.—The Commercial Director shall ensure full coordination of high-performance green building information and activities, including activities under this subtitle, within the Federal Government by working with the General Services Administration and all relevant agencies, including, at a minimum—

(1) the Environmental Protection Agency;

(2)the Office of the Federal Environmental Executive;

(3)the Office of Federal Procurement Policy;

(4)the Department of Energy, particularly the Federal Energy Management Program;

(5)the Department of Health and Human Services;

(6)the Department of Housing and Urban Development;

(7)the Department of Defense;

(8)the National Institute of Standards and Technology;

(9)the Department of Transportation;

(10)the Office of Science Technology and Policy; and

(11) such nonprofit high-performance green building rating and analysis entities as the Commercial Director determines can offer support, expertise, and review services.

(f) HIGH-PERFORMANCE GREEN BUILDING PARTNERSHIP CONSORTIUM.—

(1) RECOGNITION.—Not later than 90 days after the date Deadline. of enactment of this Act, the Commercial Director shall formally recognize one or more groups that qualify as a high-performance green building partnership consortium.

(2) REPRESENTATION TO QUALIFY.—To qualify under this section, any consortium shall include representation from—

(A) the design professions, including national associations of architects and of professional engineers;

(B) the development, construction, financial, and real estate industries;

(C) building owners and operators from the public and private sectors;

(D) academic and research organizations, including at least one national laboratory with extensive commercial building energy expertise;

(E) building code agencies and organizations, including a model energy code-setting organization;

(F) independent high-performance green building associations or councils;

(G) experts in indoor air quality and environmental factors;

(H) experts in intelligent buildings and integrated building information systems;

(I) utility energy efficiency programs;

(J) manufacturers and providers of equipment and techniques used in high-performance green buildings;

(K) public transportation industry experts; and

(L) nongovernmental energy efficiency organizations.

(3) FUNDING.—The Secretary may make payments to the Consortium pursuant to the terms of a public-private partnership for such activities of the Consortium undertaken under such a partnership as described in this subtitle directly to the Consortium or through one or more of its members.

(g) REPORT.—Not later than 2 years after the date of enactment of this Act, and biennially thereafter, the Commercial Director, in consultation with the Consortium, shall submit to Congress a report that—

(1) describes the status of the high-performance green building initiatives under this subtitle and other Federal programs affecting commercial high-performance green buildings in effect as of the date of the report, including—

(A) the extent to which the programs are being carried out in accordance with this subtitle; and

(B) the status of funding requests and appropriations for those programs; and

(1) summarizes and highlights development, at the State and local level, of high-performance green building initiatives, including executive orders, policies, or laws adopted promoting high-performance green building (including the status of implementation of those initiatives).

Sec. 422. Zero Net Energy Commercial Buildings Initiative.

(a) DEFINITIONS.—In this section:

(1) CONSORTIUM.—The term "consortium" means a High- Performance Green Building Consortium selected by the Commercial Director.

(2) INITIATIVE.—The term “initiative” means the Zero-Net- Energy Commercial Buildings Initiative established under subsection (b)(1).

(3) ZERO-NET-ENERGY COMMERCIAL BUILDING.—The term “zero-net-energy commercial building” means a high-performance commercial building that is designed, constructed, and operated—

(A) to require a greatly reduced quantity of energy to operate;

(B) to meet the balance of energy needs from sources of energy that do not produce greenhouse gases;

(C) in a manner that will result in no net emissions of greenhouse gases; and

(D) to be economically viable.

(b) ESTABLISHMENT.—

(1) IN GENERAL.—The Commercial Director shall establish an initiative, to be known as the “Zero-Net-Energy Commercial Buildings Initiative”—

(A) to reduce the quantity of energy consumed by commercial buildings located in the United States; and

(B) to achieve the development of zero net energy commercial buildings in the United States.

(2) CONSORTIUM.—

(A) IN GENERAL.—Not later than 180 days after the Deadline. date of enactment of this Act, the Commercial Director shall competitively select, and enter into an agreement with, a consortium to develop and carry out the initiative.

(B) AGREEMENTS.—In entering into an agreement with a consortium under subparagraph (A), the Commercial Director shall use the authority described in section 646(g) of the Department of Energy Organization Act (42 U.S.C. 7256(g)), to the maximum extent practicable.

(c) GOAL OF INITIATIVE.—The goal of the initiative shall be to develop and disseminate technologies, practices, and policies for the development and establishment of zero net energy commercial buildings for—

(1) any commercial building newly constructed in the United States by 2030;

(2) 50 percent of the commercial building stock of the United States by 2040; and

(3) all commercial buildings in the United States by 2050.

(d) COMPONENTS.—In carrying out the initiative, the Commercial Director, in consultation with the consortium, may—

(1) conduct research and development on building science, design, materials, components, equipment and controls, operation and other practices, integration, energy use measurement, and benchmarking;

(2) conduct pilot programs and demonstration projects to evaluate replicable approaches to achieving energy efficient commercial buildings for a variety of building types in a variety of climate zones;

(3) conduct deployment, dissemination, and technical assistance activities to encourage widespread adoption of technologies, practices, and policies to achieve energy efficient commercial buildings;

(4) conduct other research, development, demonstration, and deployment activities necessary to achieve each goal of the initiative, as determined by the Commercial Director, in consultation with the consortium;

(5) develop training materials and courses for building professionals and trades on achieving cost-effective high- performance energy efficient buildings;

(6) develop and disseminate public education materials to share information on the benefits and cost-effectiveness of high- performance energy efficient buildings;

(7) support code-setting organizations and State and local governments in developing minimum performance standards in building codes that recognize the ready availability of many technologies utilized in high-performance energy efficient buildings;

(8) develop strategies for overcoming the split incentives between builders and purchasers, and landlords and tenants, to ensure that energy efficiency and high-performance investments are made that are cost-effective on a lifecycle basis; and

(9) develop improved means of measurement and verification of energy savings and performance for public dissemination.

(e) COST SHARING.—In carrying out this section, the Commercial Director shall require cost sharing in accordance with section 988 of the Energy Policy Act of 2005 (42 U.S.C. 16352).

(f) AUTHORIZATION OF APPROPRIATIONS.—There are authorized to be appropriated to carry out this section—

(1) $20,000,000 for fiscal year 2008;

(2) $50,000,000 for each of fiscal years 2009 and 2010;

(3) $100,000,000 for each of fiscal years 2011 and 2012; and

(4) $200,000,000 for each of fiscal years 2013 through 2018.

Sec. 423. Public Outreach.

The Commercial Director and Federal Director, in coordination with the Consortium, shall carry out public outreach to inform individuals and entities of the information and services available governmentwide by—

(1) establishing and maintaining a national high-performance green building clearinghouse, including on the Internet, that—

(A) identifies existing similar efforts and coordinates activities of common interest; and

(B) provides information relating to high-performance green buildings, including hyperlinks to Internet sites that describe the activities, information, and resources of—

(i) the Federal Government;

(ii) State and local governments;

(iii) the private sector (including nongovernmental and nonprofit entities and organizations); and

(iv) international organizations;

(2) identifying and recommending educational resources for implementing high-performance green building practices, including security and emergency benefits and practices;

(3) providing access to technical assistance, tools, and resources for constructing high-performance green buildings, particularly tools to conduct life-cycle costing and life-cycle assessment;

(4) providing information on application processes for certifying a high-performance green building, including certification and commissioning;
(5) providing to the public, through the Commercial Director, technical and research information or other forms of assistance or advice that would be useful in planning and constructing high-performance green buildings;
(6) using such additional methods as are determined by the Commercial Director to be appropriate to conduct public outreach;
(7) surveying existing research and studies relating to high- performance green buildings; and
(8) coordinating activities of common interest.

Subtitle C—High-Performance Federal Buildings

Sec. 431. Energy Reduction Goals For Federal Buildings.

Section 543(a)(1) of the National Energy Conservation Policy Act (42 U.S.C. 8253(a)(1)) is amended by striking the table and inserting the following:

"Fiscal Year	Percentage Reduction
2006	2
2007	4
2008	9
2009	12
2010	15
2011	18
2012	21
2013	24
2014	27
2015	30.".

Sec. 432. Management Of Energy And Water Efficiency In Federal Buildings.

Section 543 of the National Energy Conservation Policy Act (42 U.S.C. 8253) is amended by adding at the end the following:

"(f) USE OF ENERGY AND WATER EFFICIENCY MEASURES IN FEDERAL BUILDINGS.—

"(1) DEFINITIONS.—In this subsection:

"(A) COMMISSIONING.—The term 'commissioning', with respect to a facility, means a systematic process—

"(i) of ensuring, using appropriate verification and documentation, during the period beginning on the initial day of the design phase of the facility and ending not earlier than 1 year after the date of completion of construction of the facility, that all facility systems perform interactively in accordance with—

"(I) the design documentation and intent of the facility; and

"(II) the operational needs of the owner of the facility, including preparation of operation personnel; and

"(ii) the primary goal of which is to ensure fully functional systems that can be properly operated and maintained during the useful life of the facility.

"(B) ENERGY MANAGER.—

"(i) IN GENERAL.—The term 'energy manager', with respect to a facility, means the individual who is responsible for—

"(I) ensuring compliance with this subsection by the facility; and

"(II) reducing energy use at the facility.

"(ii) INCLUSIONS.—The term 'energy manager' may include—

"(I) a contractor of a facility;

"(II) a part-time employee of a facility; and

"(III) an individual who is responsible for multiple facilities.

"(C) FACILITY.—

"(i) IN GENERAL.—The term 'facility' means any building, installation, structure, or other property (including any applicable fixtures) owned or operated by, or constructed or manufactured and leased to, the Federal Government.

"(ii) INCLUSIONS.—The term 'facility' includes—

"(I) a group of facilities at a single location or multiple locations managed as an integrated operation; and

"(II) contractor-operated facilities owned by the Federal Government.

"(iii) EXCLUSIONS.—The term 'facility' does not include any land or site for which the cost of utilities is not paid by the Federal Government.

"(D) LIFE CYCLE COST-EFFECTIVE.—The term 'life cycle cost-effective', with respect to a measure, means a measure, the estimated savings of which exceed the estimated costs over the lifespan of the measure, as determined in accordance with section 544.

"(E) PAYBACK PERIOD.—

"(i) IN GENERAL.—Subject to clause (ii), the term 'payback period', with respect to a measure, means a value equal to the quotient obtained by dividing—

"(I) the estimated initial implementation cost of the measure (other than financing costs); by

"(II) the annual cost savings resulting from the measure, including—

"(aa) net savings in estimated energy and water costs; and

"(bb) operations, maintenance, repair, replacement, and other direct costs.

"(ii) MODIFICATIONS AND EXCEPTIONS.—The Secretary, in guidelines issued pursuant to paragraph (6), may make such modifications and provide such exceptions to the calculation of the payback period of a measure as the Secretary determines to be appropriate to achieve the purposes of this Act.

"(F) RECOMMISSIONING.—The term 'recommissioning' means a process—

"(i) of commissioning a facility or system beyond the project development and warranty phases of the facility or system; and

"(ii) the primary goal of which is to ensure optimum performance of a facility, in accordance with design or current operating needs, over the useful life of the facility, while meeting building occupancy requirements.

"(G) RETROCOMMISSIONING.—The term 'retrocommis- sioning' means a process of commissioning a facility or system that was not commissioned at the time of construction of the facility or system.

"(2) FACILITY ENERGY MANAGERS.—

"(A) IN GENERAL.—Each Federal agency shall designate an energy manager responsible for implementing this subsection and reducing energy use at each facility that meets criteria under subparagraph (B).

"(B) COVERED FACILITIES.—The Secretary shall develop Criteria. criteria, after consultation with affected agencies, energy efficiency advocates, and energy and utility service providers, that cover, at a minimum, Federal facilities, including central utility plants and distribution systems and other energy intensive operations, that constitute at least 75 percent of facility energy use at each agency.

"(3) ENERGY AND WATER EVALUATIONS.—

"(A) EVALUATIONS.—Effective beginning on the date Effective date. that is 180 days after the date of enactment of this subsection and annually thereafter, energy managers shall complete, for each calendar year, a comprehensive energy and water evaluation for approximately 25 percent of the facilities of each agency that meet the criteria under paragraph (2)(B) in a manner that ensures that an evaluation of each such facility is completed at least once every 4 years.

"(B) RECOMMISSIONING AND RETROCOMMISSIONING.—As part of the evaluation under subparagraph (A), the energy manager shall identify and assess recommissioning measures (or, if the facility has never been commissioned, retrocommissioning measures) for each such facility.

"(4) IMPLEMENTATION OF IDENTIFIED ENERGY AND WATER EFFICIENCY MEASURES.—Not later than 2 years after the Deadline. completion of each evaluation under paragraph (3), each energy manager may—

"(A) implement any energy- or water-saving measure that the Federal agency identified in the evaluation conducted under paragraph (3) that is life cycle cost-effective; and

"(B) bundle individual measures of varying paybacks together into combined projects.

"(5) FOLLOW-UP ON IMPLEMENTED MEASURES.—For each measure implemented under paragraph (4), each energy manager shall ensure that—

"(A) equipment, including building and equipment controls, is fully commissioned at acceptance to be operating at design specifications;

"(B) a plan for appropriate operations, maintenance, and repair of the equipment is in place at acceptance and is followed;

"(C) equipment and system performance is measured during its entire life to ensure proper operations, maintenance, and repair; and

"(D) energy and water savings are measured and verified.

"(6) GUIDELINES.—

"(A) IN GENERAL.—The Secretary shall issue guidelines Deadlines. and necessary criteria that each Federal agency shall follow for implementation of—

"(i) paragraphs (2) and (3) not later than 180 days after the date of enactment of this subsection; and

"(ii) paragraphs (4) and (5) not later than 1 year after the date of enactment of this subsection.

"(B) RELATIONSHIP TO FUNDING SOURCE.—The guidelines issued by the Secretary under subparagraph (A) shall be appropriate and uniform for measures funded with each type of funding made available under paragraph (10), but may distinguish between different types of measures project size, and other criteria the Secretary determines are relevant.

"(7) WEB-BASED CERTIFICATION.—

"(A) IN GENERAL.—For each facility that meets the criteria established by the Secretary under paragraph (2)(B), the energy manager shall use the web-based tracking system under subparagraph (B) to certify compliance with the requirements for—

"(i) energy and water evaluations under paragraph (3);

"(ii) implementation of identified energy and water measures under paragraph (4); and

"(iii) follow-up on implemented measures under paragraph (5).

"(B) DEPLOYMENT.—

"(i) IN GENERAL.—Not later than 1 year after the date of enactment of this subsection, the Secretary shall develop and deploy a web-based tracking system required under this paragraph in a manner that tracks, at a minimum—

"(I) the covered facilities;

"(II) the status of meeting the requirements specified in subparagraph (A);

"(III) the estimated cost and savings for measures required to be implemented in a facility;

"(IV) the measured savings and persistence of savings for implemented measures; and

"(V) the benchmarking information disclosed under paragraph (8)(C).

"(ii) EASE OF COMPLIANCE.—The Secretary shall ensure that energy manager compliance with the requirements in this paragraph, to the maximum extent practicable—

"(I) can be accomplished with the use of streamlined procedures and templates that minimize the time demands on Federal employees; and

"(II) is coordinated with other applicable energy reporting requirements.

"(C) AVAILABILITY.—

"(i) IN GENERAL.—Subject to clause (ii), the Secretary shall make the web-based tracking system required under this paragraph available to Congress, other Federal agencies, and the public through the Internet.

"(ii) EXEMPTIONS.—At the request of a Federal agency, the Secretary may exempt specific data for specific facilities from disclosure under clause (i) for national security purposes.

"(8) BENCHMARKING OF FEDERAL FACILITIES.—

"(A) IN GENERAL.—The energy manager shall enter energy use data for each metered building that is (or is a part of) a facility that meets the criteria established by the Secretary under paragraph (2)(B) into a building energy use benchmarking system, such as the Energy Star Portfolio Manager.

"(B) SYSTEM AND GUIDANCE.—Not later than 1 year Deadline. after the date of enactment of this subsection, the Secretary shall—

"(i) select or develop the building energy use benchmarking system required under this paragraph for each type of building; and "(ii) issue guidance for use of the system.

"(C) PUBLIC DISCLOSURE.—Each energy manager shall post the information entered into, or generated by, a benchmarking system under this subsection, on the web- based tracking system under paragraph (7)(B). The energy manager shall update such information each year, and shall include in such reporting previous years' information to allow changes in building performance to be tracked over time.

"(9) FEDERAL AGENCY SCORECARDS.—

"(A) IN GENERAL.—The Director of the Office of Management and Budget shall issue semiannual scorecards for energy management activities carried out by each Federal agency that includes—

"(i) summaries of the status of implementing the various requirements of the agency and its energy managers under this subsection; and

"(ii) any other means of measuring performance that the Director considers appropriate.

"(B) AVAILABILITY.—The Director shall make the score Internet. cards required under this paragraph available to Congress,

other Federal agencies, and the public through the Internet.

"(10) FUNDING AND IMPLEMENTATION.—

"(A) AUTHORIZATION OF APPROPRIATIONS.—There are authorized to be appropriated such sums as are necessary to carry out this subsection.

"(B) FUNDING OPTIONS.—

"(i) IN GENERAL.—To carry out this subsection, a Federal agency may use any combination of—

"(I) appropriated funds made available under subparagraph (A); and

"(II) private financing otherwise authorized under Federal law, including financing available through energy savings performance contracts or utility energy service contracts.

"(ii) COMBINED FUNDING FOR SAME MEASURE.—A Federal agency may use any combination of appropriated funds and private financing described in clause (i) to carry out the same measure under this subsection.

"(C) IMPLEMENTATION.—Each Federal agency may implement the requirements under this subsection itself or may contract out performance of some or all of the requirements.

"(11) RULE OF CONSTRUCTION.—This subsection shall not be construed to require or to obviate any contractor savings guarantees.".

Sec. 433. Federal Building Energy Efficiency Performance Standards.

(a) STANDARDS.—Section 305(a)(3) of the Energy Conservation and Production Act (42 U.S.C. 6834(a)(3)) is amended by adding at the end the following new subparagraph:

"(D) Not later than 1 year after the date of enactment of the Energy Independence and Security Act of 2007, the Secretary shall establish, by rule, revised Federal building energy efficiency performance standards that require that:

"(i) For new Federal buildings and Federal buildings undergoing major renovations, with respect to which the Administrator of General Services is required to transmit a prospectus to Congress under section 3307 of title 40, United States Code, in the case of public buildings (as defined in section 3301 of title 40, United States Code), or of at least $2,500,000 in costs adjusted annually for inflation for other buildings:

"(I) The buildings shall be designed so that the fossil fuel-generated energy consumption of the buildings is reduced, as compared with such energy consumption by a similar building in fiscal year 2003 (as measured by Commercial Buildings Energy Consumption Survey or Residential Energy Consumption Survey data from the Energy Information Agency), by the percentage specified in the following table:

"Fiscal Year	Percentage Reduction
2010	55
2015	65
2020	80
2025	90
2030	100.

"(II) Upon petition by an agency subject to this subparagraph, the Secretary may adjust the applicable numeric requirement under subclause (I) downward with respect to a specific building, if the head of the agency designing the building certifies in writing that meeting such requirement would be technically impracticable in light of the agency's specified functional needs for that building and the Secretary concurs with the agency's conclusion. This subclause shall not apply to the General Services Administration.

"(III) Sustainable design principles shall be applied to the siting, design, and construction of such buildings. Not later than 90 days after the date of enactment of the Energy Independence and Security Act of 2007, the Secretary, after reviewing the findings of the Federal Director under section 436(h) of that Act, in consultation with the Administrator of General Services, and in consultation with the Secretary of Defense for considerations relating to those facilities under the custody and control of the Department of Defense, shall identify a certification system and level for green buildings that the Secretary determines to be the most likely to encourage a comprehensive and environmentally-sound approach to certification of green buildings. The identification of the certification system and level shall be based on a review of the Federal Director's findings under section 436(h) of the Energy Independence and Security Act of 2007 and the criteria specified in clause (iii), shall identify the highest level the Secretary determines is appropriate above the minimum level required for certification under the system selected, and shall achieve results at least comparable to the system used by and highest level referenced by the General Services Administration as of the date of enactment of the Energy Independence and Security Act of 2007.

Within 90 days of the completion of each study required by clause (iv), the Secretary, in consultation with the Administrator of General Services, and in consultation with the Secretary of Defense for considerations relating to those facilities under the custody and control of the Department of Defense, shall review and update the certification system and level, taking into account the conclusions of such study.

"(ii) In establishing criteria for identifying major renovations that are subject to the requirements of this subparagraph, the Secretary shall take into account the scope, degree, and types of renovations that are likely to provide significant opportunities for substantial improvements in energy efficiency.

"(iii) In identifying the green building certification system and level, the Secretary shall take into consideration—

"(I) the ability and availability of assessors and auditors to independently "verify the criteria and measurement of metrics at the scale necessary to implement this subparagraph;

"(II) the ability of the applicable certification organization to collect and reflect public comment;

"(III) the ability of the standard to be developed and revised through a consensus-based process;

"(IV) an evaluation of the robustness of the criteria for a high-performance green building, which shall give credit for promoting—

"(aa) efficient and sustainable use of water, energy, and other natural resources;

"(bb) use of renewable energy sources;

"(cc) improved indoor environmental quality through enhanced indoor air quality, thermal comfort, acoustics, day lighting, pollutant source control, and use of low-emission materials and building system controls; and

"(dd) such other criteria as the Secretary determines to be appropriate; and

"(I) national recognition within the building industry.

"(iv) At least once every 5 years, and in accordance with Study. section 436 of the Energy Independence and Security Act of 2007, the Administrator of General Services shall conduct a study to evaluate and compare available third-party green building certification systems and levels, taking into account the criteria listed in clause (iii).

"(v) The Secretary may by rule allow Federal agencies to develop internal certification processes, using certified professionals, in lieu of certification by the certification entity identified under clause (i)(III). The Secretary shall include in any such rule guidelines to ensure that the certification process results in buildings meeting the applicable certification system and level identified under clause (i)(III). An agency employing an internal certification process must continue to obtain external certification by the certification entity identified under clause (i)(III) for at least 5 percent of the total number of buildings certified annually by the agency.

"(vi) With respect to privatized military housing, the Secretary of Defense, after consultation with the Secretary may, through rulemaking, develop alternative criteria to those established by subclauses (I) and (III) of clause (i) that achieve an equivalent result in terms of energy savings, sustainable design, and green building performance.

"(vii) In addition to any use of water conservation technologies otherwise required by this section, water conservation technologies shall be applied to the extent that the technologies are life-cycle cost-effective.".

(b) DEFINITIONS.—Section 303(6) of the Energy Conservation and Production Act (42 U.S.C. 6832(6)) is amended by striking "which is not legally subject to State or local building codes or similar requirements." and inserting ". Such term shall include buildings built for the purpose of being leased by a Federal agency, and privatized military housing.".

(c) REVISION OF FEDERAL ACQUISITION REGULATION.—Not later than 2 years after the date of the enactment of this Act, the Federal Acquisition Regulation shall be revised to require Federal officers and employees to comply with this section and the amendments made by this section in the acquisition, construction, or major renovation of any facility. The members of the Federal Acquisition Regulatory Council (established under section 25 of the Office of Federal Procurement Policy Act (41 U.S.C. 421)) shall consult with the Federal Director and the Commercial Director before promulgating regulations to carry out this subsection.

(d) GUIDANCE.—Not later than 90 days after the date of promulgation of the revised regulations under subsection (c), the Administrator for Federal Procurement Policy shall issue guidance to all Federal procurement executives providing direction and instructions to renegotiate the design of proposed facilities and major renovations for existing facilities to incorporate improvements that are consistent with this section.

Sec. 434. Management Of Federal Building Efficiency.

(a) LARGE CAPITAL ENERGY INVESTMENTS.—Section 543 of the National Energy Conservation Policy Act (42 U.S.C. 8253) is amended by adding at the end the following:

"(f) LARGE CAPITAL ENERGY INVESTMENTS.—

"(1) IN GENERAL.—Each Federal agency shall ensure that any large capital energy investment in an existing building that is not a major renovation but involves replacement of installed equipment (such as heating and cooling systems), or involves renovation, rehabilitation, expansion, or remodeling of existing space, employs the most energy efficient designs, systems, equipment, and controls that are life-cycle cost effective.

"(A) develop a process for reviewing each decision made on a large capital energy investment described in paragraph (1) to ensure that the requirements of this subsection are met; and

"(B) report to the Director of the Office of Management Reports. and Budget on the process established.

"(3) COMPLIANCE REPORT.—Not later than 1 year after the date of enactment of this subsection, the Director of the Office of Management and Budget shall evaluate and report to Congress on the compliance of each agency with this subsection.".

(b) METERING.—Section 543(e)(1) of the National Energy Conservation Policy Act (42 U.S.C. 8253(e)(1)) is amended by inserting after the second sentence the following:

"Not later than October Deadline. 1, 2016, each agency shall provide for equivalent metering of natural gas and steam, in accordance with guidelines established by the Secretary under paragraph (2).".

Sec. 435. Leasing.

(a) IN GENERAL.—Except as provided in subsection (b), effective beginning on the date that is 3 years after the date of enactment of this Act, no Federal agency shall enter into a contract to lease space in a building that has not earned the Energy Star label in the most recent year.

(b) EXCEPTION.—

(1) APPLICATION.—This subsection applies if—

(A) no space is available in a building described in subsection (a) that meets the functional requirements of an agency, including locational needs;

(B) the agency proposes to remain in a building that the agency has occupied previously;

(C) the agency proposes to lease a building of historical, architectural, or cultural significance (as defined in section 3306(a)(4) of title 40, United States Code) or space in such a building; or

(D) the lease is for not more than 10,000 gross square feet of space.

(2) BUILDINGS WITHOUT ENERGY STAR LABEL.—If one of the conditions described in paragraph (2) is met, the agency may enter into a contract to lease space in a building that has not earned the Energy Star label in the most recent year if the lease contract includes provisions requiring that, prior to occupancy or, in the case of a contract described in paragraph (1)(B), not later than 1 year after signing the contract, the space will be renovated for all energy efficiency and conservation improvements that would be cost effective over the life of the lease, including improvements in lighting, windows, and heating, ventilation, and air conditioning systems.

(c) REVISION OF FEDERAL ACQUISITION REGULATION.—

(1) IN GENERAL.—Not later than 3 years after the date of the enactment of this Act, the Federal Acquisition Regulation described in section 6(a) of the Office of Federal Procurement Policy Act (41 U.S.C. 405(a)) shall be revised to require Federal officers and employees to comply with this section in leasing buildings.

(2) CONSULTATION.—The members of the Federal Acquisition Regulatory Council established under section 25 of the Office of Federal Procurement Policy Act (41 U.S.C. 421) shall consult with the Federal Director and the Commercial Director before promulgating regulations to carry out this subsection.

Sec. 436. High-Performance Green Federal Buildings.

(a) ESTABLISHMENT OF OFFICE.—Not later than 60 days after the date of enactment of this Act, the Administrator shall establish within the General Services Administration an Office of Federal High-Performance Green Buildings, and appoint an individual to serve as Federal Director in, a position in the career-reserved Senior Executive service, to—

(1) establish and manage the Office of Federal High- Performance Green Buildings; and

(2) carry out other duties as required under this subtitle.

(b) COMPENSATION.—The compensation of the Federal Director shall not exceed the maximum rate of basic pay for the Senior Executive Service under section 5382 of title 5, United States Code, including any applicable locality-based comparability payment that may be authorized under section 5304(h)(2)(C) of that title.

(c) DUTIES.—The Federal Director shall—

(1) coordinate the activities of the Office of Federal High- Performance Green Buildings with the activities of the Office of Commercial High-Performance Green Buildings, and the Secretary, in accordance with section 305(a)(3)(D) of the Energy Conservation and Production Act (42 U.S.C. 6834(a)(3)(D));

(2) ensure full coordination of high-performance green building information and activities within the General Services Administration and all relevant agencies, including, at a minimum—

(A) the Environmental Protection Agency;

(B) the Office of the Federal Environmental Executive;

(C) the Office of Federal Procurement Policy;

(D) the Department of Energy;

(E) the Department of Health and Human Services;

(F) the Department of Defense;

(G) the Department of Transportation;

(H) the National Institute of Standards and Technology; and

(I) the Office of Science and Technology Policy;

(3) establish a senior-level Federal Green Building Advisory Committee under section 474, which shall provide advice and recommendations in accordance with that section and subsection (d);

(4) identify and every 5 years reassess improved or higher rating standards recommended by the Advisory Committee;

(5) ensure full coordination, dissemination of information regarding, and promotion of the results of research and development information relating to Federal high-performance green building initiatives;

(6) identify and develop Federal high-performance green building standards for all types of Federal facilities, consistent with the requirements of this subtitle and section 305(a)(3)(D) of the Energy Conservation and Production Act (42 U.S.C. 6834(a)(3)(D));

(7) establish green practices that can be used throughout the life of a Federal facility;

(8) review and analyze current Federal budget practices and life-cycle costing issues, and make recommendations to Congress, in accordance with subsection (d); and

(9) identify opportunities to demonstrate innovative and emerging green building technologies and concepts.

(d) ADDITIONAL DUTIES.—The Federal Director, in consultation with the Commercial Director and the Advisory Committee, and consistent with the requirements of section 305(a)(3)(D) of the Energy Conservation and Production Act (42 U.S.C. 6834(a)(3)(D)) shall—

(1) identify, review, and analyze current budget and contracting practices that affect achievement of high-performance green buildings, including the identification of

barriers to high- performance green building life-cycle costing and budgetary issues;

(2) develop guidance and conduct training sessions with budget specialists and contracting personnel from Federal agencies and budget examiners to apply life-cycle cost criteria to actual projects;

(3) identify tools to aid life-cycle cost decisionmaking; and

(4) explore the feasibility of incorporating the benefits of high-performance green buildings, such as security benefits, into a cost-budget analysis to aid in life-cycle costing for budget and decisionmaking processes.

(e) INCENTIVES.—Within 90 days after the date of enactment of this Act, the Federal Director shall identify incentives to encourage the expedited use of high-performance green buildings and related technology in the operations of the Federal Government, in accordance with the requirements of section 305(a)(3)(D) of the Energy Conservation and Production Act (42 U.S.C. 6834(a)(3)(D)), including through—

(1) the provision of recognition awards; and

(2) the maximum feasible retention of financial savings in the annual budgets of Federal agencies for use in reinvesting in future high-performance green building initiatives.

(f) REPORT.—Not later than 2 years after the date of enactment of this Act, and biennially thereafter, the Federal Director, in consultation with the Secretary, shall submit to Congress a report that—

(1) describes the status of compliance with this subtitle, the requirements of section 305(a)(3)(D) of the Energy Conservation and Production Act (42 U.S.C. 6834(a)(3)(D)), and other Federal high-performance green building initiatives in effect as of the date of the report, including—

(A) the extent to which the programs are being carried out in accordance with this subtitle and the requirements of section 305(a)(3)(D) of that Act; and

(B) the status of funding requests and appropriations for those programs;

(2) identifies within the planning, budgeting, and construction process all types of Federal facility procedures that may affect the certification of new and existing Federal facilities as high-performance green buildings under the provisions of section 305(a)(3)(D) of that Act and the criteria established in subsection (h);

(3) identifies inconsistencies, as reported to the Advisory Committee, in Federal law with respect to product acquisition guidelines and high-performance product guidelines;

(4) recommends language for uniform standards for use by Federal agencies in environmentally responsible acquisition;

(5) in coordination with the Office of Management and Budget, reviews the budget process for capital programs with respect to alternatives for—

(A) restructuring of budgets to require the use of complete energy and environmental cost accounting;

(B) using operations expenditures in budget-related decisions while simultaneously incorporating productivity and health measures (as those measures can be quantified by the Office of Federal High-Performance Green Buildings, with the assistance of universities and national laboratories);

(C) streamlining measures for permitting Federal agencies to retain all identified savings accrued as a result of the use of life-cycle costing for future high-performance green building initiatives; and

(D) identifying short-term and long-term cost savings that accrue from high-performance green buildings, including those relating to health and productivity;

(6) identifies green, self-sustaining technologies to address the operational needs of Federal facilities in times of national security emergencies, natural disasters, or other dire emergencies;

(7) summarizes and highlights development, at the State and local level, of high-performance green building initiatives, including executive orders, policies, or laws adopted promoting high-performance green building (including the status of implementation of those initiatives); and

(8) includes, for the 2-year period covered by the report, recommendations to address each of the matters, and a plan for implementation of each recommendation, described in paragraphs (1) through (7).

(g) IMPLEMENTATION.—The Office of Federal High-Performance Green Buildings shall carry out each plan for implementation of recommendations under subsection (f)(8).

(h) IDENTIFICATION OF CERTIFICATION SYSTEM.—

(1) IN GENERAL.—For the purpose of this section, not later than 60 days after the date of enactment of this Act, the Federal Director shall identify and shall provide to the Secretary pursuant to section 305(a)(3)(D) of the Energy Conservation and Production Act (42 U.S.C. 6834(a)(3)(D)), a certification system that the Director determines to be the most likely to encourage a comprehensive and environmentally-sound approach to certification of green buildings.

(2) BASIS.—The system identified under paragraph (1) shall be based on—

(A) a study completed every 5 years and provided to the Secretary pursuant to section 305(a)(3)(D) of that Act, which shall be carried out by the Federal Director to compare and evaluate standards;

(B) the ability and availability of assessors and auditors to independently verify the criteria and measurement of metrics at the scale necessary to implement this subtitle;

(C) the ability of the applicable standard-setting organization to collect and reflect public comment;

(D) the ability of the standard to be developed and revised through a consensus-based process;

(E) an evaluation of the robustness of the criteria for a high-performance green building, which shall give credit for promoting—

(i) efficient and sustainable use of water, energy, and other natural resources;

(ii) use of renewable energy sources;

(iii) improved indoor environmental quality through enhanced indoor air quality, thermal comfort, acoustics, day lighting, pollutant source control, and use of low-emission materials and building system controls;

(iv) reduced impacts from transportation through building location and site design that promote access by public transportation; and

(v) such other criteria as the Federal Director determines to be appropriate; and

(A) national recognition within the building industry.

Sec. 437. Federal Green Building Performance.

(a) IN GENERAL.—Not later than October 31 of each of the 2 fiscal years following the fiscal year in which this Act is enacted, and at such times thereafter as the Comptroller General of the United States determines to be appropriate, the Comptroller General of the United States shall, with respect to the fiscal years that have passed since the preceding report—

(1) conduct an audit of the implementation of this subtitle, section 305(a)(3)(D) of the Energy Conservation and Production Act (42 U.S.C. 6834(a)(3)(D)), and section 435; and

(2) submit to the Federal Director, the Advisory Committee, the Administrator, and Congress a report describing the results of the audit.

(b) CONTENTS.—An audit under subsection (a) shall include a review, with respect to the period covered by the report under subsection (a)(2), of—

(1) budget, life-cycle costing, and contracting issues, using best practices identified by the Comptroller General of the United States and heads of other agencies in accordance with section 436(d);

(2) the level of coordination among the Federal Director, the Office of Management and Budget, the Department of Energy, and relevant agencies;

(3) the performance of the Federal Director and other agencies in carrying out the implementation plan;

(4) the design stage of high-performance green building measures;

(5) high-performance building data that were collected and reported to the Office; and

(6) such other matters as the Comptroller General of the United States determines to be appropriate.

(c) ENVIRONMENTAL STEWARDSHIP SCORECARD.—The Federal Director shall consult with the Advisory Committee to enhance, and assist in the implementation of, the Office of Management and Budget government efficiency reports and scorecards under section 528 and the Environmental Stewardship Scorecard announced at the White House summit on Federal sustainable buildings in January 2006, to measure the implementation by each Federal agency of sustainable design and green building initiatives.

Sec. 438. Storm Water Runoff Requirements For Federal Development Projects.

The sponsor of any development or redevelopment project involving a Federal facility with a footprint that exceeds 5,000 square feet shall use site planning, design, construction, and maintenance strategies for the property to maintain or restore, to the maximum extent technically feasible, the predevelopment hydrology of the property with regard to the temperature, rate, volume, and duration of flow.

Sec. 439. Cost-Effective Technology Acceleration Program.

(a) DEFINITION OF ADMINISTRATOR.—In this section, the term "Administrator" means the Administrator of General Services.

(b) ESTABLISHMENT.—

(1)IN GENERAL.—The Administrator shall establish a program to accelerate the use of more cost-effective technologies and practices at GSA facilities.

(2) REQUIREMENTS.—The program established under this subsection shall—

(A) ensure centralized responsibility for the coordination of cost reduction-related recommendations, practices, and activities of all relevant Federal agencies;

(B) provide technical assistance and operational guidance to applicable tenants to achieve the goal identified in subsection (c)(2)(B)(ii);

(C) establish methods to track the success of Federal departments and agencies with respect to that goal; and

(D) be fully coordinated with and no less stringent nor less energy-conserving or water-conserving than required by other provisions of this Act and other applicable law, including sections 321 through 324, 431 through 438, 461, 511 through 518, and 523 through 525 and amendments made by those sections.

(c) ACCELERATED USE OF TECHNOLOGIES.—

(1) REVIEW.—

(A) IN GENERAL.—As part of the program under this section, not later than 90 days after the date of enactment of this Act, the Administrator shall conduct a review of—

(i) current use of cost-effective lighting technologies and geothermal heat pumps in GSA facilities; and

(ii) the availability to managers of GSA facilities of cost-effective lighting technologies and geothermal heat pumps.

(B) REQUIREMENTS.—The review under subparagraph (A) shall—

(i) examine the use of cost-effective lighting technologies, geothermal heat pumps, and other cost-effective technologies and practices by Federal agencies in GSA facilities; and

(ii) as prepared in consultation with the Administrator of the Environmental Protection Agency, identify cost-effective lighting technology and geothermal heat pump technology standards that could be used for all types of GSA facilities.

(2) REPLACEMENT.—

(A) IN GENERAL.—As part of the program under this Deadline. section, not later than 180 days after the date of enactment of this Act, the Administrator shall establish, using available appropriations and programs implementing sections 432 and 525 (and amendments made by those sections), a cost-effective lighting technology and geothermal heat pump technology acceleration program to achieve maximum feasible replacement of existing lighting, heating, cooling technologies with cost-effective lighting technologies and geothermal heat pump technologies in each GSA facility. Such program shall fully comply with the requirements of sections 321 through 324, 431 through 438, 461, 511 through 518, and 523 through 525 and amendments made by those sections and any other provisions of law, which shall be applicable to the extent that they are more stringent or would achieve greater energy savings than required by this section.

(B) ACCELERATION PLAN TIMETABLE.—

(i) IN GENERAL.—To implement the program estab Deadline. lished under subparagraph (A), not later than 1 year after the date of enactment of this Act, the Administrator shall establish a timetable of actions to comply with the requirements of this section and sections 431 through 435, whichever achieves greater energy savings most expeditiously, including milestones for specific activities needed to replace existing lighting, heating, cooling technologies with cost-effective lighting technologies and geothermal heat pump technologies, to the maximum extent feasible (including at the maximum rate feasible), at each GSA facility.

(ii) GOAL.—The goal of the timetable under clause (i) shall be to complete, using available appropriations and programs implementing sections 431 through 435 (and amendments made by those sections), maximum feasible replacement of existing lighting, heating, and cooling technologies with cost-effective lighting technologies and geothermal heat pump technologies consistent with the requirements of this section and sections 431 through 435, whichever achieves greater energy savings most expeditiously. Notwithstanding any provision of this section, such program shall fully comply with the requirements of the Act including sections 321 through 324, 431 through 438, 461, 511 through 518, and 523 through 525 and amendments made by those sections and other provisions of law, which shall be applicable to the extent that they are more stringent or would achieve greater energy or water savings than required by this section.

(d) GSA FACILITY TECHNOLOGIES AND PRACTICES.—

(1) IN GENERAL.—Not later than 180 days after the date Deadline. of enactment of this Act, and annually thereafter, the Adminis-
trator shall—

(A) ensure that a manager responsible for implementing section 432 and for accelerating the use of cost- effective technologies and practices is designated for each GSA facility; and

(B) submit to Congress a plan to comply with section 432, this section, and other applicable provisions of this Act and applicable law with respect to energy and water conservation at GSA facilities.

(2) MEASURES.—The plan shall implement measures required by such other provisions of law in accordance with those provisions, and shall implement the measures required by this section to the maximum extent feasible (including at the maximum rate feasible) using available appropriations and programs implementing sections 431 through 435 and 525 (and amendments made by those sections), by not later than the date that is 5 years after the date of enactment of this Act.

(3) CONTENTS OF PLAN.—The plan shall—

(A) with respect to cost-effective technologies and practices—

(i) identify the specific activities needed to comply with sections 431 through 435;

(ii) identify the specific activities needed to achieve at least a 20-percent reduction in operational costs through the application of cost-effective technologies and practices from 2003 levels at GSA facilities by not later than 5 years after the date of enactment of this Act;

(iii) describe activities required and carried out to estimate the funds necessary to achieve the reduction described in clauses (i) and (ii);

(B) include an estimate of the funds necessary to carry out this section;

(C) describe the status of the implementation of cost- effective technologies and practices at GSA facilities, including—

(i) the extent to which programs, including the program established under subsection (b), are being carried out in accordance with this subtitle; and

(ii) the status of funding requests and appropriations for those programs;

(D) identify within the planning, budgeting, and construction processes, all types of GSA facility-related procedures that inhibit new and existing GSA facilities from implementing cost-effective technologies;

(E) recommend language for uniform standards for use by Federal agencies in implementing cost-effective technologies and practices;

(F) in coordination with the Office of Management and Budget, review the budget process for capital programs with respect to alternatives for—

(i) implementing measures that will assure that Federal agencies retain all identified savings accrued as a result of the use of cost-effective technologies, consistent with section 543(a)(1) of the National Energy Conservation Policy Act (42 U.S.C. 8253(a)(1), and other applicable law; and

(ii) identifying short- and long-term cost savings that accrue from the use of cost-effective technologies and practices;

(G) with respect to cost-effective technologies and practices, achieve substantial operational cost savings through the application of the technologies; and

(H) include recommendations to address each of the matters, and a plan for implementation of each recommendation, described in subparagraphs (A) through (G).

(4) ADMINISTRATION.—Notwithstanding any provision of this section, the program required under this section shall fully comply with the requirements of sections 321 through 324, 431 through 438, 461, 511 through 518, and 523 through 525 and amendments made by those sections, which shall be applicable to the extent that they are more stringent or would achieve greater energy or water savings than required by this section.

(e) AUTHORIZATION OF APPROPRIATIONS.—There are authorized to be appropriated such sums as are necessary to carry out this section, to remain available until expended.

Sec. 440. Authorization Of Appropriations.

There is authorized to be appropriated to carry out sections 434 through 439 and 482 $4,000,000 for each of fiscal years 2008 through 2012, to remain available until expended.

Sec. 441. Public Building Life-Cycle Costs.

Section 544(a)(1) of the National Energy Conservation Policy Act (42 U.S.C. 8254(a)(1)) is amended by striking “25” and inserting “40”.

Subtitle D—Industrial Energy Efficiency

Sec. 451. Industrial Energy Efficiency.

(a) IN GENERAL.—Title III of the Energy Policy and Conservation Act (42 U.S.C. 6291 et seq.) is amended by inserting after part D the following:

"Part E—Industrial Energy Efficiency

"Sec. 371. Definitions.

"In this part:

"(1) ADMINISTRATOR.—The term 'Administrator' means the Administrator of the Environmental Protection Agency.

"(2) COMBINED HEAT AND POWER.—The term 'combined heat and power system' means a facility that—

"(A) simultaneously and efficiently produces useful thermal energy and electricity; and

"(B) recovers not less than 60 percent of the energy value in the fuel (on a higher-heating-value basis) in the form of useful thermal energy and electricity.

"(3) NET EXCESS POWER.—The term 'net excess power' means, for any facility, recoverable waste energy recovered in the form of electricity in quantities exceeding the total consumption of electricity at the specific time of generation on the site at which the facility is located.

"(4) PROJECT.—The term 'project' means a recoverable waste energy project or a combined heat and power system project.

"(5) RECOVERABLE WASTE ENERGY.—The term 'recoverable waste energy' means waste energy from which electricity or useful thermal energy may be recovered through modification of an existing facility or addition of a new facility.

"(6) REGISTRY.—The term 'Registry' means the Registry of Recoverable Waste Energy Sources established under section 372(d).

"(7) USEFUL THERMAL ENERGY.—The term 'useful thermal energy' means energy—

"(A) in the form of direct heat, steam, hot water, or other thermal form that is used in production and beneficial measures for heating, cooling, humidity control, process use, or other valid thermal end-use energy requirements; and

"(B) for which fuel or electricity would otherwise be consumed.

"(8) WASTE ENERGY.—The term 'waste energy' means—

"(A) exhaust heat or flared gas from any industrial process;

"(B) waste gas or industrial tail gas that would otherwise be flared, incinerated, or vented;

"(C) a pressure drop in any gas, excluding any pressure drop to a condenser that subsequently vents the resulting heat; and

"(D) such other forms of waste energy as the Administrator may determine.

"(9) OTHER TERMS.—The terms 'electric utility', 'nonregulated electric utility', 'State regulated electric utility', and other terms have the meanings given those terms in title I of the Public Utility Regulatory Policies Act of 1978 (16 U.S.C. 2611 et seq.).

Sec. 372. Survey And Registry.

"(a) RECOVERABLE WASTE ENERGY INVENTORY PROGRAM.—

(1)IN GENERAL.—The Administrator, in cooperation with the Secretary and State energy offices, shall establish a recoverable waste energy inventory program.

(2) SURVEY.—The program shall include—

"(A) an ongoing survey of all major industrial and large commercial combustion sources in the United States (as defined by the Administrator) and the sites at which the sources are located; and

"(B) a review of each source for the quantity and quality of waste energy produced at the source.

"(b) CRITERIA.—

(1)IN GENERAL.—Not later than 270 days after the date of enactment of the Energy Independence and Security Act of 2007, the Administrator shall publish a rule for establishing criteria for including sites in the Registry.

(2) INCLUSIONS.—The criteria shall include—

(A) a requirement that, to be included in the Registry, a project at the site shall be determined to be economically feasible by virtue of offering a payback of invested costs not later than 5 years after the date of first full project operation (including incentives offered under this part);

(B) standards to ensure that projects proposed for inclusion in the Registry are not developed or used for the primary purpose of making sales of excess electric power under the regulatory provisions of this part; and procedures for contesting the listing of any source or site on the Registry by any State, utility, or other interested person.

"(c) TECHNICAL SUPPORT.—On the request of the owner or operator of a source or site included in the Registry, the Secretary shall—

"(1) provide to owners or operators of combustion sources technical support; and

"(2) offer partial funding (in an amount equal to not more than one-half of total costs) for feasibility studies to confirm whether or not investment in recovery of waste energy or combined heat and power at a source would offer a payback period of 5 years or less.

"(d) REGISTRY.—

"(1) ESTABLISHMENT.—

"(A) IN GENERAL.—Not later than 1 year after the date of enactment of the Energy Independence and Security Act of 2007, the Administrator shall establish a Registry of Recoverable Waste Energy Sources, and sites on which the sources are located, that meet the criteria established under subsection (b).

"(B) UPDATES; AVAILABILITY.—The Administrator shall—

"(i) update the Registry on a regular basis; and

"(ii) make the Registry available to the public on the website of the Environmental Protection Agency.

"(C) CONTESTING LISTING.—Any State, electric utility, or other interested person may contest the listing of any source or site by submitting a petition to the Administrator.

"(2) CONTENTS.—

"(A) IN GENERAL.—The Administrator shall register and include on the Registry all sites meeting the criteria established under subsection (b).

"(B) QUANTITY OF RECOVERABLE WASTE ENERGY.—The Administrator shall—

"(i) calculate the total quantities of potentially recoverable waste energy from sources at the sites, nationally and by State; and

"(ii) make public—

"(I) the total quantities described in clause (i); and

"(II) information on the criteria pollutant and greenhouse gas emissions savings that might be achieved with recovery of the waste energy from all sources and sites listed on the Registry.

"(3) AVAILABILITY OF INFORMATION.—

"(A) IN GENERAL.—The Administrator shall notify Notification. owners or operators of recoverable waste energy sources and sites listed on the Registry prior to publishing the listing.

"(B) DETAILED QUANTITATIVE INFORMATION.—

"(i) IN GENERAL.—Except as provided in clause (ii), the owner or operator of a source at a site may elect to have detailed quantitative information concerning the site not made public by notifying the Administrator of the election.

"(ii) LIMITED AVAILABILITY.—The information shall be made available to—

"(I) the applicable State energy office; and

"(II) any utility requested to support recovery of waste energy from the source pursuant to the incentives provided under section 374.

"(iii) STATE TOTALS.—Information concerning the site shall be included in the total quantity of recoverable waste energy for a State unless there are fewer than 3 sites in the State.

"(4) REMOVAL OF PROJECTS FROM REGISTRY.—

"(A) IN GENERAL.—Subject to subparagraph (B), as a project achieves successful recovery of waste energy, the Administrator shall—

"(i) remove the related sites or sources from the Registry; and

"(ii) designate the removed projects as eligible for incentives under section 374.

"(B) LIMITATION.—No project shall be removed from the Registry without the consent of the owner or operator of the project if—

"(i) the owner or operator has submitted a petition under section 374; and

"(ii) the petition has not been acted on or denied.

"(5) INELIGIBILITY OF CERTAIN SOURCES.—The Administrator shall not list any source constructed after the date of the enactment of the Energy Independence and Security Act of 2007 on the Registry if the Administrator determines that the source—

"(A) was developed for the primary purpose of making sales of excess electric power under the regulatory provisions of this part; or

"(B) does not capture at least 60 percent of the total energy value of the fuels used (on a higher-heating-value basis) in the form of useful thermal energy, electricity, mechanical energy, chemical output, or any combination thereof.

"(e) SELF-CERTIFICATION.—

"(1) IN GENERAL.—Subject to any procedures that are established by the Administrator, an owner, operator, or third-party developer of a recoverable waste energy project that qualifies under standards established by the Administrator may self- certify the sites or sources of the owner, operator, or developer to the Administrator for inclusion in the Registry.

"(2) REVIEW AND APPROVAL.—To prevent a fraudulent listing, a site or source shall be included on the Registry only if the Administrator reviews and approves the self-certification.

"(f) NEW FACILITIES.—As a new energy-consuming industrial facility is developed after the date of enactment of the Energy Independence and Security Act of 2007, to the extent the facility may constitute a site with recoverable waste energy that may qualify for inclusion on the Registry, the Administrator may elect to include the facility on the Registry, at the request of the owner, operator, or developer of the facility, on a conditional basis with the site to be removed from the Registry if the development ceases or the site fails to qualify for listing under this part.

"(g) OPTIMUM MEANS OF RECOVERY.—For each site listed in the Registry, at the request of the owner or operator of the site, the Administrator shall offer, in cooperation with Clean Energy Application Centers operated by the Secretary of Energy, suggestions for optimum means of recovery of value from waste energy stream in the form of electricity, useful thermal energy, or other energy-related products.

"(h) REVISION.—Each annual report of a State under section 548(a) of the National Energy Conservation Policy Act (42 U.S.C. 8258(a)) shall include the results of the survey for the State under this section.

"(i) AUTHORIZATION OF APPROPRIATIONS.—There are authorized to be appropriated to—

"(1) the Administrator to create and maintain the Registry and services authorized by this section, $1,000,000 for each of fiscal years 2008 through 2012; and

"(2) the Secretary—

"(A) to assist site or source owners and operators in determining the feasibility of projects authorized by this section, $2,000,000 for each of fiscal years 2008 through 2012; and

"(B) to provide funding for State energy office functions under this section, $5,000,000.

Sec. 373. Waste Energy Recovery Incentive Grant Program.

"(a) ESTABLISHMENT.—The Secretary shall establish in the Department of Energy a waste energy recovery incentive grant program to provide incentive grants to—

"(1) owners and operators of projects that successfully produce electricity or incremental useful thermal energy from waste energy recovery;

"(2) utilities purchasing or distributing the electricity; and

"(3) States that have achieved 80 percent or more of recoverable waste heat recovery opportunities.

"(b) GRANTS TO PROJECTS AND UTILITIES.—

"(1) IN GENERAL.—The Secretary shall make grants under this section—

"(A) to the owners or operators of waste energy recovery projects; and

"(B) in the case of excess power purchased or transmitted by a electric utility, to the utility.

"(2) PROOF.—Grants may only be made under this section on receipt of proof of waste energy recovery or excess electricity generation, or both, from the project in a form prescribed by the Secretary.

"(3) EXCESS ELECTRIC ENERGY.—

"(A) IN GENERAL.—In the case of waste energy recovery, a grant under this section shall be made at the rate of $10 per megawatt hour of documented electricity produced from recoverable waste energy (or by prevention of waste energy in the case of a new facility) by the project during the first 3 calendar years of production, beginning on or after the date of enactment of the Energy Independence and Security Act of 2007.

"(B) UTILITIES.—If the project produces net excess power and an electric utility purchases or transmits the excess power, 50 percent of so much of the grant as is attributable to the net excess power shall be paid to the electric utility purchasing or transporting the net excess power.

"(4) USEFUL THERMAL ENERGY.—In the case of waste energy recovery that produces useful thermal energy that is used for a purpose different from that for which the project is principally designed, a grant under this section shall be made to the owner or operator of the waste energy recovery project at the rate of $10 for each 3,412,000 Btus of the excess thermal energy used for the different purpose.

"(c) GRANTS TO STATES.—In the case of any State that has achieved 80 percent or more of waste heat recovery opportunities identified by the Secretary under this part, the Administrator shall make a 1-time grant to the State in an amount of not more than $1,000 per megawatt of waste-heat capacity recovered (or a thermal equivalent) to support State-level programs to identify and achieve additional energy efficiency.

"(d) ELIGIBILITY.—The Secretary shall—

"(1) establish rules and guidelines to establish eligibility for grants under subsection (b);

"(2) publicize the availability of the grant program known to owners or operators of recoverable waste energy sources and sites listed on the Registry; and

"(3) award grants under the program on the basis of the merits of each project in recovering or preventing waste energy throughout the United States on an impartial, objective, and not unduly discriminatory basis.

"(e) LIMITATION.—The Secretary shall not award grants to any person for a combined heat and power project or a waste heat recovery project that qualifies for specific Federal tax incentives for combined heat and power or for waste heat recovery.

"(f) AUTHORIZATION OF APPROPRIATIONS.—There are authorized to be appropriated to the Secretary—

"(1) to make grants to projects and utilities under subsection (b)—

"(A) $100,000,000 for fiscal year 2008 and $200,000,000 for each of fiscal years 2009 through 2012; and

"(B) such additional amounts for fiscal year 2008 and each fiscal year thereafter as may be necessary for administration of the waste energy recovery incentive grant program; and

"(2) to make grants to States under subsection (b), $10,000,000 for each of fiscal years 2008 through 2012, to remain available until expended.

Sec. 374. Additional Incentives for Recovery, Use, and Prevention Of Industrial Waste Energy.

"(a) CONSIDERATION OF STANDARD.—

"(1) IN GENERAL.—Not later than 180 days after the receipt by a State regulatory authority (with respect to each electric utility for which the authority has ratemaking authority), or nonregulated electric utility, of a request from a project sponsor or owner or operator, the State regulatory authority or nonregulated electric utility shall—

"(A) provide public notice and conduct a hearing respecting the standard established by subsection (b); and

"(B) on the basis of the hearing, consider and make a determination whether or not it is appropriate to implement the standard to carry out the purposes of this part.

"(2) RELATIONSHIP TO STATE LAW.—For purposes of any determination under paragraph (1) and any review of the determination in any court, the purposes of this section supplement otherwise applicable State law.

"(3) NONADOPTION OF STANDARD.—Nothing in this part prohibits any State regulatory authority or nonregulated electric utility from making any determination that it is not appropriate to adopt any standard described in paragraph (1), pursuant to authority under otherwise applicable State law.

"(b) STANDARD FOR SALES OF EXCESS POWER.—For purposes of this section, the standard referred to in subsection (a) shall provide that an owner or operator of a waste energy recovery project identified on the Registry that generates net excess power shall be eligible to benefit from at least 1 of the options described in subsection (c) for disposal of the net excess power in accordance with the rate conditions and limitations described in subsection (d).

"(b) OPTIONS.—The options referred to in subsection (b) are as follows:

"(1) SALE OF NET EXCESS POWER TO UTILITY.—The electric utility shall purchase the net excess power from the owner or operator of the eligible waste energy recovery project during the operation of the project under a contract entered into for that purpose.

"(2) TRANSPORT BY UTILITY FOR DIRECT SALE TO THIRD PARTY.—The electric utility shall transmit the net excess power on behalf of the project owner or operator to up to 3 separate locations on the system of the utility for direct sale by the owner or operator to third parties at those locations.

"(3) TRANSPORT OVER PRIVATE TRANSMISSION LINES.—The State and the electric utility shall permit, and shall waive or modify such laws as would otherwise prohibit, the construction and operation of private electric wires constructed, owned, and operated by the project owner or operator, to transport the power to up to 3 purchasers within a 3-mile radius of the project, allowing the wires to use or cross public rights- of-way, without subjecting the project to regulation as a public utility, and according the wires the same treatment for safety, zoning, land use, and other legal privileges as apply or would apply to the wires of the utility, except that—

"(A) there shall be no grant of any power of eminent domain to take or cross private property for the wires; and

"(B) the wires shall be physically segregated and not interconnected with any portion of the system of the utility, except on the customer side of the revenue meter of the utility and in a manner that precludes any possible export of the electricity onto the utility system, or disruption of the system.

"(4) AGREED ON ALTERNATIVES.—The utility and the owner or operator of the project may reach agreement on any alternate arrangement and payments or rates associated with the arrangement that is mutually satisfactory and in accord with State law.

"(d) RATE CONDITIONS AND CRITERIA.—

"(1) DEFINITIONS.—In this subsection:

"(A) PER UNIT DISTRIBUTION COSTS.—The term 'per unit distribution costs' means (in kilowatt hours) the quotient obtained by dividing—

"(i) the depreciated book-value distribution system costs of a utility; by

"(ii) the volume of utility electricity sales or transmission during the previous year at the distribution level.

"(B) PER UNIT DISTRIBUTION MARGIN.—The term 'per unit distribution margin' means—

"(i) in the case of a State-regulated electric utility, a per-unit gross pretax profit equal to the product obtained by multiplying—

"(I) the State-approved percentage rate of return for the utility for distribution system assets; by

"(II) the per unit distribution costs; and

"(ii) in the case of a nonregulated utility, a per unit contribution to net revenues determined multiplying—

"(I) the percentage (but not less than 10 percent) obtained by dividing—

"(aa) the amount of any net revenue payment or contribution to the owners or subscribers of the nonregulated utility during the prior year; by

"(bb) the gross revenues of the utility during the prior year to obtain a percentage; by

"(II) the per unit distribution costs.

"(C) PER UNIT TRANSMISSION COSTS.—The term 'per unit transmission costs' means the total cost of those transmission services purchased or provided by a utility on a per-kilowatt-hour basis as included in the retail rate of the utility.

"(2) OPTIONS.—The options described in paragraphs (1) and (2) in subsection (c) shall be offered under purchase and transport rate conditions that reflect the rate components defined under paragraph (1) as applicable under the circumstances described in paragraph (3).

"(3) APPLICABLE RATES.—

"(A) RATES APPLICABLE TO SALE OF NET EXCESS POWER.—

"(i) IN GENERAL.—Sales made by a project owner or operator of a facility under the option described in subsection (c)(1) shall be paid for on a per kilowatt hour basis that shall equal the full undiscounted retail rate paid to the utility for power purchased by the facility minus per unit distribution costs, that applies to the type of utility purchasing the power.

"(ii) VOLTAGES EXCEEDING 25 KILOVOLTS.—If the net excess power is made available for purchase at voltages that must be transformed to or from voltages exceeding 25 kilovolts to be available for resale by the utility, the purchase price shall further be reduced by per unit transmission costs.

"(B) RATES APPLICABLE TO TRANSPORT BY UTILITY FOR DIRECT SALE TO THIRD PARTIES.—

"(i) IN GENERAL.—Transportation by utilities of power on behalf of the owner or operator of a project under the option described in subsection (c)(2) shall incur a transportation rate that shall equal the per unit distribution costs and per unit distribution margin, that applies to the type of utility transporting the power.

"(ii) VOLTAGES EXCEEDING 25 KILOVOLTS.—If the net excess power is made available for transportation at voltages that must be transformed to or from voltages exceeding 25 kilovolts to be transported to the designated third-party purchasers, the transport rate shall further be increased by per unit transmission costs.

"(iii) STATES WITH COMPETITIVE RETAIL MARKETS FOR ELECTRICITY.—In a State with a competitive retail market for electricity, the applicable transportation rate for similar transportation shall be applied in lieu of any rate calculated under this paragraph.

"(4) LIMITATIONS.—

"(A) IN GENERAL.—Any rate established for sale or transportation under this section shall—

"(i) be modified over time with changes in the underlying costs or rates of the electric utility; and

"(ii) reflect the same time-sensitivity and billing periods as are established in the retail sales or transportation rates offered by the utility.

"(B) LIMITATION.—No utility shall be required to purchase or transport a quantity of net excess power under this section that exceeds the available capacity of the wires, meter, or other equipment of the electric utility serving the site unless the owner or operator of the project agrees to pay necessary and reasonable upgrade costs.

"(e) PROCEDURAL REQUIREMENTS FOR CONSIDERATION AND DETERMINATION.—

"(1) PUBLIC NOTICE AND HEARING.—

"(A) IN GENERAL.—The consideration referred to in subsection (a) shall be made after public notice and hearing.

"(B) ADMINISTRATION.—The determination referred to in subsection (a) shall be—

"(i) in writing;

"(ii) based on findings included in the determination and on the evidence presented at the hearing; and

"(iii) available to the public.

"(2) INTERVENTION BY ADMINISTRATOR.—The Administrator may intervene as a matter of right in a proceeding conducted under this section—

"(A) to calculate—

"(i) the energy and emissions likely to be saved by electing to adopt 1 or more of the options; and

"(ii) the costs and benefits to ratepayers and the utility; and

"(B) to advocate for the waste-energy recovery opportunity.

"(3) PROCEDURES.—

"(A) IN GENERAL.—Except as otherwise provided in paragraphs (1) and (2), the procedures for the consideration and determination referred to in subsection (a) shall be the procedures established by the State regulatory authority or the nonregulated electric utility.

"(B) MULTIPLE PROJECTS.—If there is more than 1 project seeking consideration simultaneously in connection with the same utility, the proceeding may encompass all such projects, if full attention is paid to individual circumstances and merits and an individual judgment is reached with respect to each project.

"(f) IMPLEMENTATION.—

"(1) IN GENERAL.—The State regulatory authority (with respect to each electric utility for which the authority has ratemaking authority) or nonregulated electric utility may, to the extent consistent with otherwise applicable State law—

"(A) implement the standard determined under this section; or

":(B) decline to implement any such standard.

"(2) NONIMPLEMENTATION OF STANDARD.—

"(A)IN GENERAL.—If a State regulatory authority (with respect to each electric utility for which the authority has ratemaking authority) or nonregulated electric utility declines to implement any standard established by this section, the authority or nonregulated electric utility shall state in writing the reasons for declining to implement the standard.

"(B) AVAILABILITY TO PUBLIC.—The statement of reasons shall be available to the public.

"(C)ANNUAL REPORT.—The Administrator shall include in an annual report submitted to Congress a description of the lost opportunities for waste-heat recovery from the project described in subparagraph (A), specifically identifying the utility and stating the quantity of lost energy and emissions savings calculated.

"(D) NEW PETITION.—If a State regulatory authority (with respect to each electric utility for which the authority has ratemaking authority) or nonregulated electric utility declines to implement the standard established by this section, the project sponsor may submit a new petition under this section with respect to the project at any time after the date that is 2 years after the date on which the State regulatory authority or nonregulated utility declined to implement the standard.

Sec. 375. Clean Energy Application Centers.

"(a) RENAMING.—

"(1) IN GENERAL.—The Combined Heat and Power Application Centers of the Department of Energy are redesignated as Clean Energy Application Centers.

"(2) REFERENCES.—Any reference in any law, rule, regulation, or publication to a Combined Heat and Power Application Center shall be treated as a reference to a Clean Energy Application Center.

"(b) RELOCATION.—

"(1) IN GENERAL.—In order to better coordinate efforts with the separate Industrial Assessment Centers and to ensure that the energy efficiency and, when applicable, the renewable nature of deploying mature clean energy technology is fully accounted for, the Secretary shall relocate the administration of the Clean Energy Application Centers to the Office of Energy Efficiency and Renewable Energy within the Department of Energy.

"(2) OFFICE OF ELECTRICITY DELIVERY AND ENERGY RELIABILITY.—The Office of Electricity Delivery and Energy Reliability shall—

"(A) continue to perform work on the role of technology described in paragraph (1) in support of the grid and the reliability and security of the technology; and

"(B) shall assist the Clean Energy Application Centers in the work of the Centers with regard to the grid and with electric utilities.

"(c) GRANTS.—

"(1) IN GENERAL.—The Secretary shall make grants to universities, research centers, and other appropriate institutions to ensure the continued operations and effectiveness of 8 Regional Clean Energy Application Centers in each of the following regions (as designated for such purposes as of the date of the enactment of the Energy Independence and Security Act of 2007):

"(A) Gulf Coast.

"(B) Intermountain.

"(C) Mid-Atlantic.

"(D) Midwest.

"(E) Northeast.

"(F) Northwest.

"(G) Pacific.

"(H) Southeast.

"(2) ESTABLISHMENT OF GOALS AND COMPLIANCE.—In making grants under this subsection, the Secretary shall ensure that sufficient goals are established and met by each Center throughout the program duration concerning outreach and technology deployment.

"(d) ACTIVITIES.—

"(1) IN GENERAL.—Each Clean Energy Application Center shall—

"(A) operate a program to encourage deployment of clean energy technologies through education and outreach to building and industrial professionals; and other individuals and organizations with an interest in efficient energy use; and

"(B) provide project specific support to building and industrial professionals through assessments and advisory activities.

"(2) TYPES OF ACTIVITIES.—Funds made available under this section may be used—

"(A) to develop and distribute informational materials on clean energy technologies, including continuation of the 8 websites in existence on the date of enactment of the Energy Independence and Security Act of 2007;

"(B) to develop and conduct target market workshops, seminars, Internet programs, and other activities to educate end users, regulators, and stakeholders in a manner that leads to the deployment of clean energy technologies;

"(C) to provide or coordinate onsite assessments for sites and enterprises that may consider deployment of clean energy technology;

"(D) to perform market research to identify high profile candidates for clean energy deployment;

"(E) to provide consulting support to sites considering deployment of clean energy technologies;

"(F) to assist organizations developing clean energy technologies to overcome barriers to deployment; and

"(G) to assist companies and organizations with performance evaluations of any clean energy technology implemented.

''(e) DURATION.—

"(1) IN GENERAL.—A grant awarded under this section shall be for a period of 5 years

"(2) ANNUAL EVALUATIONS.—Each grant shall be evaluated annually for the continuation of the grant based on the activities and results of the grant.

''(f) AUTHORIZATION.—There is authorized to be appropriated to carry out this section $10,000,000 for each of fiscal years 2008 through 2012.".

(a) TABLE OF CONTENTS.—The table of contents of the Energy Policy and Conservation Act (42 U.S.C. prec. 6201) is amended by inserting after the items relating to part D of title III the following:

Sec. 371. Definitions.

Sec. 372. Survey and Registry.

Sec. 373. Waste energy recovery incentive grant program.

Sec. 374. Additional incentives for recovery, utilization and prevention of industrial waste energy.

Sec. 375. Clean Energy Application Centers.".

Sec. 452. Energy-Intensive Industries Program.

(a) DEFINITIONS.—In this section:

(1) ELIGIBLE ENTITY.—The term ''eligible entity" means—

(A) an energy-intensive industry;

(B) a national trade association representing an energy-intensive industry; or

(C) a person acting on behalf of 1 or more energy- intensive industries or sectors, as determined by the Secretary.

(2) ENERGY-INTENSIVE INDUSTRY.—The term ''energy-intensive industry" means an industry that uses significant quantities of energy as part of its primary economic activities, including—

(A) information technology, including data centers containing electrical equipment used in processing, storing, and transmitting digital information;

(B) consumer product manufacturing;

(C) food processing;

(D) materials manufacturers, including—

(i) aluminum;

(ii) chemicals;

(iii) forest and paper products;
(iv) metal casting;
(v) glass;
(vi) petroleum refining;
(vii)mining; and
(viii) steel;

(E) other energy-intensive industries, as determined by the Secretary.

(3) FEEDSTOCK.—The term "feedstock" means the raw material supplied for use in manufacturing, chemical, and biological processes.

(4) PARTNERSHIP.—The term "partnership" means an energy efficiency partnership established under subsection (c)(1)(A).

(5) PROGRAM.—The term "program" means the energy- intensive industries program established under subsection (b).

(b) ESTABLISHMENT OF PROGRAM.—The Secretary shall establish a program under which the Secretary, in cooperation with energy-intensive industries and national industry trade associations representing the energy-intensive industries, shall support, research, develop, and promote the use of new materials processes, technologies, and techniques to optimize energy efficiency and the economic competitiveness of the United States' industrial and commercial sectors.

(c) PARTNERSHIPS.—

(1) IN GENERAL.—As part of the program, the Secretary shall establish energy efficiency partnerships between the Secretary and eligible entities to conduct research on, develop, and demonstrate new processes, technologies, and operating practices and techniques to significantly improve the energy efficiency of equipment and processes used by energy-intensive industries, including the conduct of activities to—

(A) increase the energy efficiency of industrial processes and facilities;

(B) research, develop, and demonstrate advanced technologies capable of energy intensity reductions and increased environmental performance; and

(C) promote the use of the processes, technologies, and techniques described in subparagraphs (A) and (B).

(2) ELIGIBLE ACTIVITIES.—Partnership activities eligible for funding under this subsection include—

(A) feedstock and recycling research, development, and demonstration activities to identify and promote—

(i) opportunities for meeting industry feedstock requirements with more energy efficient and flexible sources of feedstock or energy supply;

(ii) strategies to develop and deploy technologies that improve the quality and quantity of feedstocks recovered from process and waste streams; and

(iii) other methods using recycling, reuse, and improved industrial materials;

(B) research to develop and demonstrate technologies and processes that utilize alternative energy sources to supply heat, power, and new feedstocks for energy-intensive industries;

(C) research to achieve energy efficiency in steam, power, control system, and process heat technologies, and in other manufacturing processes; and

(D) industrial and commercial energy efficiency and sustainability assessments to—

(i) assist individual industrial and commercial sectors in developing tools, techniques, and methodologies to assess—
(I) the unique processes and facilities of the sectors;
(II) the energy utilization requirements of the sectors; and
(III) the application of new, more energy efficient technologies; and
(ii) conduct energy savings assessments;
(E) the incorporation of technologies and innovations that would significantly improve the energy efficiency and utilization of energy-intensive commercial applications; and
(F) any other activities that the Secretary determines to be appropriate.
(3) PROPOSALS.—
(A) IN GENERAL.—To be eligible for funding under this subsection, a partnership shall submit to the Secretary a proposal that describes the proposed research, development, or demonstration activity to be conducted by the partnership.
(B) REVIEW.—After reviewing the scientific, technical, and commercial merit of a proposals submitted under subparagraph (A), the Secretary shall approve or disapprove the proposal.
(C) COMPETITIVE AWARDS.—The provision of funding under this subsection shall be on a competitive basis.
(4) COST-SHARING REQUIREMENT.—In carrying out this section, the Secretary shall require cost sharing in accordance with section 988 of the Energy Policy Act of 2005 (42 U.S.C. 16352).
(d) GRANTS.—The Secretary may award competitive grants for innovative technology research, development and demonstrations to universities, individual inventors, and small companies, based on energy savings potential, commercial viability, and technical merit.
(e) INSTITUTION OF HIGHER EDUCATION-BASED INDUSTRIAL RESEARCH AND ASSESSMENT CENTERS.—The Secretary shall provide funding to institution of higher education-based industrial research and assessment centers, whose purpose shall be—
(1) to identify opportunities for optimizing energy efficiency and environmental performance;
(2) to promote applications of emerging concepts and technologies in small- and medium-sized manufacturers;
(3) to promote research and development for the use of alternative energy sources to supply heat, power, and new feedstocks for energy-intensive industries;
(4) to coordinate with appropriate Federal and State research offices, and provide a clearinghouse for industrial process and energy efficiency technical assistance resources; and
(5) to coordinate with State-accredited technical training centers and community colleges, while ensuring appropriate services to all regions of the United States.
(f) AUTHORIZATION OF APPROPRIATIONS.—
(1) IN GENERAL.—There are authorized to be appropriated to the Secretary to carry out this section—
(A) $184,000,000 for fiscal year 2008;
(B) $190,000,000 for fiscal year 2009;

(C) $196,000,000 for fiscal year 2010;
(D) $202,000,000 for fiscal year 2011;
(E) $208,000,000 for fiscal year 2012; and
(F) such sums as are necessary for fiscal year 2013 and each fiscal year thereafter.

(2) PARTNERSHIP ACTIVITIES.—Of the amounts made available under paragraph (1), not less than 50 percent shall be used to pay the Federal share of partnership activities under subsection (c).

(3) COORDINATION AND NONDUPLICATION.—The Secretary shall coordinate efforts under this section with other programs of the Department and other Federal agencies to avoid duplication of effort.

Sec. 453. Energy Efficiency For Data Center Buildings.

(a) DEFINITIONS.—In this section:

(1) DATA CENTER.—The term "data center" means any facility that primarily contains electronic equipment used to process, store, and transmit digital information, which may be—

(A) a free-standing structure; or
(B) a facility within a larger structure, that uses environmental control equipment to maintain the proper conditions for the operation of electronic equipment.

(2) DATA CENTER OPERATOR.—The term "data center operator" means any person or government entity that builds or operates a data center or purchases data center services, equipment, and facilities.

(b) VOLUNTARY NATIONAL INFORMATION PROGRAM.—

(1) IN GENERAL.—Not later than 90 days after the date Deadline. of enactment of this Act, the Secretary and the Administrator of the Environmental Protection Agency shall, after consulting with information technology industry and other interested parties, initiate a voluntary national information program for those types of data centers and data center equipment and facilities that are widely used and for which there is a potential for significant data center energy savings as a result of the program.

(2) REQUIREMENTS.—The program described in paragraph (1) shall—

(A) address data center efficiency holistically, reflecting the total energy consumption of data centers as whole systems, including both equipment and facilities;

(B) consider prior work and studies undertaken in this area, including by the Environmental Protection Agency and the Department of Energy;

(C) consistent with the objectives described in paragraph (1), determine the type of data center and data center equipment and facilities to be covered under the program;

(D) produce specifications, measurements, best practices, and benchmarks that will enable data center operators to make more informed decisions about the energy efficiency and costs of data centers, and that take into account—

(i) the performance and use of servers, data storage devices, and other information technology equipment;

(ii) the efficiency of heating, ventilation, and air conditioning, cooling, and power conditioning systems, provided that no modification shall be required of a standard then in effect under the Energy Policy and

Conservation Act (42 U.S.C. 6201 et seq.) for any covered heating, ventilation, air-conditioning, cooling or power-conditioning product;

(iii) energy savings from the adoption of software and data management techniques; and

(iv) other factors determined by the organization described in subsection (c);

(E) allow for creation of separate specifications, measurements, and benchmarks based on data center size and function, as well as other appropriate characteristics;

(F) advance the design and implementation of efficiency technologies to the maximum extent economically practical;

(G) provide to data center operators in the private sector and the Federal Government information about best practices and purchasing decisions that reduce the energy consumption of data centers; and

(H) publish the information described in subparagraph (G), which may be disseminated through catalogs, trade publications, the Internet, or other mechanisms, that will allow data center operators to assess the energy consumption and potential cost savings of alternative data centers and data center equipment and facilities.

(3) PROCEDURES.—The program described in paragraph (1) shall be developed in consultation with and coordinated by the organization described in subsection (c) according to commonly accepted procedures for the development of specifications, measurements, and benchmarks.

(c) DATA CENTER EFFICIENCY ORGANIZATION.—

(1) IN GENERAL.—After the establishment of the program described in subsection (b), the Secretary and the Administrator shall jointly designate an information technology industry organization to consult with and to coordinate the program.

(2) REQUIREMENTS.—The organization designated under paragraph (1), whether preexisting or formed specifically for the purposes of subsection (b), shall—

(A) consist of interested parties that have expertise in energy efficiency and in the development, operation, and functionality of computer data centers, information technology equipment, and software, as well as representatives of hardware manufacturers, data center operators, and facility managers;

(B) obtain and address input from Department of Energy National Laboratories or any college, university, research institution, industry association, company, or public interest group with applicable expertise in any of the areas listed in paragraph (1);

(C) follow commonly accepted procedures for the development of specifications and accredited standards development processes;

(D) have a mission to develop and promote energy efficiency for data centers and information technology; and

(E) have the primary responsibility to consult in the development and publishing of the information, measurements, and benchmarks described in subsection (b) and transmission of the information to the Secretary and the Administrator for consideration under subsection (d).

(d) MEASUREMENTS AND SPECIFICATIONS.—

(1) IN GENERAL.—The Secretary and the Administrator shall consider the specifications, measurements, and benchmarks described in subsection (b) for use by the Federal Energy Management Program, the Energy Star Program, and other

efficiency programs of the Department of Energy and Environmental Protection Agency, respectively.

(2) REJECTIONS.—If the Secretary or the Administrator rejects 1 or more specifications, measurements, or benchmarks described in subsection (b), the rejection shall be made consistent with section 12(d) of the National Technology Transfer and Advancement Act of 1995 (15 U.S.C. 272 note; Public Law 104–113).

(3) DETERMINATION OF IMPRACTICABILITY.—A determination that a specification, measurement, or benchmark described in subsection (b) is impractical may include consideration of the maximum efficiency that is technologically feasible and economically justified.

(e) MONITORING.—The Secretary and the Administrator shall—

(1) monitor and evaluate the efforts to develop the program described in subsection (b); and

(2) not later than 3 years after the date of enactment of this Act, make a determination as to whether the program is consistent with the objectives of subsection (b).

(f) ALTERNATIVE SYSTEM.—If the Secretary and the Administrator make a determination under subsection (e) that a voluntary national information program for data centers consistent with the objectives of subsection (b) has not been developed, the Secretary and the Administrator shall, after consultation with the National Institute of Standards and Technology and not later than 2 years after the determination, develop and implement the program under subsection (b).

(g) PROTECTION OF PROPRIETARY INFORMATION.—The Secretary, the Administrator, or the data center efficiency organization shall not disclose any proprietary information or trade secrets provided by any individual or company for the purposes of carrying out this section or the program established under this section.

Subtitle E—Healthy High-Performance Schools

Sec. 461. Healthy High-Performance Schools.

(d) AMENDMENT.—The Toxic Substances Control Act (15 U.S.C. 2601 et seq.) is amended by adding at the end the following new title:

"TITLE V—HEALTHY HIGH-PERFORMANCE SCHOOLS

Sec. 501. Grants For Healthy School Environments.

"(a) IN GENERAL.—The Administrator, in consultation with the Secretary of Education, may provide grants to States for use in—

"(1) providing technical assistance for programs of the Environmental Protection Agency (including the Tools for Schools Program and the Healthy School Environmental Assessment Tool) to schools for use in addressing environmental issues; and

"(2) development and implementation of State school environmental health programs that include—

"(A) standards for school building design, construction, and renovation; and

"(B) identification of ongoing school building environmental problems, including contaminants, hazardous substances, and pollutant emissions, in the State and recommended solutions to address those problems, including assessment of information on the exposure of children to environmental hazards in school facilities.

"(b) SUNSET.—The authority of the Administrator to carry out this section shall expire 5 years after the date of enactment of this section.

Sec. 502. Model Guidelines For Siting Of School Facilities.

"Not later than 18 months after the date of enactment of this section, the Administrator, in consultation with the Secretary of Education and the Secretary of Health and Human Services, shall issue voluntary school site selection guidelines that account for—

"(1) the special vulnerability of children to hazardous substances or pollution exposures in any case in which the potential for contamination at a potential school site exists;

"(2) modes of transportation available to students and staff;

"(3) the efficient use of energy; and

"(4) the potential use of a school at the site as an emergency shelter.

"Sec. 503. Public Outreach.

"(a) REPORTS.—The Administrator shall publish and submit to Congress an annual report on all activities carried out under this title, until the expiration of authority described in section 501(b).

"(b) PUBLIC OUTREACH.—The Federal Director appointed under section 436(a) of the Energy Independence and Security Act of 2007 (in this title referred to as the 'Federal Director') shall ensure, to the maximum extent practicable, that the public clearinghouse established under section 423(1) of the Energy Independence and Security Act of 2007 receives and makes available information on the exposure of children to environmental hazards in school facilities, as provided by the Administrator.

Sec. 504. Environmental Health Program.

"(a) IN GENERAL.—Not later than 2 years after the date of enactment of this section, the Administrator, in consultation with the Secretary of Education, the Secretary of Health and Human Services, and other relevant agencies, shall issue voluntary guidelines for use by the State in developing and implementing an environmental health program for schools that—

"(1) takes into account the status and findings of Federal initiatives established under this title or subtitle C of title IV of the Energy Independence and Security Act of 2007 and other relevant Federal law with respect to school facilities, including relevant updates on trends in the field, such as the impact of school facility environments on student and staff—

"(A) health, safety, and productivity; and

"(B) disabilities or special needs;

"(2) takes into account studies using relevant tools identified or developed in accordance with section 492 of the Energy Independence and Security Act of 2007;

"(3) takes into account, with respect to school facilities, each of—

"(A) environmental problems, contaminants, hazardous substances, and pollutant emissions, including—

"(i) lead from drinking water;

"(ii) lead from materials and products;

"(iii) asbestos;

"(iv) radon;

"(v) the presence of elemental mercury releases from products and containers;

"(vi) pollutant emissions from materials and products; and

"(vii) any other environmental problem, contaminant, hazardous substance, or pollutant emission that present or may present a risk to the health of occupants of the school facilities or environment;

"(B) natural day lighting;

"(C) ventilation choices and technologies;

"(D) heating and cooling choices and technologies;

"(E) moisture control and mold;

"(F) maintenance, cleaning, and pest control activities;

"(G) acoustics; and

"(H) other issues relating to the health, comfort, productivity, and performance of occupants of the school facilities;

"(4) provides technical assistance on siting, design, management, and operation of school facilities, including facilities used by students with disabilities or special needs;

"(5) collaborates with federally funded pediatric environmental health centers to assist in on-site school environmental investigations;

"(6) assists States and the public in better understanding and improving the environmental health of children; and

"(7) takes into account the special vulnerability of children in low-income and minority communities to exposures from contaminants, hazardous substances, and pollutant emissions.

"(b) PUBLIC OUTREACH.—The Federal Director and Commercial Director shall ensure, to the maximum extent practicable, that the public clearinghouse established under section 423 of the Energy Independence and Security Act of 2007 receives and makes available—

"(1) information from the Administrator that is contained in the report described in section 503(a); and

"(2) information on the exposure of children to environmental hazards in school facilities, as provided by the Administrator.

Sec. 505. Authorization Of Appropriations.

"There are authorized to be appropriated to carry out this title $1,000,000 for fiscal year 2009, and $1,500,000 for each of fiscal years 2010 through 2013, to remain available until expended.".

(b) TABLE OF CONTENTS AMENDMENT.—The table of contents for the Toxic Substances Control Act (15 U.S.C. 2601 et seq.) is amended by adding at the end the following:

"TITLE V—HEALTHY HIGH-PERFORMANCE SCHOOLS

"Sec. 501. Grants for healthy school environments. "Sec. 502. Model guidelines for siting of school facilities. "Sec. 503. Public outreach.

"Sec. 504. Environmental health program.

"Sec. 505. Authorization of appropriations.".

Sec. 462. Study On Indoor Environmental Quality In Schools.

(a) IN GENERAL.—The Administrator of the Environmental Protection Agency shall enter into an arrangement with the Secretary of Education and the Secretary of Energy to conduct a detailed study of how sustainable building features such as energy efficiency affect multiple perceived indoor environmental quality stressors on students in K–12 schools.

(b) CONTENTS.—The study shall—

(1) investigate the combined effect building stressors such as heating, cooling, humidity, lighting, and acoustics have on building occupants' health, productivity, and overall well-being;

(2) identify how sustainable building features, such as energy efficiency, are influencing these human outcomes singly and in concert; and

(3) ensure that the impacts of the indoor environmental quality are evaluated as a whole.

(d) AUTHORIZATION OF APPROPRIATIONS.—There are authorized to be appropriated for carrying out this section $200,000 for each of the fiscal years 2008 through 2012.

Subtitle F—Institutional Entities

Sec. 471. Energy Sustainability And Efficiency Grants And Loans For Institutions.

Part G of title III of the Energy Policy and Conservation Act is amended by inserting after section 399 (42 U.S.C. 6371h) the following:

Sec. 399a. Energy Sustainability And Efficiency Grants And Loans For Institutions.

"(a) DEFINITIONS.—In this section:

"(1) COMBINED HEAT AND POWER.—The term 'combined heat and power' means the generation of electric energy and heat in a single, integrated system, with an overall thermal efficiency of 60 percent or greater on a higher-heating-value basis.

"(2) DISTRICT ENERGY SYSTEMS.—The term 'district energy systems' means systems providing thermal energy from a renewable energy source, thermal energy source, or highly efficient technology to more than 1 building or fixed energy-consuming use from 1 or more thermal-energy production facilities through pipes or other means to provide space heating, space conditioning, hot water, steam, compression, process energy, or other end uses for that energy.

"(3) ENERGY SUSTAINABILITY.—The term 'energy sustain- ability' includes using a renewable energy source, thermal energy source, or a highly efficient technology for transportation, electricity generation, heating, cooling, lighting, or other energy services in fixed installations.

"(4) INSTITUTION OF HIGHER EDUCATION.—The term 'institution of higher education' has the meaning given the term in section 2 of the Energy Policy Act of 2005 (42 U.S.C. 15801).

"(5) INSTITUTIONAL ENTITY.—The term 'institutional entity' means an institution of higher education, a public school district, a local government, a municipal utility, or a designee of 1 of those entities.

"(6) RENEWABLE ENERGY SOURCE.—The term 'renewable energy source' has the meaning given the term in section 609 of the Public Utility Regulatory Policies Act of 1978 (7 U.S.C. 918c).

"(7) SUSTAINABLE ENERGY INFRASTRUCTURE.—The term 'sustainable energy infrastructure' means—facilities for production of energy from renewable energy sources, thermal energy sources, or highly efficient technologies, including combined heat and power or other waste heat use; and district energy systems.

"(8) THERMAL ENERGY SOURCE.—The term 'thermal energy source' means—

"(A) a natural source of cooling or heating from lake or ocean water; and

"(B) recovery of useful energy that would otherwise be wasted from ongoing energy uses.

"(b) TECHNICAL ASSISTANCE GRANTS.—

"(1) IN GENERAL.—Subject to the availability of appropriated funds, the Secretary shall implement a program of information dissemination and technical assistance to institutional entities to assist the institutional entities in identifying, evaluating, designing, and implementing sustainable energy infrastructure projects in energy sustainability.

"(2) ASSISTANCE.—The Secretary shall support institutional entities in—

"(A) identification of opportunities for sustainable energy infrastructure;

"(B) understanding the technical and economic characteristics of sustainable energy infrastructure;

"(C) utility interconnection and negotiation of power and fuel contracts;

"(D) understanding financing alternatives;

"(E) permitting and siting issues;

"(F) obtaining case studies of similar and successful sustainable energy infrastructure systems; and

"(G) reviewing and obtaining computer software for assessment, design, and operation and maintenance of sustainable energy infrastructure systems.

"(3) ELIGIBLE COSTS FOR TECHNICAL ASSISTANCE GRANTS.—

On receipt of an application of an institutional entity, the Secretary may make grants to the institutional entity to fund a portion of the cost of—

"(A) feasibility studies to assess the potential for implementation or improvement of sustainable energy infrastructure;

"(B) analysis and implementation of strategies to overcome barriers to project implementation, including financial, contracting, siting, and permitting barriers; and

"(C) detailed engineering of sustainable energy infrastructure.

"(c) GRANTS FOR ENERGY EFFICIENCY IMPROVEMENT AND ENERGY SUSTAINABILITY.—

"(1) GRANTS.—

"(A) IN GENERAL.—The Secretary shall award grants to institutional entities to carry out projects to improve energy efficiency on the grounds and facilities of the institutional entity.

"(B) REQUIREMENT.—To the extent that applications have been submitted, grants under subparagraph (A) shall include not less than 1 grant each year to an institution of higher education in each State.

"(C) MINIMUM FUNDING.—Not less than 50 percent of the total funding for all grants under this subsection shall be awarded in grants to institutions of higher education.

"(2) CRITERIA.—Evaluation of projects for grant funding shall be based on criteria established by the Secretary, including criteria relating to—

"(A) improvement in energy efficiency;

"(B) reduction in greenhouse gas emissions and other air emissions, including criteria air pollutants and ozone- depleting refrigerants;

"(C) increased use of renewable energy sources or thermal energy sources;

"(D) reduction in consumption of fossil fuels;

"(E) active student participation; and

"(F) need for funding assistance.

"(3) CONDITION.—As a condition of receiving a grant under this subsection, an institutional entity shall agree—

"(A) to implement a public awareness campaign concerning the project in the community in which the institutional entity is located; and

"(B) to submit to the Secretary, and make available to the public, reports on any efficiency improvements, energy cost savings, and environmental benefits achieved as part of a project carried out under paragraph (1), including quantification of the results relative to the criteria described under paragraph (2).

"(d) GRANTS FOR INNOVATION IN ENERGY SUSTAINABILITY.—

"(1) GRANTS.—

"(A) IN GENERAL.—The Secretary shall award grants to institutional entities to engage in innovative energy sustainability projects.

"(B) REQUIREMENT.—To the extent that applications have been submitted, grants under subparagraph (A) shall include not less than 2 grants each year to institutions of higher education in each State.

"(C) MINIMUM FUNDING.—Not less than 50 percent of the total funding for all grants under this subsection shall be awarded in grants to institutions of higher education.

"(2) INNOVATION PROJECTS.—An innovation project carried out with a grant under this subsection shall—

"(A) involve—

"(i) an innovative technology that is not yet commercially available; or

"(ii) available technology in an innovative application that maximizes energy efficiency and sustain- ability;

"(B) have the greatest potential for testing or demonstrating new technologies or processes; and

"(C) to the extent undertaken by an institution of higher education, ensure active student participation in the project, including the planning, implementation, evaluation, and other phases of projects.

"(3) CONDITION.—As a condition of receiving a grant under this subsection, an institutional entity shall agree to submit to the Secretary, and make available to the public, reports that describe the results of the projects carried out using grant funds.

"(e) ALLOCATION TO INSTITUTIONS OF HIGHER EDUCATION WITH SMALL ENDOWMENTS.—

"(1) IN GENERAL.—Of the total amount of grants provided to institutions of higher education for a fiscal year under this section, the Secretary shall provide not less than 50 percent of the amount to institutions of higher education that have an endowment of not more than $100,000,000.

"(2) REQUIREMENT.—To the extent that applications have been submitted, at least 50 percent of the amount described in paragraph (1) shall be provided to institutions of higher education that have an endowment of not more than $50,000,000.

"(f) GRANT AMOUNTS.—

"(1) IN GENERAL.—If the Secretary determines that cost sharing is appropriate, the amounts of grants provided under this section shall be limited as provided in this subsection.

"(2) TECHNICAL ASSISTANCE GRANTS.—In the case of grants for technical assistance under subsection (b), grant funds shall be available for not more than—

"(A) an amount equal to the lesser of—

"(i)$50,000; or

"(ii) 75 percent of the cost of feasibility studies to assess the potential for implementation or improvement of sustainable energy infrastructure;

"(B) an amount equal to the lesser of—

"(i) $90,000; or

"(ii) 60 percent of the cost of guidance on overcoming barriers to project implementation, including financial, contracting, siting, and permitting barriers; and

"(C) an amount equal to the lesser of—

"(i) $250,000; or

"(ii) 40 percent of the cost of detailed engineering and design of sustainable energy infrastructure.

"(3) GRANTS FOR EFFICIENCY IMPROVEMENT AND ENERGY SUSTAINABILITY.—In the case of grants for efficiency improvement and energy sustainability under subsection (c), grant funds shall be available for not more than an amount equal to the lesser of—

"(A) $1,000,000; or

"(B) 60 percent of the total cost.

"(4) GRANTS FOR INNOVATION IN ENERGY SUSTAINABILITY.—

In the case of grants for innovation in energy sustainability under subsection (d), grant funds shall be available for not more than an amount equal to the lesser of—

"(A) $500,000; or

"(B) 75 percent of the total cost.

"(g) LOANS FOR ENERGY EFFICIENCY IMPROVEMENT AND ENERGY SUSTAINABILITY.—

"(1) IN GENERAL.—Subject to the availability of appropriated funds, the Secretary shall provide loans to institutional entities for the purpose of implementing energy efficiency improvements and sustainable energy infrastructure.

"(2) TERMS AND CONDITIONS.—

"(A) IN GENERAL.—Except as otherwise provided in this paragraph, loans made under this subsection shall be on such terms and conditions as the Secretary may prescribe.

"(B) MATURITY.—The final maturity of loans made within a period shall be the lesser of, as determined by the Secretary—

"(i) 20 years; or

"(ii) 90 percent of the useful life of the principal physical asset to be financed by the loan.

"(C) DEFAULT.—No loan made under this subsection may be subordinated to another debt contracted by the institutional entity or to any other claims against the institutional entity in the case of default.

"(D) BENCHMARK INTEREST RATE.—

"(i) IN GENERAL.—Loans under this subsection shall be at an interest rate that is set by reference to a benchmark interest rate (yield) on marketable Treasury securities with a similar maturity to the direct loans being made.

"(ii) MINIMUM.—The minimum interest rate of loans under this subsection shall be at the interest rate of the benchmark financial instrument.

"(iii) NEW LOANS.—The minimum interest rate of new loans shall be adjusted each quarter to take account of changes in the interest rate of the benchmark financial instrument.

"(E) CREDIT RISK.—The Secretary shall—

"(i) prescribe explicit standards for use in periodically assessing the credit risk of making direct loans under this subsection; and

"(ii) find that there is a reasonable assurance of repayment before making a loan.

"(F) ADVANCE BUDGET AUTHORITY REQUIRED.—New direct loans may not be obligated under this subsection except to the extent that appropriations of budget authority to cover the costs of the new direct loans are made in advance, as required by section 504 of the Federal Credit Reform Act of 1990 (2 U.S.C. 661c).

"(3) CRITERIA.—Evaluation of projects for potential loan funding shall be based on criteria established by the Secretary, including criteria relating to—

"(A) improvement in energy efficiency;

"(B) reduction in greenhouse gas emissions and other air emissions, including criteria air pollutants and ozone- depleting refrigerants;

"(C) increased use of renewable electric energy sources or renewable thermal energy sources;

"(D) reduction in consumption of fossil fuels; and

"(E) need for funding assistance, including consideration of the size of endowment or other financial resources available to the institutional entity.

"(4) LABOR STANDARDS.—

"(A) IN GENERAL.—All laborers and mechanics employed by contractors or subcontractors in the performance of construction, repair, or alteration work funded in whole or in part under this section shall be paid wages at rates not less than those prevailing on projects of a character similar in the locality as determined by the Secretary of Labor in accordance with sections 3141 through 3144, 3146, and 3147 of title 40, United States Code. The Secretary shall not approve any such funding without first obtaining adequate assurance that required labor standards will be maintained upon the construction work.

"(B) AUTHORITY AND FUNCTIONS.—The Secretary of Labor shall have, with respect to the labor standards specified in paragraph (1), the authority and functions set forth in Reorganization Plan Number 14 of 1950 (15 Fed. Reg. 3176; 64 Stat. 1267) and section 3145 of title 40, United States Code.

"(h) PROGRAM PROCEDURES.—Not later than 180 days after the date of enactment of this section, the Secretary shall establish procedures for the solicitation and evaluation of potential projects for grant and loan funding and administration of the grant and loan programs.

"(i) AUTHORIZATION.—

"(1) GRANTS.—There is authorized to be appropriated for the cost of grants authorized in subsections (b), (c), and (d) $250,000,000 for each of fiscal years 2009 through 2013, of which not more than 5 percent may be used for administrative expenses.

"(2) LOANS.—There is authorized to be appropriated for the initial cost of direct loans authorized in subsection (g) $500,000,000 for each of fiscal years 2009 through 2013, of which not more than 5 percent may be used for administrative expenses.".

Sec. 481. Application Of International Energy Conservation Code To Public And Assisted Housing.

Section 109 of the Cranston-Gonzalez National Affordable Housing Act (42 U.S.C. 12709) is amended—

(1) in subsection (a)—

(A) in paragraph (1)(C), by striking ", where such standards are determined to be cost effective by the Secretary of Housing and Urban Development"; and

(B) in the first sentence of paragraph (2)—

(i)by striking "Council of American Building Officials Model Energy Code, 1992" and inserting "2006 International Energy Conservation Code"; and

(ii) by striking ", and, with respect to rehabilitation and new construction of public and assisted housing funded by HOPE VI revitalization grants under section 24 of the United States Housing Act of 1937 (42 U.S.C. 1437v), the 2003 International Energy Conservation Code";

(2) in subsection (b)—

(A) in the heading, by striking "MODEL ENERGY CODE.—" and inserting "INTERNATIONAL ENERGY CONSERVATION CODE.—";

(B) by inserting "and rehabilitation" after "all new construction"; and

(C) by striking ", and, with respect to rehabilitation and new construction of public and assisted housing funded by HOPE VI revitalization grants under section 24 of the United States Housing Act of 1937 (42 U.S.C. 1437v), the 2003 International Energy Conservation Code";

(3) in subsection (c)—

(A) in the heading, by striking "MODEL ENERGY CODE AND"; and

(B) by striking ", or, with respect to rehabilitation and new construction of public and assisted housing funded by HOPE VI revitalization grants under section 24 of the United States Housing Act of 1937 (42 U.S.C. 1437v), the 2003 International Energy Conservation Code";

(4) by adding at the end the following:

"(d) FAILURE TO AMEND THE STANDARDS.—If the Secretary of Housing and Urban Development and the Secretary of Agriculture have not, within 1 year after the requirements of the 2006 IECC or the ASHRAE Standard 90.1–2004 are revised, amended the standards or made a determination under subsection (c), all new construction and rehabilitation of housing specified in subsection (a) shall meet the requirements of the revised code or standard if—

"(1) the Secretary of Housing and Urban Development or the Secretary of Agriculture make a determination that the revised codes do not negatively affect the availability or affordability of new construction of assisted housing and single family and multifamily residential housing (other than manufactured homes) subject to mortgages insured under the National Housing Act (12 U.S.C. 1701 et seq.) or insured, guaranteed, or made by the Secretary of Agriculture under title V of the Housing Act of 1949 (42 U.S.C. 1471 et seq.), respectively; and

"(2) the Secretary of Energy has made a determination under section 304 of the Energy Conservation and Production Act (42 U.S.C. 6833) that the revised code or standard would improve energy efficiency.";

(5) by striking "CABO Model Energy Code, 1992" each place it appears and inserting "the 2006 IECC"; and

(6) by striking "1989" each place it appears and inserting "2004".

Subtitle H—General Provisions

Sec. 491. Demonstration Project.

(a) IN GENERAL.—The Federal Director and the Commercial Director shall establish guidelines to implement a demonstration project to contribute to the research goals of the Office of Commercial High-Performance Green Buildings and the Office of Federal High-Performance Green Buildings.

(b) PROJECTS.—In accordance with guidelines established by the Federal Director and the Commercial Director under subsection (a) and the duties of the Federal Director and the Commercial Director described in this title, the Federal Director or the Commercial Director shall carry out—

(1) for each of fiscal years 2009 through 2014, 1 demonstration project per year of green features in a Federal building selected by the Federal Director in accordance with relevant agencies and described in subsection (c)(1), that—

(A)provides for instrumentation, monitoring, and data collection related to the green features, for study of the impact of the features on overall energy use and operational costs, and for the evaluation of the information obtained through the conduct of projects and activities under this title; and

(B) achieves the highest rating offered by the high performance green building system identified pursuant to section 436(h);

(2) no fewer than 4 demonstration projects at 4 universities, that, as competitively selected by the Commercial Director in accordance with subsection (c)(2), have—
(A) appropriate research resources and relevant projects to meet the goals of the demonstration project established by the Office of Commercial High-Performance Green Buildings; and
(B) the ability—
(i) to serve as a model for high-performance green building initiatives, including research and education by achieving the highest rating offered by the high performance green building system identified pursuant to section 436(h);
(ii) to identify the most effective ways to use high- performance green building and landscape technologies to engage and educate undergraduate and graduate students;
(iii) to effectively implement a high-performance green building education program for students and occupants;
(iv) to demonstrate the effectiveness of various high-performance technologies, including their impacts on energy use and operational costs, in each of the 4 climatic regions of the United States described in subsection (c)(2)(B); and
(v) to explore quantifiable and nonquantifiable beneficial impacts on public health and employee and student performance;
(3) demonstration projects to evaluate replicable approaches of achieving high performance in actual building operation in various types of commercial buildings in various climates; and
(4) deployment activities to disseminate information on and encourage widespread adoption of technologies, practices, and policies to achieve zero-net-energy commercial buildings or low energy use and effective monitoring of energy use in commercial buildings.

(c) CRITERIA.—
(1) FEDERAL FACILITIES.—With respect to the existing or proposed Federal facility at which a demonstration project under this section is conducted, the Federal facility shall—
(A) be an appropriate model for a project relating to—
(i) the effectiveness of high-performance technologies;
(ii) analysis of materials, components, systems, and emergency operations in the building, and the impact of those materials, components, and systems, including the impact on the health of building occupants;
(iii) life-cycle costing and life-cycle assessment of building materials and systems; and
(iv) location and design that promote access to the Federal facility through walking, biking, and mass transit; and
(B) possess sufficient technological and organizational adaptability.
(2) UNIVERSITIES.—With respect to the 4 universities at which a demonstration project under this section is conducted—
(A) the universities should be selected, after careful review of all applications received containing the required information, as determined by the

Commercial Director, based on—

(i) successful and established public-private research and development partnerships;

(ii) demonstrated capabilities to construct or renovate buildings that meet high indoor environmental quality standards;

(iii) organizational flexibility;

(iv)technological adaptability;

(v) the demonstrated capacity of at least 1 university to replicate lessons learned among nearby or sister universities, preferably by participation in groups or consortia that promote sustainability;

(vi) the demonstrated capacity of at least 1 university to have officially-adopted, institution-wide "high- performance green building" guidelines for all campus building projects; and

(vii) the demonstrated capacity of at least 1 university to have been recognized by similar institutions as a national leader in sustainability education and curriculum for students of the university; and

(B) each university shall be located in a different climatic region of the United States, each of which regions shall have, as determined by the Office of Commercial High-Performance Green Buildings—

(i) a hot, dry climate;

(ii)a hot, humid climate;

(iii) a cold climate; or

(iv) a temperate climate (including a climate with cold winters and humid summers).

(d) APPLICATIONS.—To receive a grant under subsection (b), an eligible applicant shall submit to the Federal Director or the Commercial Director an application at such time, in such manner, and containing such information as the Director may require, including a written assurance that all laborers and mechanics employed by contractors or subcontractors during construction, alteration, or repair that is financed, in whole or in part, by a grant under this section shall be paid wages at rates not less than those prevailing on similar construction in the locality, as determined by the Secretary of Labor in accordance with sections 3141 through 3144, 3146, and 3147 of title 40, United States Code. The Secretary of Labor shall, with respect to the labor standards described in this subsection, have the authority and functions set forth in Reorganization Plan Numbered 14 of 1950 (5 U.S.C. App.) and section 3145 of title 40, United States Code.

(e) REPORT.—Not later than 1 year after the date of enactment of this Act, and annually thereafter through September 30, 20 14—

(1)the Federal Director and the Commercial Director shall submit to the Secretary a report that describes the status of the demonstration projects; and

(2) each University at which a demonstration project under this section is conducted shall submit to the Secretary a report that describes the status of the demonstration projects under this section.

(f) AUTHORIZATION OF APPROPRIATIONS.—There is authorized to be appropriated to carry out the demonstration project described in section (b)(1), $10,000,000 for the period of fiscal years 2008 through 2012, and to carry out the demonstration project described in section (b)(2), $10,000,000 for the period of fiscal years 2008 through 2012, to remain available until expended.

Sec. 492. Research And Development.

(a) ESTABLISHMENT.—The Federal Director and the Commercial Director, jointly and in coordination with the Advisory Committee, shall—

(1)(A) survey existing research and studies relating to high- performance green buildings; and

(B) coordinate activities of common interest;

(2) develop and recommend a high-performance green building research plan that—

(A) identifies information and research needs, including the relationships between human health, occupant productivity, safety, security, and accessibility and each of—

(i) emissions from materials and products in the building;

(ii) natural day lighting;

(iii) ventilation choices and technologies;

(iv) heating, cooling, and system control choices and technologies;

(v) moisture control and mold;

(vi) maintenance, cleaning, and pest control activities;

(vii) acoustics;

(viii) access to public transportation; and

(ix) other issues relating to the health, comfort, productivity, and performance of occupants of the building;

(B) promotes the development and dissemination of high-performance green building measurement tools that, at a minimum, may be used—

(i) to monitor and assess the life-cycle performance of facilities (including demonstration projects) built as high-performance green buildings; and

(ii) to perform life-cycle assessments; and

(C) identifies and tests new and emerging technologies for high-performance green buildings;

(3) assist the budget and life-cycle costing functions of the Directors' Offices under section 436(d);

(4) study and identify potential benefits of green buildings relating to security, natural disaster, and emergency needs of the Federal Government; and

(5) support other research initiatives determined by the Directors' Offices.

(b) INDOOR AIR QUALITY.—The Federal Director, in consultation with the Administrator of the Environmental Protection Agency and the Advisory Committee, shall develop and carry out a comprehensive indoor air quality program for all Federal facilities to ensure the safety of Federal workers and facility occupants—

(1) during new construction and renovation of facilities; and

(2) in existing facilities.

Sec. 493. Environmental Protection Agency DemonstrationGrant Program For Local Governments.

Title III of the Clean Air Act (42 U.S.C. 7601 et seq.) is amended by adding at the end the following:

Sec. 329. Demonstration Grant Program For Local Governments.

"(a) GRANT PROGRAM.—

"(1) IN GENERAL.—The Administrator shall establish a demonstration program under which the Administrator shall provide competitive grants to assist local governments (such as municipalities and counties), with respect to local government buildings—

"(A) to deploy cost-effective technologies and practices;

"(B) to achieve operational cost savings, through the application of cost-effective technologies and practices, as verified by the Administrator.

"(2) COST SHARING.—

"(A) IN GENERAL.—The Federal share of the cost of an activity carried out using a grant provided under this section shall be 40 percent.

"(B) WAIVER OF NON-FEDERAL SHARE.—The Administrator may waive up to 100 percent of the local share of the cost of any grant under this section should the Administrator determine that the community is economically distressed, pursuant to objective economic criteria established by the Administrator in published guidelines.

"(3) MAXIMUM AMOUNT.—The amount of a grant provided under this subsection shall not exceed $1,000,000.

"(b) GUIDELINES.—

"(1) IN GENERAL.—Not later than 1 year after the date Deadline. of enactment of this section, the Administrator shall issue guidelines to implement the grant program established under subsection (a).

"(2) REQUIREMENTS.—The guidelines under paragraph (1) shall establish—

"(A) standards for monitoring and verification of operational cost savings through the application of cost-effective technologies and practices reported by grantees under this section;

"(B) standards for grantees to implement training programs, and to provide technical assistance and education, relating to the retrofit of buildings using cost-effective technologies and practices; and

"(C) a requirement that each local government that receives a grant under this section shall achieve facility- wide cost savings, through renovation of existing local government buildings using cost-effective technologies and practices, of at least 40 percent as compared to the baseline operational costs of the buildings before the renovation (as calculated assuming a 3-year, weather-normalized average).

"(c) COMPLIANCE WITH STATE AND LOCAL LAW.—Nothing in this section or any program carried out using a grant provided under this section supersedes or otherwise affects any State or local law, to the extent that the State or local law contains a requirement that is more stringent than the relevant requirement of this section.

"(d) AUTHORIZATION OF APPROPRIATIONS.—There is authorized to be appropriated to carry out this section $20,000,000 for each of fiscal years 2007 through 2012.

"(e) REPORTS.—

"(1) IN GENERAL.—The Administrator shall provide annual reports to Congress on cost savings achieved and actions taken and recommendations made under this section, and any recommendations for further action.

"(2) FINAL REPORT.—The Administrator shall issue a final report at the conclusion of the program, including findings, a summary of total cost savings achieved, and recommendations for further action.

(f) TERMINATION.—The program under this section shall terminate on September 30, 2012.

(g) DEFINITIONS.—In this section, the terms 'cost-effective technologies and practices' and 'operating cost savings' shall have the meanings defined in section 401 of the Energy Independence and Security Act of 2007.".

Sec. 494. Green Building Advisory Committee.

(a) ESTABLISHMENT.—Not later than 180 days after the date of enactment of this Act, the Federal Director, in coordination with the Commercial Director, shall establish an advisory committee, to be known as the "Green Building Advisory Committee".

(b) MEMBERSHIP.—

(1) IN GENERAL.—The Committee shall be composed of representatives of, at a minimum—

(A) each agency referred to in section 421(e); and

(B) other relevant agencies and entities, as determined by the Federal Director, including at least 1 representative of each of—

(i) State and local governmental green building programs;

(ii) independent green building associations or councils;

(iii) building experts, including architects, material suppliers, and construction contractors;

(iv) security advisors focusing on national security needs, natural disasters, and other dire emergency situations;

(v) public transportation industry experts; and

(vi) environmental health experts, including those with experience in children's health.

(2) NON-FEDERAL MEMBERS.—The total number of non-Federal members on the Committee at any time shall not exceed 15.

(c) MEETINGS.—The Federal Director shall establish a regular schedule of meetings for the Committee.

(d) DUTIES.—The Committee shall provide advice and expertise for use by the Federal Director in carrying out the duties under this subtitle, including such recommendations relating to Federal activities carried out under sections 434 through 436 as are agreed to by a majority of the members of the Committee.

(e) FACA EXEMPTION.—The Committee shall not be subject to section 14 of the Federal Advisory Committee Act (5 U.S.C. App.).

Sec. 495. Advisory Committee on Energy Efficiency Finance

(a) ESTABLISHMENT.—The Secretary, acting through the Assistant Secretary of Energy for Energy Efficiency and Renewable Energy, shall establish an Advisory Committee on Energy Efficiency Finance to provide advice and recommendations to the Department on energy efficiency finance and investment issues, options, ideas, and trends, and to assist the energy community in identifying practical ways of lowering costs and increasing investments in energy efficiency technologies.

(b) MEMBERSHIP.—The advisory committee established under this section shall have a balanced membership that shall include members with expertise in—
(1) availability of seed capital;
(2) availability of venture capital;
(3) availability of other sources of private equity;
(4) investment banking with respect to corporate finance;
(5) investment banking with respect to mergers and acquisitions;
(6) equity capital markets;
(7) debt capital markets;
(8) research analysis;
(9) sales and trading;
(10) commercial lending; and
(11) residential lending.

(c) TERMINATION.—The Advisory Committee on Energy Efficiency Finance shall terminate on the date that is 10 years after the date of enactment of this Act.

(d) AUTHORIZATION OF APPROPRIATIONS.—There are authorized to be appropriated such sums as are necessary to the Secretary for carrying out this section.

TITLE V—ENERGY SAVINGS IN GOVERNMENT AND PUBLIC INSTITUTIONS

Subtitle A — United States Capitol Complex

Sec. 501. Capitol Complex Photovoltaic Roof Feasibility Studies.

(a) STUDIES.—The Architect of the Capitol may conduct feasibility studies regarding construction of photovoltaic roofs for the Rayburn House Office Building and the Hart Senate Office Building.

(b) REPORT.—Not later than 6 months after the date of enactment of this Act, the Architect of the Capitol shall transmit to the Committee on Transportation and Infrastructure of the House of Representatives and the Committee on Rules and Administration of the Senate, a report on the results of the feasibility studies and recommendations regarding construction of photovoltaic roofs for the buildings referred to in subsection (a).

(c) AUTHORIZATION OF APPROPRIATIONS.—There is authorized to be appropriated to carry out this section $500,000.

Sec. 502. Capitol Complex E–85 Refueling Station.

(a) CONSTRUCTION.—The Architect of the Capitol may construct a fuel tank and pumping system for E–85 fuel at or within close proximity to the Capitol Grounds Fuel Station.

(b) USE.—The E–85 fuel tank and pumping system shall be available for use by all legislative branch vehicles capable of operating with E–85 fuel, subject to such other legislative branch agencies reimbursing the Architect of the Capitol for the costs of E–85 fuel used by such other legislative branch vehicles.

(c) AUTHORIZATION OF APPROPRIATIONS.—There is authorized to be appropriated to carry out this section $640,000 for fiscal year 2008.

Sec. 503. Energy And Environmental Measures In Capitol Complex Master Plan.

(a) IN GENERAL.—To the maximum extent practicable, the Architect of the Capitol shall include energy efficiency and conservation measures, greenhouse gas emission reduction measures, and other appropriate environmental measures in the Capitol Complex Master Plan.

(b) REPORT.—Not later than 6 months after the date of enactment of this Act, the Architect of the Capitol shall submit to the Committee on Transportation and Infrastructure of the House of Representatives and the Committee on Rules and Administration of the Senate, a report on the energy efficiency and conservation measures, greenhouse gas emission reduction measures, and other appropriate environmental measures included in the Capitol Complex Master Plan pursuant to subsection (a).

Sec. 504. Promoting Maximum Efficiency In Operation Of Capitol Power Plant.

(a) STEAM BOILERS.—

(1) IN GENERAL.—The Architect of the Capitol shall take such steps as may be necessary to operate the steam boilers at the Capitol Power Plant in the most energy efficient manner possible to minimize carbon emissions and operating costs, including adjusting steam pressures and adjusting the operation of the boilers to take into account variations in demand, including seasonality, for the use of the system.

(2) EFFECTIVE DATE.—The Architect shall implement the steps required under paragraph (1) not later than 30 days after the date of the enactment of this Act.

(b) CHILLER PLANT.—

(1) IN GENERAL.—The Architect of the Capitol shall take such steps as may be necessary to operate the chiller plant at the Capitol Power Plant in the most energy efficient manner possible to minimize carbon emissions and operating costs, including adjusting water temperatures and adjusting the operation of the chillers to take into account variations in demand, including seasonality, for the use of the system.

(2) EFFECTIVE DATE.—The Architect shall implement the steps required under paragraph (1) not later than 30 days after the date of the enactment of this Act.

(c) METERS.—Not later than 90 days after the date of the enactment of this Act, the Architect of the Capitol shall evaluate the accuracy of the meters in use at the Capitol Power Plant and correct them as necessary.

(d) REPORT ON IMPLEMENTATION.—Not later than 180 days after the date of the enactment of this Act, the Architect of the Capitol shall complete the implementation of the requirements of this section and submit a report describing the actions taken and the energy efficiencies achieved to the Committee on Transportation and Infrastructure of the House of Representatives, the Committee on Commerce, Science, and Transportation of the Senate, the Committee on House Administration of the House of Representatives, and the Committee on Rules and Administration of the Senate.

Sec. 505. Capitol Power Plant Carbon Dioxide Emissions Feasibility Study And Demonstration Projects.

The first section of the Act of March 4, 1911 (2 U.S.C. 2162; 36 Stat. 1414, chapter 285) is amended in the seventh undesignated paragraph (relating to the Capitol Power Plant) under the heading "Public Buildings", under the heading "Under the Department of Interior"—

(1) by striking "ninety thousand dollars:" and inserting $90,000."; and

(2) by striking "Provided, That hereafter the" and all that follows through the end of the proviso and inserting the following:

"(a) DESIGNATION.—The heating, lighting, and power plant constructed under the terms of the Act approved April 28, 1904 (33 Stat. 479, chapter 1762) shall be known as the 'Capitol Power Plant'.

"(b) DEFINITION.—In this section, the term 'carbon dioxide energy efficiency' means the quantity of electricity used to power equipment for carbon dioxide capture and storage or use.

"(c) FEASIBILITY STUDY.—The Architect of the Capitol shall conduct a feasibility study evaluating the available methods to capture, store, and use carbon dioxide emitted from the Capitol Power Plant as a result of burning fossil fuels. In carrying out the feasibility study, the Architect of the Capitol is encouraged to consult with individuals with expertise in carbon capture and storage or use, including experts with the Environmental Protection Agency, Department of Energy, academic institutions, non-profit organizations, and industry, as appropriate. The study shall consider—

"(1) the availability of technologies to capture and store or use Capitol Power Plant carbon dioxide emissions;

"(2) strategies to conserve energy and reduce carbon dioxide emissions at the Capitol Power Plant; and

"(3) other factors as determined by the Architect of the Capitol.

"(d) DEMONSTRATION PROJECTS.—

"(1) IN GENERAL.—If the feasibility study determines that a demonstration project to capture and store or use Capitol Power Plant carbon dioxide emissions is technologically feasible and economically justified (including direct and indirect economic and environmental benefits), the Architect of the Capitol may conduct 1 or more demonstration projects to capture and store or use carbon dioxide emitted from the Capitol Power Plant as a result of burning fossil fuels.

"(2) FACTORS FOR CONSIDERATION.—In carrying out such demonstration projects, the Architect of the Capitol shall consider—

"(A) the amount of Capitol Power Plant carbon dioxide emissions to be captured and stored or used;

"(B) whether the proposed project is able to reduce air pollutants other than carbon dioxide;

"(C) the carbon dioxide energy efficiency of the proposed project;

"(D) whether the proposed project is able to use carbon dioxide emissions;

"(E) whether the proposed project could be expanded to significantly increase the amount of Capitol Power Plant carbon dioxide emissions to be captured and stored or used;

"(F) the potential environmental, energy, and educational benefits of demonstrating the capture and storage or use of carbon dioxide at the U.S. Capitol; and

"(G) other factors as determined by the Architect of the Capitol.

"(3) TERMS AND CONDITIONS.—A demonstration project funded under this section shall be subject to such terms and conditions as the Architect of the Capitol may prescribe.

"(e) AUTHORIZATION OF APPROPRIATIONS.—There is authorized to be appropriated to carry out the feasibility study and demonstration project $3,000,000. Such sums shall remain available until expended.".

Subtitle B — Energy Savings Performance Contracting

Sec. 511. Authority To Enter Into Contracts; Reports.

(a) IN GENERAL.—Section 801(a)(2)(D) of the National Energy Conservation Policy Act (42 U.S.C. 8287(a)(2)(D)) is amended—

(1) in clause (ii), by inserting "and" after the semicolon at the end;

(2) by striking clause (iii); and

(3) by redesignating clause (iv) as clause (iii).

(b) REPORTS.—Section 548(a)(2) of the National Energy Conservation Policy Act (42 U.S.C. 8258(a)(2)) is amended by inserting "and any termination penalty exposure" after "the energy and cost savings that have resulted from such contracts".

(c) CONFORMING AMENDMENT.—Section 2913 of title 10, United States Code, is amended by striking subsection (e).

Sec. 512. Financing Flexibility.

Section 801(a)(2) of the National Energy Conservation Policy Act (42 U.S.C. 8287(a)(2)) is amended by adding at the end the following:

"(E) FUNDING OPTIONS.—In carrying out a contract under this title, a Federal agency may use any combination of—

"(i) appropriated funds; and

"(ii) private financing under an energy savings performance contract.".

Sec. 513. Promoting Long-Term Energy Savings Performance Contracts And Verifying Savings.

Section 801(a)(2) of the National Energy Conservation Policy Act (42 U.S.C. 8287(a)(2)) (as amended by section 512) is amended—

(1) in subparagraph (D), by inserting "beginning on the date of the delivery order" after "25 years"; and

(2) by adding at the end the following:

"(F) PROMOTION OF CONTRACTS.—In carrying out this section, a Federal agency shall not—

"(i) establish a Federal agency policy that limits the maximum contract term under subparagraph (D) to a period shorter than 25 years; or

"(ii) limit the total amount of obligations under energy savings performance contracts or other private financing of energy savings measures.

"(G) MEASUREMENT AND VERIFICATION REQUIREMENTS FOR PRIVATE FINANCING.—

"(i) IN GENERAL.—In the case of energy savings performance contracts, the evaluations and savings measurement and verification required under paragraphs (2) and (4) of section 543(f) shall be used by a Federal agency to meet the requirements for the need for energy audits, calculation of energy savings, and any other evaluation of costs and savings needed to implement the guarantee of savings under this section.

"(ii) MODIFICATION OF EXISTING CONTRACTS.—Not Deadline. Later than 18 months after the date of enactment of this subparagraph, each Federal agency shall, to the maximum extent practicable, modify any indefinite delivery and indefinite quantity energy savings performance contracts, and other indefinite delivery and indefinite quantity contracts using private financing, to conform to the amendments made by subtitle B of title V of the Energy Independence and Security Act of 2007.".

Sec. 514. Permanent Reauthorization.

Section 801 of the National Energy Conservation Policy Act (42 U.S.C. 8287) is amended by striking subsection (c).

Sec. 515. Definition Of Energy Savings.

Section 804(2) of the National Energy Conservation Policy Act (42 U.S.C. 8287c(2)) is amended—

(1) by redesignating subparagraphs (A), (B), and (C) as clauses (i), (ii), and (iii), respectively, and indenting appropriately;

(2) by striking "means a reduction" and inserting "means—

"(A) a reduction";

(3) by striking the period at the end and inserting a semicolon; and

(4) by adding at the end the following:

"(B) the increased efficient use of an existing energy source by cogeneration or heat recovery;

"(C) if otherwise authorized by Federal or State law (including regulations), the sale or transfer of electrical or thermal energy generated on-site from renewable energy sources or cogeneration, but in excess of Federal needs, to utilities or non-Federal energy users; and

"(D) the increased efficient use of existing water sources in interior or exterior applications.".

Sec. 516. Retention Of Savings.

Section 546(c) of the National Energy Conservation Policy Act (42 U.S.C. 8256(c)) is amended by striking paragraph (5).

Sec. 517. Training Federal Contracting Officers To Nego 42 Usc 17131. Tiate Energy Efficiency Contracts.

(a) PROGRAM.—The Secretary shall create and administer in the Federal Energy Management Program a training program to educate Federal contract negotiation and contract management personnel so that the contract officers are prepared to—

(1) negotiate energy savings performance contracts;

(2) conclude effective and timely contracts for energy efficiency services with all companies offering energy efficiency services; and

(3) review Federal contracts for all products and services for the potential energy efficiency opportunities and implications of the contracts.

(b) SCHEDULE.—Not later than 1 year after the date of enactment of this Act, the Secretary shall plan, staff, announce, and begin training under the Federal Energy Management Program.

(c) PERSONNEL TO BE TRAINED.—Personnel appropriate to receive training under the Federal Energy Management Program shall be selected by and sent for the training from—

(1) the Department of Defense;

(2) the Department of Veterans Affairs;

(3) the Department;

(4) the General Services Administration;

(5) the Department of Housing and Urban Development;

(6) the United States Postal Service; and

(7) all other Federal agencies and departments that enter contracts for buildings, building services, electricity and electricity services, natural gas and natural gas services, heating and air conditioning services, building fuel purchases, and other types of procurement or service contracts determined by the Secretary, in carrying out the Federal Energy Management Program, to offer the potential for energy savings and greenhouse gas emission reductions if negotiated with taking into account those goals.

(d) TRAINERS.—Training under the Federal Energy Management Program may be conducted by—

(1) attorneys or contract officers with experience in negotiating and managing contracts described in subsection (c)(7) from any agency, except that the Secretary shall reimburse the related salaries and expenses of the attorneys or contract officers from amounts made available for carrying out this section to the extent the attorneys or contract officers are not employees of the Department; and

(2) private experts hired by the Secretary for the purposes of this section, except that the Secretary may not hire experts who are simultaneously employed by any company under contract to provide energy efficiency services to the Federal Government.

(a) AUTHORIZATION OF APPROPRIATIONS.—There are authorized to be appropriated to the Secretary to carry out this section $750,000 for each of fiscal years 2008 through 2012.

Sec. 518. Study Of Energy And Cost Savings In Nonbuilding Applications.

(a) DEFINITIONS.—In this section:

(1) NONBUILDING APPLICATION.—The term "nonbuilding application" means—

(A) any class of vehicles, devices, or equipment that is transportable under the power of the applicable vehicle, device, or equipment by land, sea, or air and that consumes energy from any fuel source for the purpose of—

(i) that transportation; or

(ii) maintaining a controlled environment within the vehicle, device, or equipment; and

(B) any federally-owned equipment used to generate electricity or transport water.

(2) SECONDARY SAVINGS.—

(A) IN GENERAL.—The term "secondary savings" means additional energy or cost savings that are a direct consequence of the energy savings that result from the energy efficiency improvements that were financed and implemented pursuant to an energy savings performance contract.

(B) INCLUSIONS.—The term "secondary savings" includes—

(i) energy and cost savings that result from a reduction in the need for fuel delivery and logistical support;

(ii) personnel cost savings and environmental benefits; and

(iii) in the case of electric generation equipment, the benefits of increased efficiency in the production of electricity, including revenues received by the Federal Government from the sale of electricity so produced.

(b) STUDY.—

(1) IN GENERAL.—As soon as practicable after the date Reports. of enactment of this Act, the Secretary and the Secretary of Defense shall jointly conduct, and submit to Congress and the President, a report of, a study of the potential for the use of energy savings performance contracts to reduce energy consumption and provide energy and cost savings in non- building applications.

(2) REQUIREMENTS.—The study under this subsection shall include—

(A) an estimate of the potential energy and cost savings to the Federal Government, including secondary savings and benefits, from increased efficiency in nonbuilding applications;

(B) an assessment of the feasibility of extending the use of energy savings performance contracts to nonbuilding applications, including an identification of any regulatory or statutory barriers to that use; and

(C) such recommendations as the Secretary and the Secretary of Defense determine to be appropriate.

Subtitle C—Energy Efficiency in Federal Agencies

Sec. 521. Installation Of Photovoltaic System At Department Of Energy Headquarters Building.

(a) IN GENERAL.—The Administrator of General Services shall install a photovoltaic system, as set forth in the Sun Wall Design Project, for the headquarters building of the Department located at 1000 Independence Avenue, SW., Washington, DC, commonly known as the Forrestal Building.

(b) FUNDING.—There shall be available from the Federal Buildings Fund established by section 592 of title 40, United States Code, $30,000,000 to carry out this section. Such sums shall be derived from the unobligated balance of amounts made available from the Fund for fiscal year 2007, and prior fiscal years, for repairs and alternations and other activities (excluding amounts made available for the energy program). Such sums shall remain available until expended.

Sec. 522. Prohibition On Incandescent Lamps By Coast Guard.

(a) PROHIBITION.—Except as provided by subsection (b), on and after January 1, 2009, a general service incandescent lamp shall not be purchased or installed in a Coast Guard facility by or on behalf of the Coast Guard.

(b) EXCEPTION.—A general service incandescent lamp may be purchased, installed, and used in a Coast Guard facility whenever the application of a general service incandescent lamp is—

(1) necessary due to purpose or design, including medical, security, and industrial applications;

(2) reasonable due to the architectural or historical value of a light fixture installed before January 1, 2009; or

(3) the Commandant of the Coast Guard determines that operational requirements necessitate the use of a general service incandescent lamp.

(c) LIMITATION.—In this section, the term "facility" does not include a vessel or aircraft of the Coast Guard.

Sec. 523. Standard Relating To Solar Hot Water Heaters.

Section 305(a)(3)(A) of the Energy Conservation and Production Act (42 U.S.C. 6834(a)(3)(A)) is amended—

(1) in clause (i)(II), by striking "and" at the end;

(2) in clause (ii), by striking the period at the end and inserting "; and"; and

(3) by adding at the end the following:

"(iii) if lifecycle cost-effective, as compared to other reasonably available technologies, not less than 30 percent of the hot water demand for each new Federal building or Federal building undergoing a major renovation be met through the installation and use of solar hot water heaters.".

Sec. 524. Federally-Procured Appliances With Standby Power.

Section 553 of the National Energy Conservation Policy Act (42 U.S.C. 8259b) is amended—

(1) by redesignating subsection (e) as subsection (f); and

(2) by inserting after subsection (d) the following:

"(e) FEDERALLY-PROCURED APPLIANCES WITH STANDBY POWER.—

(1) DEFINITION OF ELIGIBLE PRODUCT.—In this subsection, the term 'eligible product' means a commercially available, off-the-shelf product that—

"(A)(i) uses external standby power devices; or

"(ii) contains an internal standby power function; and "(B) is included on the list compiled under paragraph (4).

(2) FEDERAL PURCHASING REQUIREMENT.—Subject to paragraph (3), if an agency purchases an eligible product, the agency shall purchase—

(A) an eligible product that uses not more than 1 watt in the standby power consuming mode of the eligible product; or

(B) if an eligible product described in subparagraph (A) is not available, the eligible product with the lowest available standby power wattage in the standby power consuming mode of the eligible product.

"(3) LIMITATION.—The requirements of paragraph (2) shall apply to a purchase by an agency only if—

"(A) the lower-wattage eligible product is—

"(i) lifecycle cost-effective; and

"(ii) practicable; and

"(B) the utility and performance of the eligible product is not compromised by the lower wattage requirement.

"(4) ELIGIBLE PRODUCTS.—The Secretary, in consultation. with the Secretary of Defense, the Administrator of the Environmental Protection Agency, and the Administrator of General Services, shall compile a publicly accessible list of cost-effective eligible products that shall be subject to the purchasing requirements of paragraph (2).".

Sec. 525. Federal Procurement of Energy Efficient Products.

(a) AMENDMENTS.—Section 553 of the National Energy Conservation Policy Act (42 U.S.C. 8259b) is amended—

(1) in subsection (b)(1), by inserting "in a product category covered by the Energy Star program or the Federal Energy Management Program for designated products" after "energy consuming product"; and

(2) in the second sentence of subsection (c)—

(A) by inserting "list in their catalogues, represent as available, and" after "Logistics Agency shall"; and

(B) by striking "where the agency" and inserting "in which the head of the agency".

(a) CATALOGUE LISTING DEADLINE.—Not later than 9 months after the date of enactment of this Act, the General Services note. Administration and the Defense Logistics Agency shall ensure that the requirement established by the amendment made by subsection (a)(2)(A) has been fully complied with.

Sec. 526. Procurement And Acquisition Of Alternative Fuels.

No Federal agency shall enter into a contract for procurement of an alternative or synthetic fuel, including a fuel produced from nonconventional petroleum sources, for any mobility-related use, other than for research or testing, unless the contract specifies that the lifecycle greenhouse gas emissions associated with the production and combustion of the fuel supplied under the contract must, on an ongoing basis, be less than or equal to such emissions from the equivalent conventional fuel produced from conventional petroleum sources.

Sec. 527. Government Efficiency Status Reports.

(a) IN GENERAL.—Each Federal agency subject to any of the requirements of this title or the amendments made by this title shall compile and submit to the Director of the Office of Management and Budget an annual Government efficiency status report on—

(1) compliance by the agency with each of the requirements of this title and the amendments made by this title;

(2) the status of the implementation by the agency of initiatives to improve energy efficiency, reduce energy costs, and reduce emissions of greenhouse gases; and

(3) savings to the taxpayers of the United States resulting from mandated improvements under this title and the amendments made by this title.

(b) SUBMISSION.—The report shall be submitted—

(1) to the Director at such time as the Director requires;

(2) in electronic, not paper, format; and

(3) consistent with related reporting requirements.

Sec. 528. Omb Government Efficiency Reports And Scorecards.

(a) REPORTS.—Not later than April 1 of each year, the Director of the Office of Management and Budget shall submit an annual Government efficiency report to the Committee on Oversight and Government Reform of the House of Representatives and the Committee on Governmental Affairs of the Senate, which shall contain—

(1) a summary of the information reported by agencies under section 527;

(2) an evaluation of the overall progress of the Federal Government toward achieving the goals of this title and the amendments made by this title; and

(3) recommendations for additional actions necessary to meet the goals of this title and the amendments made by this title.

(b) SCORECARDS.—The Director of the Office of Management and Budget shall include in any annual energy scorecard the Director is otherwise required to submit a description of the compliance of each agency with the requirements of this title and the amendments made by this title.

Sec. 529. Electricity Sector Demand Response.

(a) IN GENERAL.—Title V of the National Energy Conservation Policy Act (42 U.S.C. 8241 et seq.) is amended by adding at the end the following:

Part 5—Peak Demand Reduction

Sec. 571. National Action Plan For Demand Response.

"(a) NATIONAL ASSESSMENT AND REPORT.—The Federal Energy Regulatory Commission ('Commission') shall conduct a National Assessment of Demand Response. The Commission shall, within 18 months of the date of enactment of this part, submit a report to Congress that includes each of the following:

"(1) Estimation of nationwide demand response potential in 5 and 10 year horizons, including data on a State-by-State basis, and a methodology for updates of such estimates on an annual basis.

"(2) Estimation of how much of this potential can be achieved within 5 and 10 years after the enactment of this part accompanied by specific policy recommendations that if implemented can achieve the estimated potential. Such recommendations shall include options for funding and/or incentives for the development of demand response resources.

"(3) The Commission shall further note any barriers to demand response programs offering flexible, non-discriminatory, and fairly compensatory terms for the services and benefits made available, and shall provide recommendations for overcoming such barriers.

"(4) The Commission shall seek to take advantage of preexisting research and ongoing work, and shall insure that there is no duplication of effort.

"(b) NATIONAL ACTION PLAN ON DEMAND RESPONSE.—The Commission shall further develop a National Action Plan on Demand Response, soliciting and accepting input and participation from a broad range of industry stakeholders, State regulatory utility commissioners, and non-governmental groups. The Commission shall seek consensus where possible, and decide on optimum solutions to issues that defy consensus. Such Plan shall be completed within 1 year after the completion of the National Assessment of Demand Response, and shall meet each of the following objectives:

"(1) Identification of requirements for technical assistance to States to allow them to maximize the amount of demand response resources that can be developed and deployed.

"(2) Design and identification of requirements for implementation of a national communications program that includes broad-based customer education and support.

"(3) Development or identification of analytical tools, information, model regulatory provisions, model contracts, and other support materials for use by customers, States, utilities and demand response providers.

"(c) Upon completion, the National Action Plan on Demand Response shall be published, together with any favorable and dissenting comments submitted by participants in its preparation. Six months after publication, the Commission, together with the Secretary of Energy, shall submit to Congress a proposal to implement the Action Plan, including specific proposed assignments of responsibility, proposed budget amounts, and any agreements secured for participation from State and other participants.

"(d) AUTHORIZATION.—There are authorized to be appropriated to the Commission to carry out this section not more than $10,000,000 for each of the fiscal years 2008, 2009, and 2010.". (b) TABLE OF CONTENTS.—The table of contents for the National Energy Conservation Policy Act (42 U.S.C. 8201 note) is amended by adding after the items relating to part 4 of title V the following:

"PART 5—PEAK DEMAND REDUCTION

"Sec. 571. National Action Plan for Demand Response.".

Subtitle D—Energy Efficiency of Public Institutions

Sec. 531. Reauthorization Of State Energy Programs.

Section 365(f) of the Energy Policy and Conservation Act (42 U.S.C. 6325(f)) is amended by striking "$100,000,000 for each of the fiscal years 2006 and 2007 and $125,000,000 for fiscal year 2008" and inserting "$125,000,000 for each of fiscal years 2007 through 2012".

Sec. 532. Utility Energy Efficiency Programs.

(a) ELECTRIC UTILITIES.—Section 111(d) of the Public Utility Regulatory Policies Act of 1978 (16 U.S.C. 2621(d)) is amended by adding at the end the following:

"(16) INTEGRATED RESOURCE PLANNING.—Each electric utility shall—

"(A) integrate energy efficiency resources into utility, State, and regional plans; and

"(B) adopt policies establishing cost-effective energy efficiency as a priority resource.

"(17) RATE DESIGN MODIFICATIONS TO PROMOTE ENERGY EFFICIENCY INVESTMENTS.—

"(A) IN GENERAL.—The rates allowed to be charged by any electric utility shall—

"(i) align utility incentives with the delivery of cost-effective energy efficiency; and

"(ii) promote energy efficiency investments.

"(B) POLICY OPTIONS.—In complying with subparagraph (A), each State regulatory authority and each non- regulated utility shall consider—

"(i) removing the throughput incentive and other regulatory and management disincentives to energy efficiency;

"(ii) providing utility incentives for the successful management of energy efficiency programs;

"(iii) including the impact on adoption of energy efficiency as 1 of the goals of retail rate design, recognizing that energy efficiency must be balanced with other objectives;

"(iv) adopting rate designs that encourage energy efficiency for each customer class;

"(v) allowing timely recovery of energy efficiency- related costs; and

"(vi) offering home energy audits, offering demand response programs, publicizing the financial and environmental benefits associated with making home energy efficiency improvements, and educating home-owners about all existing Federal and State incentives, including the availability of low-cost loans, that make energy efficiency improvements more affordable.".

(b) NATURAL GAS UTILITIES.—Section 303(b) of the Public Utility Regulatory Policies Act of 1978 (15 U.S.C. 3203(b)) is amended by adding at the end the following:

"(5) ENERGY EFFICIENCY.—Each natural gas utility shall—

"(A) integrate energy efficiency resources into the plans and planning processes of the natural gas utility; and

"(B) adopt policies that establish energy efficiency as a priority resource in the plans and planning processes of the natural gas utility.

"(6) RATE DESIGN MODIFICATIONS TO PROMOTE ENERGY EFFICIENCY INVESTMENTS.—

"(A) IN GENERAL.—The rates allowed to be charged by a natural gas utility shall align utility incentives with the deployment of cost-effective energy efficiency.

"(B) POLICY OPTIONS.—In complying with subparagraph (A), each State regulatory authority and each non- regulated utility shall consider—

"(i)separating fixed-cost revenue recovery from the volume of transportation or sales service provided to the customer;

"(ii) providing to utilities incentives for the successful management of energy efficiency programs, such

"(iii) as allowing utilities to retain a portion of the cost- reducing benefits accruing from the programs;

"(iv) promoting the impact on adoption of energy efficiency as 1 of the goals of retail rate design, recognizing that energy efficiency must be balanced with other objectives; and

"(v) adopting rate designs that encourage energy efficiency for each customer class.

For purposes of applying the provisions of this subtitle to this paragraph, any reference in this subtitle to the date of enactment of this Act shall be treated as a reference to the date of enactment of this paragraph.".

(c) CONFORMING AMENDMENT.—Section 303(a) of the Public Utility Regulatory Policies Act of 1978 (15 U.S.C. 3203(a)) is amended by striking "and (4)" inserting "(4), (5), and (6)".

Subtitle E—Energy Efficiency and Conservation Block Grants

Sec. 541. Definitions.

In this subtitle:

(1) ELIGIBLE ENTITY.—The term "eligible entity" means—

(A) a State;

(B) an eligible unit of local government; and

(C) an Indian tribe.

(2) ELIGIBLE UNIT OF LOCAL GOVERNMENT.—The term "eligible unit of local government" means—

(A) an eligible unit of local government-alternative 1;
and

(B) an eligible unit of local government-alternative 2.

(3)(A) ELIGIBLE UNIT OF LOCAL GOVERNMENT-ALTERNATIVE 1.—The term "eligible unit of local government-alternative 1" means—

(i) a city with a population—

(I) of at least 35,000; or

(II) that causes the city to be 1 of the 10 highest- populated cities of the State in which the city is located; and

(ii) a county with a population—

(I) of at least 200,000; or

(II) that causes the county to be 1 of the 10 highest- populated counties of the State in which the county is located.

(B) ELIGIBLE UNIT OF LOCAL GOVERNMENT-ALTERNATIVE 2.— The term "eligible unit of local government-alternative 2" means—

(i) a city with a population of at least 50,000; or

(ii) a county with a population of at least 200,000.

(4) INDIAN TRIBE.—The term "Indian tribe" has the meaning given the term in section 4 of the Indian Self-Determination and Education Assistance Act (25 U.S.C. 450b).

(5) PROGRAM.—The term "program" means the Energy Efficiency and Conservation Block Grant Program established under section 542(a).

(6) STATE.—The term "State" means—

(A) a State;

(B) the District of Columbia;

(C) the Commonwealth of Puerto Rico; and

(D) any other territory or possession of the United States.

Sec. 542. Energy Efficiency And Conservation Block Grant Program.

(a) ESTABLISHMENT.—The Secretary shall establish a program, to be known as the "Energy Efficiency and Conservation Block Grant Program", under which the Secretary shall provide grants to eligible entities in accordance with this subtitle.

(b) PURPOSE.—The purpose of the program shall be to assist eligible entities in implementing strategies—

(1) to reduce fossil fuel emissions created as a result of activities within the jurisdictions of eligible entities in a manner that—

(A) is environmentally sustainable; and

(B) to the maximum extent practicable, maximizes benefits for local and regional communities;

(2) to reduce the total energy use of the eligible entities;

and

(3) to improve energy efficiency in—

(A) the transportation sector;

(B) the building sector; and

(C) other appropriate sectors.

Sec. 543. Allocation of Funds.

(a) IN GENERAL.—Of amounts made available to provide grants under this subtitle for each fiscal year, the Secretary shall allocate—

(1) 68 percent to eligible units of local government in accordance with subsection (b);

(2) 28 percent to States in accordance with subsection (c);

(3) 2 percent to Indian tribes in accordance with subsection (d); and

(4) 2 percent for competitive grants under section 546.

(b) ELIGIBLE UNITS OF LOCAL GOVERNMENT.—Of amounts available for distribution to eligible units of local government under subsection (a)(1), the Secretary shall provide grants to eligible units of local government under this section based on a formula established by the Secretary according to—

(1) the populations served by the eligible units of local government, according to the latest available decennial census; and

(2) the daytime populations of the eligible units of local government and other similar factors (such as square footage of commercial, office, and industrial space), as determined by the Secretary.

(c) STATES.—Of amounts available for distribution to States under subsection (a)(2), the Secretary shall provide—

(1) not less than 1.25 percent to each State; and

(2) the remainder among the States, based on a formula to be established by the Secretary that takes into account—

(A) the population of each State; and

(B) any other criteria that the Secretary determines to be appropriate.

(d) INDIAN TRIBES.—Of amounts available for distribution to Indian tribes under subsection (a)(3), the Secretary shall establish a formula for allocation of the amounts to Indian tribes, taking into account any factors that the Secretary determines to be appropriate.

(e) PUBLICATION OF ALLOCATION FORMULAS.—Not later than 90 days before the beginning of each fiscal year for which grants are provided under this subtitle, the Secretary shall publish in the Federal Register the formulas for allocation established under this section.

(f) STATE AND LOCAL ADVISORY COMMITTEE.—The Secretary shall establish a State and local advisory committee to advise the Secretary regarding administration, implementation, and evaluation of the program.

Sec. 544. Use Of Funds.

An eligible entity may use a grant received under this subtitle to carry out activities to achieve the purposes of the program, including—

(1) development and implementation of an energy efficiency and conservation strategy under section 545(b);

(2) retaining technical consultant services to assist the eligible entity in the development of such a strategy, including—

(A) formulation of energy efficiency, energy conservation, and energy usage goals;

(B) identification of strategies to achieve those goals—

(i) through efforts to increase energy efficiency and reduce energy consumption; and

(ii) by encouraging behavioral changes among the population served by the eligible entity;

(C) development of methods to measure progress in achieving the goals;

(D) development and publication of annual reports to the population served by the eligible entity describing—

(i) the strategies and goals; and

(ii) the progress made in achieving the strategies and goals during the preceding calendar year; and

(E) other services to assist in the implementation of the energy efficiency and conservation strategy;

(3) conducting residential and commercial building energy audits;

(4) establishment of financial incentive programs for energy efficiency improvements;

(5) the provision of grants to nonprofit organizations and governmental agencies for the purpose of performing energy efficiency retrofits;

(6) development and implementation of energy efficiency and conservation programs for buildings and facilities within the jurisdiction of the eligible entity, including—

(A) design and operation of the programs;

(B) identifying the most effective methods for achieving maximum participation and efficiency rates;

(C) public education;

(D) measurement and verification protocols; and
(E) identification of energy efficient technologies;

(7) development and implementation of programs to conserve energy used in transportation, including—
(A) use of flex time by employers;
(B) satellite work centers;
(C) development and promotion of zoning guidelines or requirements that promote energy efficient development;
(D) development of infrastructure, such as bike lanes and pathways and pedestrian walkways;
(E) synchronization of traffic signals; and
(F) other measures that increase energy efficiency and decrease energy consumption;

(8) development and implementation of building codes and inspection services to promote building energy efficiency;

(9) application and implementation of energy distribution technologies that significantly increase energy efficiency, including—
(A) distributed resources; and
(B) district heating and cooling systems;

(10) activities to increase participation and efficiency rates for material conservation programs, including source reduction, recycling, and recycled content procurement programs that lead to increases in energy efficiency;

(11) the purchase and implementation of technologies to reduce, capture, and, to the maximum extent practicable, use methane and other greenhouse gases generated by landfills or similar sources;

(12) replacement of traffic signals and street lighting with energy efficient lighting technologies, including light emitting diodes; and any other technology of equal or greater energy efficiency;

(13) development, implementation, and installation on or in any government building of the eligible entity of onsite renewable energy technology that generates electricity from renewable resources, including—
(A) solar energy;
(B) wind energy;
(C) fuel cells; and
(D) biomass; and

(14) any other appropriate activity, as determined by the Secretary, in consultation with—
(A) the Administrator of the Environmental Protection Agency;
(B) the Secretary of Transportation; and
(C) the Secretary of Housing and Urban Development.

Sec. 545. Requirements For Eligible Entities.

(a) CONSTRUCTION REQUIREMENT.—

(1) IN GENERAL.—To be eligible to receive a grant under the program, each eligible applicant shall submit to the Secretary a written assurance that all laborers and mechanics employed by any contractor or subcontractor of the eligible entity during any construction, alteration, or repair activity funded, in whole or in part, by the grant shall be paid wages at rates not less than the prevailing wages for

similar construction activities in the locality, as determined by the Secretary of Labor, in accordance with sections 3141 through 3144, 3146, and 3147 of title 40, United States Code.

(2) SECRETARY OF LABOR.—With respect to the labor standards referred to in paragraph (1), the Secretary of Labor shall have the authority and functions described in—

(A) Reorganization Plan Numbered 14 of 1950 (5 U.S.C. 903 note); and

(B) section 3145 of title 40, United States Code.

(b) ELIGIBLE UNITS OF LOCAL GOVERNMENT AND INDIAN TRIBES.—

(1) PROPOSED STRATEGY.—

(A) IN GENERAL.—Not later than 1 year after the date on which an eligible unit of local government or Indian tribe receives a grant under this subtitle, the eligible unit of local government or Indian tribe shall submit to the Secretary a proposed energy efficiency and conservation strategy in accordance with this paragraph.

(B) INCLUSIONS.—The proposed strategy under subparagraph (A) shall include—

(i) a description of the goals of the eligible unit of local government or Indian tribe, in accordance with the purposes of this subtitle, for increased energy efficiency and conservation in the jurisdiction of the eligible unit of local government or Indian tribe; and

(ii) a plan for the use of the grant to assist the eligible unit of local government or Indian tribe in achieving those goals, in accordance with section 544.

(C) REQUIREMENTS FOR ELIGIBLE UNITS OF LOCAL GOVERNMENT.—In developing the strategy under subparagraph (A), an eligible unit of local government shall—

(i) take into account any plans for the use of funds by adjacent eligible units of local governments that receive grants under the program; and

(ii) coordinate and share information with the State in which the eligible unit of local government is located regarding activities carried out using the grant to maximize the energy efficiency and conservation benefits under this subtitle.

(2) APPROVAL BY SECRETARY.—

(A) IN GENERAL.—The Secretary shall approve or disapprove a proposed strategy under paragraph (1) by not later than 120 days after the date of submission of the proposed strategy.

(B) DISAPPROVAL.—If the Secretary disapproves a proposed strategy under subparagraph (A)—

(i) the Secretary shall provide to the eligible unit of local government or Indian tribe the reasons for the disapproval; and

(ii) the eligible unit of local government or Indian tribe may revise and resubmit the proposed strategy as many times as necessary until the Secretary approves a proposed strategy.

(C) REQUIREMENT.—The Secretary shall not provide to an eligible unit of local government or Indian tribe any grant under the program until a proposed strategy of the eligible unit of local government or Indian tribe is approved by the Secretary under this paragraph.

(3) LIMITATIONS ON USE OF FUNDS.—Of amounts provided to an eligible unit of local government or Indian tribe under
the program, an eligible unit of local government or Indian tribe may use—
(A) for administrative expenses, excluding the cost of meeting the reporting requirements of this subtitle, an amount equal to the greater of—
(i) 10 percent; and
(ii) $75,000;
(B) for the establishment of revolving loan funds, an amount equal to the greater of—
(i) 20 percent; and
(ii) $250,000; and
(C) for the provision of subgrants to nongovernmental organizations for the purpose of assisting in the implementation of the energy efficiency and conservation strategy of the eligible unit of local government or Indian tribe, an amount equal to the greater of—
(i) 20 percent; and
(ii) $250,000.
(4) ANNUAL REPORT.—Not later than 2 years after the date on which funds are initially provided to an eligible unit of local government or Indian tribe under the program, and annually thereafter, the eligible unit of local government or Indian tribe shall submit to the Secretary a report describing—
(A) the status of development and implementation of the energy efficiency and conservation strategy of the eligible unit of local government or Indian tribe; and
(B) as practicable, an assessment of energy efficiency gains within the jurisdiction of the eligible unit of local government or Indian tribe.
(c) STATES.—
(1) DISTRIBUTION OF FUNDS.—
"(A) IN GENERAL.—A State that receives a grant under the program shall use not less than 60 percent of the amount received to provide subgrants to units of local government in the State that are not eligible units of local government.
"(B) DEADLINE.—The State shall provide the subgrants required under subparagraph (A) by not later than 180 days after the date on which the Secretary approves a proposed energy efficiency and conservation strategy of the State under paragraph (3).
(2) REVISION OF CONSERVATION PLAN; PROPOSED STRATEGY.—Not later than 120 days after the date of enactment of this Act, each State shall—
(A) modify the State energy conservation plan of the State under section 362 of the Energy Policy and Conservation Act (42 U.S.C. 6322) to establish additional goals for increased energy efficiency and conservation in the State; and
(B) submit to the Secretary a proposed energy efficiency and conservation strategy that—
(i) establishes a process for providing subgrants as required under paragraph (1); and

(ii) includes a plan of the State for the use of funds received under the program to assist the State in achieving the goals established under subparagraph (A), in accordance with sections 542(b) and 544.

(3) APPROVAL BY SECRETARY.—

(A) IN GENERAL.—The Secretary shall approve or dis Deadline. approve a proposed strategy under paragraph (2)(B) by not later than 120 days after the date of submission of the proposed strategy.

(B) DISAPPROVAL.—If the Secretary disapproves a proposed strategy under subparagraph (A)—

(i) the Secretary shall provide to the State the reasons for the disapproval; and

(ii) the State may revise and resubmit the proposed strategy as many times as necessary until the Secretary approves a proposed strategy.

(C) REQUIREMENT.—The Secretary shall not provide to a State any grant under the program until a proposed strategy of the State is approved by the Secretary under this paragraph.

(4) LIMITATIONS ON USE OF FUNDS.—A State may use not more than 10 percent of amounts provided under the program for administrative expenses.

(5) ANNUAL REPORTS.—Each State that receives a grant under the program shall submit to the Secretary an annual report that describes—

(A) the status of development and implementation of the energy efficiency and conservation strategy of the State during the preceding calendar year;

(B) the status of the subgrant program of the State under paragraph (1);

(C) the energy efficiency gains achieved through the energy efficiency and conservation strategy of the State during the preceding calendar year; and

(D) specific energy efficiency and conservation goals of the State for subsequent calendar years.

Sec. 546. Competitive Grants.

(a) IN GENERAL.—Of the total amount made available for each fiscal year to carry out this subtitle, the Secretary shall use not less than 2 percent to provide grants under this section, on a competitive basis, to—

(1) units of local government (including Indian tribes) that are not eligible entities; and

(2) consortia of units of local government described in paragraph (1).

(b) APPLICATIONS.—To be eligible to receive a grant under this section, a unit of local government or consortia shall submit to the Secretary an application at such time, in such manner, and containing such information as the Secretary may require, including a plan of the unit of local government to carry out an activity described in section 544.

(c) PRIORITY.—In providing grants under this section, the Secretary shall give priority to units of local government—

(1) located in States with populations of less than 2,000,000; or

(2) that plan to carry out projects that would result in significant energy efficiency improvements or reductions in fossil fuel use.

Sec. 547. Review And Evaluation.

(a) IN GENERAL.—The Secretary may review and evaluate the performance of any eligible entity that receives a grant under the program, including by conducting an audit, as the Secretary determines to be appropriate.

(b) WITHHOLDING OF FUNDS.—The Secretary may withhold from an eligible entity any portion of a grant to be provided to the eligible entity under the program if the Secretary determines that the eligible entity has failed to achieve compliance with—
(1) any applicable guideline or regulation of the Secretary relating to the program, including the misuse or misappropriation of funds provided under the program; or
(2) the energy efficiency and conservation strategy of the eligible entity.

Sec. 548. Funding.

(a) AUTHORIZATION OF APPROPRIATIONS.—
(1) GRANTS.—There is authorized to be appropriated to the Secretary for the provision of grants under the program $2,000,000,000 for each of fiscal years 2008 through 2012; provided that 49 percent of the appropriated funds shall be distributed using the definition of eligible unit of local government-alternative 1 in section 541(3)(A) and 49 percent of the appropriated funds shall be distributed using the definition of eligible unit of local government-alternative 2 in section 541(3)(B).
(2) ADMINISTRATIVE COSTS.—There are authorized to be appropriated to the Secretary for administrative expenses of the program—
(A) $20,000,000 for each of fiscal years 2008 and 2009;
(B) $25,000,000 for each of fiscal years 2010 and 2011; and
(C) $30,000,000 for fiscal year 2012.
(b) MAINTENANCE OF FUNDING.—The funding provided under this section shall supplement (and not supplant) other Federal funding provided under—
(1) a State energy conservation plan established under part D of title III of the Energy Policy and Conservation Act (42 U.S.C. 6321 et seq.); or
(2) the Weatherization Assistance Program for Low-Income Persons established under part A of title IV of the Energy Conservation and Production Act (42 U.S.C. 6861 et seq.).

TITLE VI — ACCELERATED RESEARCH AND DEVELOPMENT

Subtitle A—Solar Energy

Sec. 601. Short Title

This subtitle may be cited as the “Solar Energy Research and Advancement Act of 2007”.

Sec. 602. Thermal Energy Storage Research And Development Program.

(a) ESTABLISHMENT.—The Secretary shall establish a program of research and development to provide lower cost and more viable thermal energy storage technologies to enable the shifting of electric power loads on demand and extend the operating time of concentrating solar power electric generating plants.
(b) AUTHORIZATION OF APPROPRIATIONS.—There are authorized to be appropriated to the Secretary for carrying out this section $5,000,000 for fiscal year 2008, $7,000,000 for fiscal year 2009, $9,000,000 for fiscal year 2010, $10,000,000 for fiscal year 2011, and $12,000,000 for fiscal year 2012.

Sec. 603. Concentrating Solar Power Commercial Application Studies.

(a) INTEGRATION.—The Secretary shall conduct a study on methods to integrate concentrating solar power and utility-scale photovoltaic systems into regional electricity transmission systems, and to identify new transmission or transmission upgrades needed to bring electricity from high concentrating solar power resource areas to growing electric power load centers throughout the United States. The study shall analyze and assess cost-effective approaches for management and large-scale integration of concentrating solar power and utility-scale photovoltaic systems into regional electric transmission grids to improve electric reliability, to efficiently manage load, and to reduce demand on the natural gas transmission system for electric power. The Secretary shall submit a report to Congress on the results of this study not later than 12 months after the date of enactment of this Act.

(b) WATER CONSUMPTION.—Not later than 6 months after the date of the enactment of this Act, the Secretary of Energy shall transmit to Congress a report on the results of a study on methods to reduce the amount of water consumed by concentrating solar power systems.

Sec. 604. Solar Energy Curriculum Development And Certification Grants.

(a) ESTABLISHMENT.—The Secretary shall establish in the Office of Solar Energy Technologies a competitive grant program to create and strengthen solar industry workforce training and internship programs in installation, operation, and maintenance of solar energy products. The goal of this program is to ensure a supply of well-trained individuals to support the expansion of the solar energy industry.

(b) AUTHORIZED ACTIVITIES.—Grant funds may be used to support the following activities:

(1) Creation and development of a solar energy curriculum appropriate for the local educational, entrepreneurial, and environmental conditions, including curriculum for community colleges.

(2) Support of certification programs for individual solar energy system installers, instructors, and training programs.

(3) Internship programs that provide hands-on participation by students in commercial applications.

(4) Activities required to obtain certification of training programs and facilities by an industry-accepted quality-control certification program.

(5) Incorporation of solar-specific learning modules into traditional occupational training and internship programs for construction-related trades.

(6) The purchase of equipment necessary to carry out activities under this section.

(7) Support of programs that provide guidance and updates to solar energy curriculum instructors.

(c) ADMINISTRATION OF GRANTS.—Grants may be awarded under this section for up to 3 years. The Secretary shall award grants to ensure sufficient geographic distribution of training programs nationally. Grants shall only be awarded for programs certified by an industry-accepted quality-control certification institution, or for new and growing programs with a credible path to certification. Due consideration shall be given to women, underrepresented minorities, and persons with disabilities.

(d) REPORT.—The Secretary shall make public, on the website of the Department or upon request, information on the name and institution for all grants awarded under this section, including a brief description of the project as well as the grant award amount.

(e) AUTHORIZATION OF APPROPRIATIONS.—There are authorized to be appropriated to the Secretary for carrying out this section $10,000,000 for each of the fiscal years 2008 through 2012.

Sec. 605. Daylighting Systems And Direct Solar Light Pipe Technology.

(a) ESTABLISHMENT.—The Secretary shall establish a program of research and development to provide assistance in the demonstration and commercial application of direct solar renewable energy sources to provide alternatives to traditional power generation for lighting and illumination, including light pipe technology, and to promote greater energy conservation and improved efficiency. All direct solar renewable energy devices supported under this program shall have the capability to provide measurable data on the amount of kilowatt-hours saved over the traditionally powered light sources they have replaced.

(b) REPORTING.—The Secretary shall transmit to Congress an annual report assessing the measurable data derived from each project in the direct solar renewable energy sources program and the energy savings resulting from its use.

(c) DEFINITIONS.—For purposes of this section—

(1) the term "direct solar renewable energy" means energy from a device that converts sunlight into useable light within a building, tunnel, or other enclosed structure, replacing artificial light generated by a light fixture and doing so without the conversion of the sunlight into another form of energy; and

(2) the term "light pipe" means a device designed to transport visible solar radiation from its collection point to the interior of a building while excluding interior heat gain in the nonheating season.

(d) AUTHORIZATION OF APPROPRIATIONS.—There are authorized to be appropriated to the Secretary for carrying out this section $3,500,000 for each of the fiscal years 2008 through 2012.

Sec. 606. Solar Air Conditioning Research And Development Program.

(a) ESTABLISHMENT.—The Secretary shall establish a research, development, and demonstration program to promote less costly and more reliable decentralized distributed solar-powered air conditioning for individuals and businesses.

(b) AUTHORIZED ACTIVITIES.—Grants made available under this section may be used to support the following activities:

(1) Advancing solar thermal collectors, including concentrating solar thermal and electric systems, flat plate and evacuated tube collector performance.

(2) Achieving technical and economic integration of solar- powered distributed air-conditioning systems with existing hot water and storage systems for residential applications.

(3) Designing and demonstrating mass manufacturing capability to reduce costs of modular standardized solar-powered distributed air conditioning systems and components.

(4) Improving the efficiency of solar-powered distributed air-conditioning to increase the effectiveness of solar-powered absorption chillers, solar-driven compressors and condensors, and cost-effective precooling approaches.

(5) Researching and comparing performance of solar-powered distributed air conditioning systems in different regions of the country, including potential integration with other onsite systems, such as solar, biogas, geothermal heat pumps, and propane assist or combined propane fuel cells, with a goal to develop site-specific energy production and management systems that ease fuel and peak utility loading.

(c) COST SHARING.—Section 988 of the Energy Policy Act of 2005 (42 U.S.C. 16352) shall apply to a project carried out under this section.

(d) AUTHORIZATION OF APPROPRIATIONS.—There are authorized to be appropriated to the Secretary for carrying out this section $2,500,000 for each of the fiscal years 2008 through 2012.

Sec. 607. Photovoltaic Demonstration Program.

(a) IN GENERAL.—The Secretary shall establish a program of grants to States to demonstrate advanced photovoltaic technology.

(b) REQUIREMENTS.—

(1) ABILITY TO MEET REQUIREMENTS.—To receive funding under the program under this section, a State must submit a proposal that demonstrates, to the satisfaction of the Secretary, that the State will meet the requirements of subsection (f).

(2) COMPLIANCE WITH REQUIREMENTS.—If a State has received funding under this section for the preceding year, the State must demonstrate, to the satisfaction of the Secretary, that it complied with the requirements of subsection (f) in carrying out the program during that preceding year, and that it will do so in the future, before it can receive further funding under this section.

(c) COMPETITION.—The Secretary shall award grants on a Grants. competitive basis to the States with the proposals the Secretary considers most likely to encourage the widespread adoption of photo-

voltaic technologies. The Secretary shall take into consideration the geographic distribution of awards.

(d) PROPOSALS.—Not later than 6 months after the date of Deadline. enactment of this Act, and in each subsequent fiscal year for the life of the program, the Secretary shall solicit proposals from the States to participate in the program under this section.

(e) COMPETITIVE CRITERIA.—In awarding funds in a competitive allocation under subsection (c), the Secretary shall consideration of, or lower the costs of, advanced photovoltaic technologies;

(1) the likelihood of a proposal to encourage the demonstration

and

(2) the extent to which a proposal is likely to—

(A) maximize the amount of photovoltaics demonstrated;

(B) maximize the proportion of non-Federal cost share; and

(C) limit State administrative costs.

(f) STATE PROGRAM.—A program operated by a State with funding under this section shall provide competitive awards for the demonstration of advanced photovoltaic technologies. Each State program shall—

(1) require a contribution of at least 60 percent per award from non-Federal sources, which may include any combination of State, local, and private funds, except that at least 10 percent of the funding must be supplied by the State;

(2) endeavor to fund recipients in the commercial, industrial, institutional, governmental, and residential sectors;

(3) limit State administrative costs to no more than 10 percent of the grant;

(4) report annually to the Secretary on—

(A) the amount of funds disbursed;

(B) the amount of photovoltaics purchased; and

(C) the results of the monitoring under paragraph (5);

(5) provide for measurement and verification of the output of a representative sample of the photovoltaics systems demonstrated throughout the average working life of the systems, or at least 20 years; and

(6) require that applicant buildings must have received an independent energy efficiency audit during the 6-month period preceding the filing of the application.

(g) UNEXPENDED FUNDS.—If a State fails to expend any funds received under this section within 3 years of receipt, such remaining funds shall be returned to the Treasury.

(h) REPORTS.—The Secretary shall report to Congress 5 years after funds are first distributed to the States under this section—

(1) the amount of photovoltaics demonstrated;

(2) the number of projects undertaken;

(3) the administrative costs of the program;

(4) the results of the monitoring under subsection (f)(5); and

(5) the total amount of funds distributed, including a breakdown by State.

(i) AUTHORIZATION OF APPROPRIATIONS.—There are authorized to be appropriated to the Secretary for the purposes of carrying out this section—

(1) $15,000,000 for fiscal year 2008;

(2) $30,000,000 for fiscal year 2009;

(3) $45,000,000 for fiscal year 2010;

(4) $60,000,000 for fiscal year 2011; and

(5) $70,000,000 for fiscal year 2012.

Subtitle B—Geothermal Energy

Sec. 611. Short Title.

This subtitle may be cited as the "Advanced Geothermal Energy Research and Development Act of 2007".

Sec. 612. Definitions.

For purposes of this subtitle:

(1) ENGINEERED.—When referring to enhanced geothermal systems, the term "engineered" means subjected to intervention, including intervention to address one or more of the following issues:

(A) Lack of effective permeability or porosity or open fracture connectivity within the reservoir.

(B) Insufficient contained geofluid in the reservoir.

(C) A low average geothermal gradient, which necessitates deeper drilling.

(2) ENHANCED GEOTHERMAL SYSTEMS.—The term "enhanced geothermal systems" means geothermal reservoir systems that are engineered, as opposed to occurring naturally.

(3) GEOFLUID.—The term "geofluid" means any fluid used to extract thermal energy from the Earth which is transported to the surface for direct use or electric power generation, except that such term shall not include oil or natural gas.

(4) GEOPRESSURED RESOURCES.—The term "geopressured resources" mean geothermal deposits found in sedimentary rocks under higher than normal pressure and saturated with gas or methane.

(5) GEOTHERMAL.—The term "geothermal" refers to heat energy stored in the Earth's crust that can be accessed for direct use or electric power generation.

(6) HYDROTHERMAL.—The term "hydrothermal" refers to naturally occurring subsurface reservoirs of hot water or steam.

(7) SYSTEMS APPROACH.—The term "systems approach" means an approach to solving problems or designing systems that attempts to optimize the performance of the overall system, rather than a particular component of the system.

Sec. 613. Hydrothermal Research And Development.

(a) IN GENERAL.—The Secretary shall support programs of research, development, demonstration, and commercial application to expand the use of geothermal energy production from hydro- thermal systems, including the programs described in subsection (b).

(b) PROGRAMS.—

(1) ADVANCED HYDROTHERMAL RESOURCE TOOLS.—The Secretary, in consultation with other appropriate agencies, shall support a program to develop advanced geophysical, geochemical, and geologic tools to assist in locating hidden hydro- thermal resources, and to increase the reliability of site characterization before, during, and after initial drilling. The program shall develop new prospecting techniques to assist in prioritization of targets for characterization. The program shall include a field component.

(2) INDUSTRY COUPLED EXPLORATORY DRILLING.—The Secretary shall support a program of cost-shared field demonstration programs, to be pursued, simultaneously and independently, in collaboration with industry partners, for the demonstration of advanced technologies and techniques of siting and exploratory drilling for undiscovered resources in a variety of geologic settings. The program shall include incentives to encourage the use of advanced technologies and techniques.

Sec. 614. General Geothermal Systems Research And Development.

(a) SUBSURFACE COMPONENTS AND SYSTEMS.—The Secretary shall support a program of research, development, demonstration, and commercial application of components and systems capable of withstanding extreme geothermal environments and necessary to cost-effectively develop, produce, and monitor geothermal reservoirs and produce geothermal energy. These components and systems shall include advanced casing systems (expandable tubular casing, low-clearance casing designs, and others), high-temperature cements, high-temperature submersible pumps, and high-temperature packers, as well as technologies for under-reaming, multilateral completions, high-temperature and high-pressure logging, logging while drilling, deep fracture stimulation, and reservoir system diagnostics.

(b) RESERVOIR PERFORMANCE MODELING.—The Secretary shall support a program of research, development, demonstration, and commercial application of models of geothermal reservoir performance, with an emphasis on accurately modeling performance over time. Models shall be developed to assist both in the development of geothermal reservoirs and to more accurately account for stress-related effects in stimulated hydrothermal and enhanced geothermal systems production environments.

(c) ENVIRONMENTAL IMPACTS.—The Secretary shall—

(1) support a program of research, development, demonstration, and commercial application of technologies and practices designed to mitigate or preclude potential adverse environmental impacts of geothermal energy development, production or use, and seek to ensure that geothermal energy development is consistent with the highest practicable standards of environmental stewardship;

(2) in conjunction with the Assistant Administrator for Research and Development at the Environmental Protection Agency, support a research program to identify potential environmental impacts of geothermal energy development, production, and use, and ensure that the program described in paragraph (1) addresses such impacts, including effects on groundwater and local hydrology; and

(3) support a program of research to compare the potential environmental impacts identified as part of the development, production, and use of geothermal energy with the potential emission reductions of greenhouse gases gained by geothermal energy development, production, and use.

Sec. 615. Enhanced Geothermal Systems Research And Development.

(a) IN GENERAL.—The Secretary shall support a program of research, development, demonstration, and commercial application for enhanced geothermal systems, including the programs described in subsection (b).

(b) PROGRAMS.—

(1) ENHANCED GEOTHERMAL SYSTEMS TECHNOLOGIES.—The Secretary shall support a program of research, development, demonstration, and commercial application of the technologies and knowledge necessary for enhanced geothermal systems to advance to a state of commercial readiness, including advances in—

(A) reservoir stimulation;

(B) reservoir characterization, monitoring, and modeling;

(C) stress mapping;

(D) tracer development;
(E) three-dimensional tomography; and
(F) understanding seismic effects of reservoir engineering and stimulation.

(2) ENHANCED GEOTHERMAL SYSTEMS RESERVOIR STIMULATION.—

(A) PROGRAM.—In collaboration with industry partners, the Secretary shall support a program of research, development, and demonstration of enhanced geothermal systems reservoir stimulation technologies and techniques. A minimum of 4 sites shall be selected in locations that show particular promise for enhanced geothermal systems development. Each site shall—

(i) represent a different class of subsurface geologic environments; and
(ii) take advantage of an existing site where subsurface characterization has been conducted or existing drill holes can be utilized, if possible.

(B) CONSIDERATION OF EXISTING SITE.—The Desert Peak, Nevada, site, where a Department of Energy and industry cooperative enhanced geothermal systems project is already underway, may be considered for inclusion among the sites selected under subparagraph (A).

Sec. 616. Geothermal Energy Production From Oil And Gas Fields And Recovery And Production Of Geopressured Gas Resources.

(a) IN GENERAL.—The Secretary shall establish a program of research, development, demonstration, and commercial application to support development of geothermal energy production from oil and gas fields and production and recovery of energy, including electricity, from geopressured resources. In addition, the Secretary shall conduct such supporting activities including research, resource characterization, and technology development as necessary.

(b) GEOTHERMAL ENERGY PRODUCTION FROM OIL AND GAS FIELDS.—The Secretary shall implement a grant program in support of geothermal energy production from oil and gas fields. The program shall include grants for a total of not less than three demonstration projects of the use of geothermal techniques such as advanced organic rankine cycle systems at marginal, unproductive, and productive oil and gas wells. The Secretary shall, to the extent Grants. practicable and in the public interest, make awards that—

(1) include not less than five oil or gas well sites per project award;
(2) use a range of oil or gas well hot water source temperatures from 150 degrees Fahrenheit to 300 degrees Fahrenheit;
(3) cover a range of sizes up to one megawatt;
(4) are located at a range of sites;
(5) can be replicated at a wide range of sites;
(6) facilitate identification of optimum techniques among competing alternatives;
(7) include business commercialization plans that have the potential for production of equipment at high volumes and operation and support at a large number of sites; and
(8) satisfy other criteria that the Secretary determines are necessary to carry out the program and collect necessary data and information.

The Secretary shall give preference to assessments that address multiple elements contained in paragraphs (1) through (8).

(c) GRANT AWARDS.—Each grant award for demonstration of geothermal technology such as advanced organic rankine cycle systems at oil and gas wells made by the Secretary under subsection (b) shall include—

(1) necessary and appropriate site engineering study;

(2) detailed economic assessment of site specific conditions;

(3) appropriate feasibility studies to determine whether the demonstration can be replicated;

(4) design or adaptation of existing technology for site specific circumstances or conditions;

(5) installation of equipment, service, and support;

(6) operation for a minimum of 1 year and monitoring for the duration of the demonstration; and

(7) validation of technical and economic assumptions and documentation of lessons learned.

(d) GEOPRESSURED GAS RESOURCE RECOVERY AND PRODUCTION.—

(1) The Secretary shall implement a program to support the research, development, demonstration, and commercial application of cost-effective techniques to produce energy from geopressured resources.

(2) The Secretary shall solicit preliminary engineering designs for geopressured resources production and recovery facilities.

(3) Based upon a review of the preliminary designs, the Secretary shall award grants, which may be cost-shared, to support the detailed development and completion of engineering, architectural and technical plans needed to support construction of new designs.

(4) Based upon a review of the final design plans above, the Secretary shall award cost-shared development and construction grants for demonstration geopressured production facilities that show potential for economic recovery of the heat, kinetic energy and gas resources from geopressured resources.

(e) COMPETITIVE GRANT SELECTION.—Not less than 90 days after the date of the enactment of this Act, the Secretary shall conduct a national solicitation for applications for grants under the programs outlined in subsections (b) and (d). Grant recipients shall be selected on a competitive basis based on criteria in the respective subsection.

(f) WELL DRILLING.—No funds may be used under this section for the purpose of drilling new wells.

Sec. 617. Cost Sharing And Proposal Evaluation.

(a) FEDERAL SHARE.—The Federal share of costs of projects funded under this subtitle shall be in accordance with section 988 of the Energy Policy Act of 2005.

(b) ORGANIZATION AND ADMINISTRATION OF PROGRAMS.—Programs under this subtitle shall incorporate the following elements:

(1) The Secretary shall coordinate with, and where appropriate may provide funds in furtherance of the purposes of

(2) this subtitle to, other Department of Energy research and development programs focused on drilling, subsurface characterization, and other related technologies.

(3) In evaluating proposals, the Secretary shall give priority to proposals that demonstrate clear evidence of employing a systems approach.

(4) The Secretary shall coordinate and consult with the appropriate Federal land management agencies in selecting proposals for funding under this subtitle.

Nothing in this subtitle shall be construed to alter or affect any law relating to the management or protection of Federal lands.

Sec. 618. Center For Geothermal Technology Transfer.

(a) IN GENERAL.—The Secretary shall award to an institution of higher education (or consortium thereof) a grant to establish a Center for Geothermal Technology Transfer (referred to in this section as the "Center").

(b) DUTIES.—The Center shall—

(1) serve as an information clearinghouse for the geothermal industry by collecting and disseminating information on best practices in all areas relating to developing and utilizing geothermal resources;

(2) make data collected by the Center available to the public; and

(3) seek opportunities to coordinate efforts and share information with domestic and international partners engaged in research and development of geothermal systems and related technology.

(c) SELECTION CRITERIA.—In awarding the grant under subsection (a) the Secretary shall select an institution of higher education (or consortium thereof) best suited to provide national leadership on geothermal related issues and perform the duties enumerated under subsection (b).

(d) DURATION OF GRANT.—A grant made under subsection (a)—

(1) shall be for an initial period of 5 years; and

(2) may be renewed for additional 5-year periods on the basis of—

(A) satisfactory performance in meeting the duties outlined in subsection (b); and

(B) any other requirements specified by the Secretary.

Sec. 619. Geopowering America.

The Secretary shall expand the Department of Energy's GeoPowering the West program to extend its geothermal technology transfer activities throughout the entire United States. The program shall be renamed "GeoPowering America". The program shall continue to be based in the Department of Energy office in Golden, Colorado.

Sec. 620. Educational Pilot Program.

The Secretary shall seek to award grant funding, on a competitive basis, to an institution of higher education for a geothermal- powered energy generation facility on the institution's campus. The purpose of the facility shall be to provide electricity and space heating. The facility shall also serve as an educational resource to students in relevant fields of study, and the data generated by the facility shall be available to students and the general public. The total funding award shall not exceed $2,000,000.

Sec. 621. Reports.

(a) REPORTS ON ADVANCED USES OF GEOTHERMAL ENERGY.— Not later than 3 years and 5 years after the date of enactment of this Act, the Secretary shall report to the Committee on Science and Technology of the House of Representatives and the Committee on Energy and Natural Resources of the Senate on advanced concepts and technologies to maximize the geothermal resource potential of the United States. The reports shall include—

(1) the use of carbon dioxide as an alternative geofluid with potential carbon sequestration benefits;
(2) mineral recovery from geofluids;
(3) use of geothermal energy to produce hydrogen;
(4) use of geothermal energy to produce biofuels;
(5) use of geothermal heat for oil recovery from oil shales and tar sands; and
(6) other advanced geothermal technologies, including advanced drilling technologies and advanced power conversion technologies.

(b) PROGRESS REPORTS.—

(1) Not later than 36 months after the date of enactment of this Act, the Secretary shall submit to the Committee on Science and Technology of the House of Representatives and the Committee on Energy and Natural Resources of the Senate an interim report describing the progress made under this subtitle. At the end of 60 months, the Secretary shall submit to Congress a report on the results of projects undertaken under this subtitle and other such information the Secretary considers appropriate.
(2) As necessary, the Secretary shall report to the Congress on any legal, regulatory, or other barriers encountered that hinder economic development of these resources, and provide recommendations on legislative or other actions needed to address such impediments.

Sec. 622. Applicability Of Other Laws.

Nothing in this subtitle shall be construed as waiving, modifying, or superseding the applicability of any requirement under any environmental or other Federal or State law. To the extent that activities authorized in this subtitle take place in coastal and ocean areas, the Secretary shall consult with the Secretary of Commerce, acting through the Under Secretary of Commerce for Oceans and Atmosphere, regarding the potential marine environmental impacts and measures to address such impacts.

Sec. 623. Authorization Of Appropriations.

There are authorized to be appropriated to the Secretary to carry out this subtitle $90,000,000 for each of the fiscal years 2008 through 2012, of which $10,000,000 for each fiscal year shall be for carrying out section 616. There are also authorized to be appropriated to the Secretary for the Intermountain West Geothermal Consortium $5,000,000 for each of the fiscal years 2008 through 2012.

Sec. 624. International Geothermal Energy Development.

(a) IN GENERAL.—The Secretary of Energy, in coordination with other appropriate Federal and multilateral agencies (including the United States Agency for International Development) shall support international collaborative efforts to promote the research, development, and deployment of geothermal technologies used to develop hydrothermal and enhanced geothermal system resources, including as partners (as appropriate) the African Rift Geothermal Development Facility, Australia, China, France, the Republic of Iceland, India, Japan, and the United Kingdom.

(b) UNITED STATES TRADE AND DEVELOPMENT AGENCY.—The Director of the United States Trade and Development Agency may—

(1) encourage participation by United States firms in actions taken to carry out subsection (a); and

(2) provide grants and other financial support for feasibility and resource assessment studies conducted in, or intended to benefit, less developed countries.

(c) AUTHORIZATION OF APPROPRIATIONS.—There are authorized to be appropriated to carry out this section $5,000,000 for each of fiscal years 2008 through 2012.

Sec. 625. High Cost Region Geothermal Energy Grant Program.

(a) DEFINITIONS.—In this section:

(1) ELIGIBLE ENTITY.—The term "eligible entity" means—

(A) a utility;

(B) an electric cooperative;

(C) a State;

(D) a political subdivision of a State;

(E) an Indian tribe; or

(F) a Native corporation.

(2) HIGH-COST REGION.—The term "high-cost region" means a region in which the average cost of electrical power exceeds 150 percent of the national average retail cost, as determined by the Secretary.

(b) PROGRAM.—The Secretary shall use amounts made available to carry out this section to make grants to eligible entities for activities described in subsection (c).

(c) ELIGIBLE ACTIVITIES.—An eligible entity may use grant funds under this section, with respect to a geothermal energy project in a high-cost region, only—

(1) to conduct a feasibility study, including a study of exploration, geochemical testing, geomagnetic surveys, geologic information gathering, baseline environmental studies, well drilling, resource characterization, permitting, and economic analysis;

(2) for design and engineering costs, relating to the project; and

(3) to demonstrate and promote commercial application of technologies related to geothermal energy as part of the project.

(d) COST SHARING.—The cost-sharing requirements of section 988 of the Energy Policy Act of 2005 (42 U.S.C. 16352) shall apply to any project carried out under this section.

(e) AUTHORIZATION OF APPROPRIATIONS.—There are authorized to be appropriated such sums as are necessary to carry out this section.

Subtitle C—Marine and Hydrokinetic Renewable Energy Technologies

Sec. 631. Short Title.

This subtitle may be cited as the "Marine and Hydrokinetic Renewable Energy Research and Development Act".

Sec. 632. Definition.

For purposes of this subtitle, the term "marine and hydrokinetic renewable energy" means electrical energy from—

(1) waves, tides, and currents in oceans, estuaries, and tidal areas;

(2) free flowing water in rivers, lakes, and streams;

(3) free flowing water in man-made channels; and

(4) differentials in ocean temperature (ocean thermal energy conversion).

The term "marine and hydrokinetic renewable energy" does not include energy from any source that uses a dam, diversionary structure, or impoundment for electric power purposes.

Sec. 633. Marine And Hydrokinetic Renewable Energy Research And Development.

(a) IN GENERAL.—The Secretary, in consultation with the Secretary of the Interior and the Secretary of Commerce, acting through the Under Secretary of Commerce for Oceans and Atmosphere, shall establish a program of research, development, demonstration, and commercial application to expand marine and hydrokinetic renewable energy production, including programs to—

(1) study and compare existing marine and hydrokinetic renewable energy technologies;

(2) research, develop, and demonstrate marine and hydrokinetic renewable energy systems and technologies;

(3) reduce the manufacturing and operation costs of marine and hydrokinetic renewable energy technologies;

(4) investigate efficient and reliable integration with the utility grid and intermittency issues;

(5) advance wave forecasting technologies;

(6) conduct experimental and numerical modeling for optimization of marine energy conversion devices and arrays;

(7) increase the reliability and survivability of marine and hydrokinetic renewable energy technologies, including development of corrosive-resistant materials;

(8) identify, in conjunction with the Secretary of Commerce, acting through the Under Secretary of Commerce for Oceans and Atmosphere, and other Federal agencies as appropriate, the potential environmental impacts, including potential impacts on fisheries and other marine resources, of marine and hydrokinetic renewable energy technologies, measures to prevent adverse impacts, and technologies and other means available for monitoring and determining environmental impacts;

(9) identify, in conjunction with the Secretary of the Department in which the United States Coast Guard is operating, acting through the Commandant of the United States Coast Guard, the potential navigational impacts of marine and hydrokinetic renewable energy technologies and measures to prevent adverse impacts on navigation;

(10) develop power measurement standards for marine and hydrokinetic renewable energy;

(11) develop identification standards for marine and hydrokinetic renewable energy devices;

(12) address standards development, demonstration, and technology transfer for advanced systems engineering and system integration methods to identify critical interfaces;

(13) identifying opportunities for cross fertilization and development of economies of scale between other renewable sources and marine and hydrokinetic renewable energy sources; and

(14) providing public information and opportunity for public comment concerning all technologies.

(b) REPORT.—Not later than 18 months after the date of enactment of this Act, the Secretary, in conjunction with the Secretary of Commerce, acting through the Undersecretary of Commerce for Oceans and Atmosphere, and the Secretary of the Interior, shall provide to the Congress a report that addresses—

(1) the potential environmental impacts, including impacts to fisheries and marine resources, of marine and hydrokinetic renewable energy technologies;

(2) options to prevent adverse environmental impacts;

(3) the potential role of monitoring and adaptive management in identifying and addressing any adverse environmental impacts; and

(4) the necessary components of such an adaptive management program.

Sec. 634. National Marine Renewable Energy Research, Development, and Demonstration Centers.

(a) CENTERS.—The Secretary shall award grants to institutions of higher education (or consortia thereof) for the establishment of 1 or more National Marine Renewable Energy Research, Development, and Demonstration Centers. In selecting locations for Centers, the Secretary shall consider sites that meet one of the following criteria:

(1) Hosts an existing marine renewable energy research and development program in coordination with an engineering program at an institution of higher education.

(2) Has proven expertise to support environmental and policy-related issues associated with harnessing of energy in the marine environment.

(3) Has access to and utilizes the marine resources in the Gulf of Mexico, the Atlantic Ocean, or the Pacific Ocean. The Secretary may give special consideration to historically black colleges and universities and land grant universities that also meet one of these criteria. In establishing criteria for the selection of the Centers, the Secretary shall consult with the Secretary of Commerce, acting through the Under Secretary of Commerce for Oceans and Atmosphere, on the criteria related to ocean waves, tides, and currents including those for advancing wave forecasting technologies, ocean temperature differences, and studying the compatibility of marine renewable energy technologies and systems with the environment, fisheries, and other marine resources.

(b) PURPOSES.—The Centers shall advance research, development, demonstration, and commercial application of marine renewable energy, and shall serve as an information clearinghouse for the marine renewable energy industry, collecting and disseminating information on best practices in all areas related to developing and managing enhanced marine renewable energy systems resources.

(c) DEMONSTRATION OF NEED.—When applying for a grant under this section, an applicant shall include a description of why Federal support is necessary for the

Center, including evidence that the research of the Center will not be conducted in the absence of Federal support.

Sec. 635. Applicability Of Other Laws.

Nothing in this subtitle shall be construed as waiving, modifying, or superseding the applicability of any requirement under any environmental or other Federal or State law.

Sec. 636. Authorization Of Appropriations.

There are authorized to be appropriated to the Secretary to carry out this subtitle $50,000,000 for each of the fiscal years 2008 through 2012, except that no funds shall be appropriated under this section for activities that are receiving funds under section 931(a)(2)(E)(i) of the Energy Policy Act of 2005 (42 U.S.C. 1623 1(a)(2)(E)(i)).

Subtitle D—Energy Storage for Transportation and Electric Power

Sec. 641. Energy Storage Competitiveness.

(a) SHORT TITLE.—This section may be cited as the "United States Energy Storage Competitiveness Act of 2007".

(b) DEFINITIONS.—In this section:

(1) COUNCIL.—The term "Council" means the Energy Storage Advisory Council established under subsection (e).

(2) COMPRESSED AIR ENERGY STORAGE.—The term "compressed air energy storage" means, in the case of an electricity grid application, the storage of energy through the compression of air.

(3) ELECTRIC DRIVE VEHICLE.—The term "electric drive vehicle" means—

(A) a vehicle that uses an electric motor for all or part of the motive power of the vehicle, including battery electric, hybrid electric, plug-in hybrid electric, fuel cell, and plug-in fuel cell vehicles and rail transportation vehicles; or

(B) mobile equipment that uses an electric motor to replace an internal combustion engine for all or part of the work of the equipment.

(4) ISLANDING.—The term "islanding" means a distributed generator or energy storage device continuing to power a location in the absence of electric power from the primary source.

(5) FLYWHEEL.—The term "flywheel" means, in the case of an electricity grid application, a device used to store rotational kinetic energy.

(6) MICROGRID.—The term "microgrid" means an integrated energy system consisting of interconnected loads and distributed energy resources (including generators and energy storage devices), which as an integrated system can operate in parallel with the utility grid or in an intentional islanding mode.

(7) SELF-HEALING GRID.—The term "self-healing grid" means a grid that is capable of automatically anticipating and responding to power system disturbances (including the isolation of failed sections and components), while optimizing the performance and service of the grid to customers.

(8) SPINNING RESERVE SERVICES.—The term "spinning reserve services" means a quantity of electric generating capacity in excess of the quantity needed to meet peak electric demand.

(9) ULTRACAPACITOR.—The term "ultracapacitor" means an energy storage device that has a power density comparable to a conventional capacitor but is capable of exceeding the energy density of a conventional capacitor by several orders of magnitude.

(c) PROGRAM.—The Secretary shall carry out a research, development, and demonstration program to support the ability of the United States to remain globally competitive in energy storage systems for electric drive vehicles, stationary applications, and electricity transmission and distribution.

(d) COORDINATION.—In carrying out the activities of this section, the Secretary shall coordinate relevant efforts with appropriate Federal agencies, including the Department of Transportation.

(e) ENERGY STORAGE ADVISORY COUNCIL.—

(1) ESTABLISHMENT.—Not later than 90 days after the date Deadline. of enactment of this Act, the Secretary shall establish an Energy Storage Advisory Council.

(2) COMPOSITION.—

(A) IN GENERAL.—Subject to subparagraph (B), the Council shall consist of not less than 15 individuals appointed by the Secretary, based on recommendations of the National Academy of Sciences.

(B) ENERGY STORAGE INDUSTRY.—The Council shall consist primarily of representatives of the energy storage industry of the United States.

(C) CHAIRPERSON.—The Secretary shall select a Chairperson for the Council from among the members appointed under subparagraph (A).

(3) MEETINGS.—

(A) IN GENERAL.—The Council shall meet not less than once a year.

(B) FEDERAL ADVISORY COMMITTEE ACT.—The Federal Advisory Committee Act (5 U.S.C. App.) shall apply to a meeting of the Council.

(4) PLANS.—No later than 1 year after the date of enact Deadlines. ment of this Act and every 5 years thereafter, the Council, in conjunction with the Secretary, shall develop a 5-year plan for integrating basic and applied research so that the United States retains a globally competitive domestic energy storage industry for electric drive vehicles, stationary applications, and electricity transmission and distribution.

(5) REVIEW.—The Council shall—

(A) assess, every 2 years, the performance of the Department in meeting the goals of the plans developed under paragraph (4); and

(B) make specific recommendations to the Secretary on programs or activities that should be established or terminated to meet those goals.

(f) BASIC RESEARCH PROGRAM.—

(1) BASIC RESEARCH.—The Secretary shall conduct a basic research program on energy storage systems to support electric drive vehicles, stationary applications, and electricity transmission and distribution, including—

(A) materials design;

(B) materials synthesis and characterization;

(C) electrode-active materials, including electrolytes and bioelectrolytes;
(D) surface and interface dynamics;
(E) modeling and simulation; and
(F) thermal behavior and life degradation mechanisms.

(2) NANOSCIENCE CENTERS.—The Secretary, in cooperation with the Council, shall coordinate the activities of the nanoscience centers of the Department to help the energy storage research centers of the Department maintain a globally competitive posture in energy storage systems for electric drive vehicles, stationary applications, and electricity transmission and distribution.

(3) FUNDING.—For activities carried out under this subsection, in addition to funding activities at National Laboratories, the Secretary shall award funds to, and coordinate activities with, a range of stakeholders including the public, private, and academic sectors.

(g) APPLIED RESEARCH PROGRAM.—

(1) IN GENERAL.—The Secretary shall conduct an applied research program on energy storage systems to support electric drive vehicles, stationary applications, and electricity transmission and distribution technologies, including—
(A) ultracapacitors;
(B) flywheels;
(C) batteries and battery systems (including flow batteries);
(D) compressed air energy systems;
(E) power conditioning electronics;
(F) manufacturing technologies for energy storage systems;
(G) thermal management systems; and
(H) hydrogen as an energy storage medium.

(2) FUNDING.—For activities carried out under this subsection, in addition to funding activities at National Laboratories, the Secretary shall provide funds to, and coordinate activities with, a range of stakeholders, including the public, private, and academic sectors.

(h) ENERGY STORAGE RESEARCH CENTERS.—

(1) IN GENERAL.—The Secretary shall establish, through competitive bids, not more than 4 energy storage research centers to translate basic research into applied technologies to advance the capability of the United States to maintain a globally competitive posture in energy storage systems for electric drive vehicles, stationary applications, and electricity transmission and distribution.

(2) PROGRAM MANAGEMENT.—The centers shall be managed by the Under Secretary for Science of the Department.

(3) PARTICIPATION AGREEMENTS.—As a condition of participating in a center, a participant shall enter into a participation agreement with the center that requires that activities conducted by the participant for the center promote the goal of enabling the United States to compete successfully in global energy storage markets.

(4) PLANS.—A center shall conduct activities that promote the achievement of the goals of the plans of the Council under subsection (e)(4).

(5) NATIONAL LABORATORIES.—A national laboratory (as defined in section 2 of the Energy Policy Act of 2005 (42 U.S.C. 15801)) may participate in a center established under this subsection, including a cooperative research and develop-

ment agreement (as defined in section 12(d) of the Stevenson- Wydler Technology Innovation Act of 1980 (15 U.S.C. 3710a(d))).

(6) DISCLOSURE.—Section 623 of the Energy Policy Act of 1992 (42 U.S.C. 13293) may apply to any project carried out through a grant, contract, or cooperative agreement under this subsection.

(7) INTELLECTUAL PROPERTY.—In accordance with section 202(a)(ii) of title 35, United States Code, section 152 of the Atomic Energy Act of 1954 (42 U.S.C. 2182), and section 9 of the Federal Nonnuclear Energy Research and Development Act of 1974 (42 U.S.C. 5908), the Secretary may require, for any new invention developed under this subsection, that—

(A) if an industrial participant is active in a energy storage research center established under this subsection relating to the advancement of energy storage technologies carried out, in whole or in part, with Federal funding, the industrial participant be granted the first option to negotiate with the invention owner, at least in the field of energy storage technologies, nonexclusive licenses, and royalties on terms that are reasonable, as determined by the Secretary;

(B) if 1 or more industry participants are active in a center, during a 2-year period beginning on the date on which an invention is made—

(i) the patent holder shall not negotiate any license or royalty agreement with any entity that is not an industrial participant under this subsection; and

(ii)the patent holder shall negotiate nonexclusive licenses and royalties in good faith with any interested industrial participant under this subsection; and

(C) the new invention be developed under such other terms as the Secretary determines to be necessary to promote the accelerated commercialization of inventions made under this subsection to advance the capability of the United States to successfully compete in global energy storage markets.

(i) ENERGY STORAGE SYSTEMS DEMONSTRATIONS.—

(1) IN GENERAL.—The Secretary shall carry out a program of new demonstrations of advanced energy storage systems.

(2) SCOPE.—The demonstrations shall—

(A) be regionally diversified; and

(B) expand on the existing technology demonstration program of the Department.

(3) STAKEHOLDERS.—In carrying out the demonstrations, the Secretary shall, to the maximum extent practicable, include the participation of a range of stakeholders, including—

(A) rural electric cooperatives;

(B) investor owned utilities;

(C) municipally owned electric utilities;

(D) energy storage systems manufacturers;

(E) electric drive vehicle manufacturers;

(F) the renewable energy production industry;

(G) State or local energy offices;

(H) the fuel cell industry; and

(I) institutions of higher education.

(4) OBJECTIVES.—Each of the demonstrations shall include 1 or more of the following:

(A) Energy storage to improve the feasibility of microgrids or islanding, or transmission and distribution capability, to improve reliability in rural areas.

(B) Integration of an energy storage system with a self-healing grid.

(C) Use of energy storage to improve security to emergency response infrastructure and ensure availability of emergency backup power for consumers.

(D) Integration with a renewable energy production source, at the source or away from the source.

(E) Use of energy storage to provide ancillary services, such as spinning reserve services, for grid management.

(F) Advancement of power conversion systems to make the systems smarter, more efficient, able to communicate with other inverters, and able to control voltage.

(G) Use of energy storage to optimize transmission and distribution operation and power quality, which could address overloaded lines and maintenance of transformers and substations.

(H) Use of advanced energy storage for peak load management of homes, businesses, and the grid.

(I) Use of energy storage devices to store energy during nonpeak generation periods to make better use of existing grid assets.

(j) VEHICLE ENERGY STORAGE DEMONSTRATION.—

(1) IN GENERAL.—The Secretary shall carry out a program of electric drive vehicle energy storage technology demonstrations.

(2) CONSORTIA.—The technology demonstrations shall be conducted through consortia, which may include—

(A) energy storage systems manufacturers and suppliers of the manufacturers;

(B) electric drive vehicle manufacturers;

(C) rural electric cooperatives;

(D) investor owned utilities;

(E) municipal and rural electric utilities;

(F) State and local governments;

(G) metropolitan transportation authorities; and

(H) institutions of higher education.

(3) OBJECTIVES.—The program shall demonstrate 1 or more of the following:

(A) Novel, high capacity, high efficiency energy storage, charging, and control systems, along with the collection of data on performance characteristics, such as battery life, energy storage capacity, and power delivery capacity.

(B) Advanced onboard energy management systems and highly efficient battery cooling systems.

(C) Integration of those systems on a prototype vehicular platform, including with drivetrain systems for passenger, commercial, and nonroad electric drive vehicles.

(D) New technologies and processes that reduce manufacturing costs.

(E) Integration of advanced vehicle technologies with electricity distribution system and smart metering technology.

(F) Control systems that minimize emissions profiles in cases in which clean diesel engines are part of a plug- in hybrid drive system.

(k) SECONDARY APPLICATIONS AND DISPOSAL OF ELECTRIC DRIVE VEHICLE BATTERIES.—The Secretary shall carry out a program of research, development, and demonstration of—

(1) secondary applications of energy storage devices following service in electric drive vehicles; and

(2) technologies and processes for final recycling and disposal of the devices.

(l) COST SHARING.—The Secretary shall carry out the programs established under this section in accordance with section 988 of the Energy Policy Act of 2005 (42 U.S.C. 16352).

(m) MERIT REVIEW OF PROPOSALS.—The Secretary shall carry out the programs established under subsections (i), (j), and (k) in accordance with section 989 of the Energy Policy Act of 2005 (42 U.S.C. 16353).

(n) COORDINATION AND NONDUPLICATION.—To the maximum extent practicable, the Secretary shall coordinate activities under this section with other programs and laboratories of the Department and other Federal research programs.

(o) REVIEW BY NATIONAL ACADEMY OF SCIENCES.—On the business day that is 5 years after the date of enactment of this Act, the Secretary shall offer to enter into an arrangement with the National Academy of Sciences to assess the performance of the Department in carrying out this section.

(p) AUTHORIZATION OF APPROPRIATIONS.—There are authorized to be appropriated to carry out—

(1) the basic research program under subsection (f) $50,000,000 for each of fiscal years 2009 through 2018;

(2) the applied research program under subsection (g) $80,000,000 for each of fiscal years 2009 through 2018; and;

(3) the energy storage research center program under subsection (h) $100,000,000 for each of fiscal years 2009 through 2018;

(4) the energy storage systems demonstration program under subsection (i) $30,000,000 for each of fiscal years 2009 through 2018;

(5) the vehicle energy storage demonstration program under subsection (j) $30,000,000 for each of fiscal years 2009 through 2018; and

(6) the secondary applications and disposal of electric drive vehicle batteries program under subsection (k) $5,000,000 for each of fiscal years 2009 through 2018.

Subtitle E—Miscellaneous Provisions

Sec. 651. Lightweight Materials Research And Development.

(a) IN GENERAL.—As soon as practicable after the date of enactment of this Act, the Secretary of Energy shall establish a program to determine ways in which the weight of motor vehicles could be reduced to improve fuel efficiency without compromising passenger safety by conducting research, development, and demonstration relating to—

(1) the development of new materials (including cast metal composite materials formed by autocombustion synthesis) and material processes that yield a higher strength-to-weight ratio or other properties that reduce vehicle weight; and

(2) reducing the cost of—

(A) lightweight materials (including high-strength steel alloys, aluminum, magnesium, metal composites, and carbon fiber reinforced polymer composites) with the properties required for construction of lighter-weight vehicles; and

(B) materials processing, automated manufacturing, joining, and recycling lightweight materials for high-volume applications.

(b) AUTHORIZATION OF APPROPRIATIONS.—There is authorized to be appropriated to carry out this section $80,000,000 for the period of fiscal years 2008 through 2012.

Sec. 652. Commercial Insulation Demonstration Program.

(a) DEFINITIONS.—In this section:

(1) ADVANCED INSULATION.—The term "advanced insulation" means insulation that has an R value of not less than R35 per inch.

(2) COVERED REFRIGERATION UNIT.—The term "covered refrigeration unit" means any—

(A) commercial refrigerated truck;

(B) commercial refrigerated trailer; or

(C) commercial refrigerator, freezer, or refrigerator- freezer described in section 342(c) of the Energy Policy and Conservation Act (42 U.S.C. 63 13(c)).

(b) REPORT.—Not later than 90 days after the date of enactment of this Act, the Secretary shall submit to Congress a report that includes an evaluation of—

(1) the state of technological advancement of advanced insulation; and

(2) the projected amount of cost savings that would be generated by implementing advanced insulation into covered refrigeration units.

(c) DEMONSTRATION PROGRAM.—

(1) ESTABLISHMENT.—If the Secretary determines in the report described in subsection (b) that the implementation of advanced insulation into covered refrigeration units would generate an economically justifiable amount of cost savings, the Secretary, in cooperation with manufacturers of covered refrigeration units, shall establish a demonstration program under which the Secretary shall demonstrate the cost-effectiveness of advanced insulation.

(2) DISCLOSURE.—The Secretary may, for a period of up to 5 years after an award is granted under the demonstration program, exempt from mandatory disclosure under section 552 of title 5, United States Code (popularly known as the Freedom of Information Act) information that the Secretary determines would be a privileged or confidential trade secret or commercial or financial information under subsection (b)(4) of such section if the information had been obtained from a non-Government party.

(3) COST-SHARING.—Section 988 of the Energy Policy Act of 2005 (42 U.S.C. 16352) shall apply to any project carried out under this subsection.

(d) AUTHORIZATION OF APPROPRIATIONS.—There is authorized to be appropriated to carry out this section $8,000,000 for the period of fiscal years 2009 through 2014.

Sec. 653. Technical Criteria For Clean Coal Power Initiative.

Section 402(b)(1)(B)(ii) of the Energy Policy Act of 2005 (42 U.S.C. 15962(b)(1)(B)(ii)) is amended by striking subclause (I) and inserting the following:

"(I)(aa) to remove at least 99 percent of sulfur dioxide; or

"(bb) to emit not more than 0.04 pound SO_2 per million Btu, based on a 30-day average;".

Sec. 654. H-Prize.

Section 1008 of the Energy Policy Act of 2005 (42 U.S.C. 16396) is amended by adding at the end the following new subsection: "(f) H-PRIZE.—

"(1) PRIZE AUTHORITY.—

"(A) IN GENERAL.—As part of the program under this section, the Secretary shall carry out a program to competitively award cash prizes in conformity with this subsection to advance the research, development, demonstration, and commercial application of hydrogen energy technologies.

"(B) ADVERTISING AND SOLICITATION OF COMPETITORS.—

"(i) ADVERTISING.—The Secretary shall widely advertise prize competitions under this subsection to encourage broad participation, including by individuals, universities (including historically Black colleges and universities and other minority serving institutions), and large and small businesses (including businesses owned or controlled by socially and economically disadvantaged persons).

"(ii) ANNOUNCEMENT THROUGH FEDERAL REGISTER NOTICE.—The Secretary shall announce each prize competition under this subsection by publishing a notice in the Federal Register. This notice shall include essential elements of the competition such as the subject of the competition, the duration of the competition, the eligibility requirements for participation in the competition, the process for participants to register for the competition, the amount of the prize, and the criteria for awarding the prize.

"(C) ADMINISTERING THE COMPETITIONS.—The Secretary shall enter into an agreement with a private, nonprofit entity to administer the prize competitions under this subsection, subject to the provisions of this subsection (in this subsection referred to as the administering entity'). The duties of the administering entity under the agreement shall include—

"(i) advertising prize competitions under this subsection and their results;

"(ii) raising funds from private entities and individuals to pay for administrative costs and to contribute to cash prizes, including funds provided in exchange for the right to name a prize awarded under this subsection;

"(iii) developing, in consultation with and subject to the final approval of the Secretary, the criteria for selecting winners in prize competitions under this subsection, based on goals provided by the Secretary;

"(iv) determining, in consultation with the Secretary, the appropriate amount and funding sources for each prize to be awarded under this subsection, subject to the final approval of the Secretary with respect to Federal funding;

"(v) providing advice and consultation to the Secretary on the selection of judges in accordance with paragraph (2)(D), using criteria developed in consultation with and subject to the final approval of the Secretary; and

"(vi) protecting against the administering entity's unauthorized use or disclosure of a registered participant's trade secrets and confidential business information. Any information properly identified as trade secrets or confidential business information that is submitted by a participant as part of a competitive program under this subsection may be withheld from public disclosure.

"(D) FUNDING SOURCES.—Prizes under this subsection shall consist of Federal appropriated funds and any funds provided by the administering entity (including funds raised pursuant to subparagraph (C)(ii)) for such cash prize programs. The Secretary may accept funds from other Federal agencies for such cash prizes and, notwithstanding section 3302(b) of title 31, United States Code, may use such funds for the cash prize program under this subsection. Other than publication of the names of prize sponsors, the Secretary may not give any special consideration to any private sector entity or individual in return for a donation to the Secretary or administering entity.

"(E) ANNOUNCEMENT OF PRIZES.—The Secretary may not issue a notice required by subparagraph (B)(ii) until all the funds needed to pay out the announced amount of the prize have been appropriated or committed in writing by the administering entity. The Secretary may increase the amount of a prize after an initial announcement is made under subparagraph (B)(ii) if—

"(i) notice of the increase is provided in the same manner as the initial notice of the prize; and

"(ii) the funds needed to pay out the announced amount of the increase have been appropriated or committed in writing by the administering entity.

"(F) SUNSET.—The authority to announce prize competitions under this subsection shall terminate on September 30, 2018.

"(2) PRIZE CATEGORIES.—

"(A) CATEGORIES.—The Secretary shall establish prizes under this subsection for—

"(i) advancements in technologies, components, or systems related to—

"(I) hydrogen production;

"(II) hydrogen storage;

"(III) hydrogen distribution; and

"(IV) hydrogen utilization;

"(ii) prototypes of hydrogen-powered vehicles or other hydrogen-based products that best meet or exceed objective performance criteria, such as completion of a race over a certain distance or terrain or generation of energy at certain levels of efficiency; and

"(iii) transformational changes in technologies for the distribution or production of hydrogen that meet or exceed far-reaching objective criteria, which shall include minimal carbon emissions and which may include cost criteria designed to facilitate the eventual market success of a winning technology.

"(B) AWARDS.—

"(i) ADVANCEMENTS.—To the extent permitted under paragraph (1)(E), the prizes authorized under subparagraph (A)(i) shall be awarded biennially to the most significant advance made in each of the four subcategories described in subclauses (I) through (IV) of subparagraph (A)(i) since the submission deadline of the previous prize competition in the same category under subparagraph (A)(i) or the date of enactment of this subsection, whichever is later, unless no such advance is significant enough to merit an award. No one such prize may exceed $1,000,000. If less than $4,000,000 is available for a prize competition under subparagraph (A)(i), the Secretary may omit one or more subcategories, reduce the amount of the prizes, or not hold a prize competition.

"(ii) PROTOTYPES.—To the extent permitted under paragraph (1)(E), prizes authorized under subparagraph (A)(ii) shall be awarded biennially in alternate years from the prizes authorized under subparagraph (A)(i). The Secretary is authorized to award up to one prize in this category in each 2-year period. No such prize may exceed $4,000,000. If no registered participants meet the objective performance criteria established pursuant to subparagraph (C) for a competition under this clause, the Secretary shall not award a prize.

"(iii) TRANSFORMATIONAL TECHNOLOGIES.—To the extent permitted under paragraph (1)(E), the Secretary shall announce one prize competition authorized under subparagraph (A)(iii) as soon after the date of enactment of this subsection as is practicable. A prize offered under this clause shall be not less than $10,000,000, paid to the winner in a lump sum, and an additional amount paid to the winner as a match for each dollar of private funding raised by the winner for the hydrogen technology beginning on the date the winner was named. The match shall be provided for 3 years after the date the prize winner is named or until the full amount of the prize has been paid out, whichever occurs first. A prize winner may elect to have the match amount paid to another entity that is continuing the development of the winning technology. The Secretary shall announce the rules for receiving the match in the notice required by paragraph (1)(B)(ii). The Secretary shall award a prize under this clause only when a registered participant has met the objective criteria established for the prize pursuant to subparagraph (C) and announced pursuant to paragraph (1)(B)(ii). Not more than $10,000,000 in Federal funds may be used for the prize award under this clause. The administering entity shall seek to raise $40,000,000 toward the matching award under this clause.

"(C) CRITERIA.—In establishing the criteria required by this subsection, the Secretary—

"(i) shall consult with the Department's Hydrogen Technical and Fuel Cell Advisory Committee;

"(ii) shall consult with other Federal agencies, including the National Science Foundation; and

"(iii) may consult with other experts such as private organizations, including professional societies, industry associations, and the National Academy of Sciences and the National Academy of Engineering.

"(D) JUDGES.—For each prize competition under this subsection, the Secretary in consultation with the administering entity shall assemble a panel of qualified judges to select the winner or winners on the basis of the criteria established under subparagraph (C). Judges for each prize competition shall include individuals from outside the Department, including from the private sector. A judge, spouse, minor children, and members of the judge's household may not—

"(i) have personal or financial interests in, or be an employee, officer, director, or agent of, any entity that is a registered participant in the prize competition for which he or she will serve as a judge; or

"(ii) have a familial or financial relationship with an individual who is a registered participant in the prize competition for which he or she will serve as a judge.

"(3) ELIGIBILITY.—To be eligible to win a prize under this subsection, an individual or entity—

"(A) shall have complied with all the requirements in accordance with the Federal Register notice required under paragraph (1)(B)(ii);

"(B) in the case of a private entity, shall be incorporated in and maintain a primary place of business in the United States, and in the case of an individual, whether participating singly or in a group, shall be a citizen of, or an alien lawfully admitted for permanent residence in, the United States; and

"(C) shall not be a Federal entity, a Federal employee acting within the scope of his employment, or an employee of a national laboratory acting within the scope of his employment.

"(4) INTELLECTUAL PROPERTY.—The Federal Government shall not, by virtue of offering or awarding a prize under this subsection, be entitled to any intellectual property rights derived as a consequence of, or direct relation to, the participation by a registered participant in a competition authorized by this subsection. This paragraph shall not be construed to prevent the Federal Government from negotiating a license for the use of intellectual property developed for a prize competition under this subsection.

"(5) LIABILITY.—

"(A) WAIVER OF LIABILITY.—The Secretary may require registered participants to waive claims against the Federal Government and the administering entity (except claims for willful misconduct) for any injury, death, damage, or loss of property, revenue, or profits arising from the registered participants' participation in a competition under this subsection. The Secretary shall give notice of any waiver required under this subparagraph in the notice required by paragraph (1)(B)(ii). The Secretary may not require a registered participant to waive claims against the administering entity arising out of the unauthorized use or disclosure by the administering entity of the registered participant's trade secrets or confidential business information.

"(B) LIABILITY INSURANCE.—

"(i) REQUIREMENTS.—Registered participants in a prize competition under this subsection shall be required to obtain liability insurance or demonstrate financial responsibility, in amounts determined by the Secretary, for claims by—

"(I) a third party for death, bodily injury, or property damage or loss resulting from an activity carried out in connection with participation in a competition under this subsection; and

"(II) the Federal Government for damage or loss to Government property resulting from such an activity.

"(ii) FEDERAL GOVERNMENT INSURED.—The Federal Government shall be named as an additional insured under a registered participant's insurance policy required under clause (i)(I), and registered participants shall be required to agree to indemnify the Federal Government against third party claims for damages arising from or related to competition activities under this subsection.

"(6) REPORT TO CONGRESS.—Not later than 60 days after the awarding of the first prize under this subsection, and annually thereafter, the Secretary shall transmit to the Congress a report that—

"(A) identifies each award recipient;

"(B) describes the technologies developed by each award recipient; and

"(C) specifies actions being taken toward commercial application of all technologies with respect to which a prize has been awarded under this subsection.

"(7) AUTHORIZATION OF APPROPRIATIONS.—

"(A) IN GENERAL.—

"(i) AWARDS.—There are authorized to be appropriated to the Secretary for the period encompassing fiscal years 2008 through 2017 for carrying out this subsection—

"(I) $20,000,000 for awards described in paragraph (2)(A)(i);

"(II) $20,000,000 for awards described in paragraph (2)(A)(ii); and

"(III) $10,000,000 for the award described in paragraph (2)(A)(iii).

"(ii) ADMINISTRATION.—In addition to the amounts authorized in clause (i), there are authorized to be appropriated to the Secretary for each of fiscal years 2008 and 2009 $2,000,000 for the administrative costs of carrying out this subsection.

"(B) CARRYOVER OF FUNDS.—Funds appropriated for prize awards under this subsection shall remain available until expended, and may be transferred, reprogrammed, or expended for other purposes only after the expiration of 10 fiscal years after the fiscal year for which the funds were originally appropriated. No provision in this subsection permits obligation or payment of funds in violation of section 1341 of title 31 of the United States Code (commonly referred to as the Anti-Deficiency Act).

"(8) NONSUBSTITUTION.—The programs created under this subsection shall not be considered a substitute for Federal research and development programs.".

Sec. 655. Bright Tomorrow Lighting Prizes.

(a) ESTABLISHMENT.—Not later than 1 year after the date of enactment of this Act, as part of the program carried out under section 1008 of the Energy Policy Act of 2005 (42 U.S.C. 16396), the Secretary shall establish and award Bright Tomorrow Lighting Prizes for solid state lighting in accordance with this section.

(b) PRIZE SPECIFICATIONS.— (1) 60-WATT INCANDESCENT REPLACEMENT LAMP PRIZE.—

The Secretary shall award a 60-Watt Incandescent Replacement Lamp Prize to an entrant that produces a solid-state-light package simultaneously capable of—

(A) producing a luminous flux greater than 900 lumens;

(B) consuming less than or equal to 10 watts;

(C) having an efficiency greater than 90 lumens per watt;

(D) having a color rendering index greater than 90;

(E) having a correlated color temperature of not less than 2,750, and not more than 3,000, degrees Kelvin;

(F) having 70 percent of the lumen value under subparagraph (A) exceeding 25,000 hours under typical conditions expected in residential use;

(G) having a light distribution pattern similar to a soft 60-watt incandescent A19 bulb;

(H) having a size and shape that fits within the maximum dimensions of an A19 bulb in accordance with American National Standards Institute standard C78.20–2003, figure C78.20–211;

(I) using a single contact medium screw socket; and

(J) mass production for a competitive sales commercial market satisfied by producing commercially accepted quality control lots of such units equal to or exceeding the criteria described in subparagraphs (A) through (I).

(2) PAR TYPE 38 HALOGEN REPLACEMENT LAMP PRIZE.—

The Secretary shall award a Parabolic Aluminized Reflector Type 38 Halogen Replacement Lamp Prize (referred to in this section as the "PAR Type 38 Halogen Replacement Lamp Prize") to an entrant that produces a solid-state-light package simultaneously capable of—

(A) producing a luminous flux greater than or equal to 1,350 lumens;

(B) consuming less than or equal to 11 watts;

(C) having an efficiency greater than 123 lumens per watt;

(D) having a color rendering index greater than or equal to 90;

(E) having a correlated color coordinate temperature of not less than 2,750, and not more than 3,000, degrees Kelvin;

(F) having 70 percent of the lumen value under subparagraph (A) exceeding 25,000 hours under typical conditions expected in residential use;

(G) having a light distribution pattern similar to a PAR 38 halogen lamp;

(H) having a size and shape that fits within the maximum dimensions of a PAR 38 halogen lamp in accordance with American National Standards Institute standard C78– 21–2003, figure C78.21–238;

(I) using a single contact medium screw socket; and

(J) mass production for a competitive sales commercial market satisfied by producing commercially accepted quality control lots of such units equal to or exceeding the criteria described in subparagraphs (A) through (I).

(3) TWENTY-FIRST CENTURY LAMP PRIZE.—The Secretary shall award a Twenty-First Century Lamp Prize to an entrant that produces a solid-state-light-light capable of—

(A) producing a light output greater than 1,200 lumens;

(B) having an efficiency greater than 150 lumens per watt;

(C) having a color rendering index greater than 90;

(D) having a color coordinate temperature between 2,800 and 3,000 degrees Kelvin; and

(E) having a lifetime exceeding 25,000 hours.

(c) PRIVATE FUNDS.—

(1) IN GENERAL.—Subject to paragraph (2), and notwithstanding section 3302 of title 31, United States Code, the Secretary may accept, retain, and use funds contributed by any person, government entity, or organization for purposes of carrying out this subsection—

(A) without further appropriation; and

(B) without fiscal year limitation.

(2) PRIZE COMPETITION.—A private source of funding may not participate in the competition for prizes awarded under this section.

(d) TECHNICAL REVIEW.—The Secretary shall establish a technical review committee composed of non-Federal officers to review entrant data submitted under this section to determine whether the data meets the prize specifications described in subsection (b).

(e) THIRD PARTY ADMINISTRATION.—The Secretary may competitively select a third party to administer awards under this section.

(f) ELIGIBILITY FOR PRIZES.—To be eligible to be awarded a prize under this section—

(1) in the case of a private entity, the entity shall be incorporated in and maintain a primary place of business in the United States; and

(2) in the case of an individual (whether participating as a single individual or in a group), the individual shall be a citizen or lawful permanent resident of the United States.

(g) AWARD AMOUNTS.—Subject to the availability of funds to carry out this section, the amount of—

(1) the 60-Watt Incandescent Replacement Lamp Prize described in subsection (b)(1) shall be $10,000,000;

(2) the PAR Type 38 Halogen Replacement Lamp Prize described in subsection (b)(2) shall be $5,000,000; and

(3) the Twenty-First Century Lamp Prize described in subsection (b)(3) shall be $5,000,000.

(h) FEDERAL PROCUREMENT OF SOLID-STATE-LIGHTS.—

(1) 60-WATT INCANDESCENT REPLACEMENT.—Subject to paragraph (3), as soon as practicable after the successful award of the 60-Watt Incandescent Replacement Lamp Prize under subsection (b)(1), the Secretary (in consultation with the Administrator of General Services) shall develop government- wide Federal purchase guidelines with a goal of replacing the use of 60-watt incandescent lamps in Federal Government buildings with a solid-state-light package described in subsection (b)(1) by not later than the date that is 5 years after the date the award is made.

(2) PAR 38 HALOGEN REPLACEMENT LAMP REPLACEMENT.—

Subject to paragraph (3), as soon as practicable after the successful award of the PAR Type 38 Halogen Replacement Lamp Prize under subsection (b)(2), the Secretary (in consultation with the Administrator of General Services) shall develop governmentwide Federal purchase guidelines with the goal of replacing the use of PAR 38 halogen lamps in Federal Government buildings with a solid-state-light package described in subsection (b)(2) by not later than the date that is 5 years after the date the award is made.

(3) WAIVERS.—

(A) IN GENERAL.—The Secretary or the Administrator of General Services may waive the application of paragraph (1) or (2) if the Secretary or Administrator determines that the return on investment from the purchase of a solid-state-light package described in paragraph (1) or (2) of subsection (b), respectively, is cost prohibitive.

(B) REPORT OF WAIVER.—If the Secretary or Administrator waives the application of paragraph (1) or (2), the Secretary or Administrator, respectively, shall submit to Congress an annual report that describes the waiver and provides a detailed justification for the waiver.

(i) REPORT.—Not later than 2 years after the date of enactment of this Act, and annually thereafter, the Administrator of General Services shall submit to the Energy Information Agency a report describing the quantity, type, and cost of each lighting product purchased by the Federal Government.

(j) BRIGHT TOMORROW LIGHTING AWARD FUND.—

(1) ESTABLISHMENT.—There is established in the United States Treasury a Bright Tomorrow Lighting permanent fund without fiscal year limitation to award prizes under paragraphs (1), (2), and (3) of subsection (b).

(2) SOURCES OF FUNDING.—The fund established under paragraph (1) shall accept—

(A) fiscal year appropriations; and

(B) private contributions authorized under subsection (c).

(k) AUTHORIZATION OF APPROPRIATIONS.—There are authorized to be appropriated such sums as are necessary to carry out this section.

Sec. 656. Renewable Energy Innovation Manufacturing Partnership.

(a) ESTABLISHMENT.—The Secretary shall carry out a program, to be known as the Renewable Energy Innovation Manufacturing Partnership Program (referred to in this section as the "Program"), to make assistance awards to eligible entities for use in carrying out research, development, and demonstration relating to the manufacturing of renewable energy technologies.

(b) SOLICITATION.—To carry out the Program, the Secretary shall annually conduct a competitive solicitation for assistance awards for an eligible project described in subsection (e).

(c) PROGRAM PURPOSES.—The purposes of the Program are—

(1) to develop, or aid in the development of, advanced manufacturing processes, materials, and infrastructure;

(2) to increase the domestic production of renewable energy technology and components; and

(3) to better coordinate Federal, State, and private resources to meet regional and national renewable energy goals through advanced manufacturing partnerships.

(d) ELIGIBLE ENTITIES.—An entity shall be eligible to receive an assistance award under the Program to carry out an eligible project described in subsection (e) if the entity is composed of—

(1) 1 or more public or private nonprofit institutions or national laboratories engaged in research, development, demonstration, or technology transfer, that would participate substantially in the project; and

(2) 1 or more private entities engaged in the manufacturing or development of renewable energy system components (including solar energy, wind energy, biomass, geothermal energy, energy storage, or fuel cells).

(e) ELIGIBLE PROJECTS.—An eligible entity may use an assistance award provided under this section to carry out a project relating to—

(1) the conduct of studies of market opportunities for component manufacturing of renewable energy systems;

(2) the conduct of multiyear applied research, development, demonstration, and deployment projects for advanced manufacturing processes, materials, and infrastructure for renewable energy systems; and

(3) other similar ventures, as approved by the Secretary, that promote advanced manufacturing of renewable technologies.

(f) CRITERIA AND GUIDELINES.—The Secretary shall establish criteria and guidelines for the submission, evaluation, and funding of proposed projects under the Program.

(g) COST SHARING.—Section 988 of the Energy Policy Act of 2005 (42 U.S.C. 16352) shall apply to a project carried out under this section.

(h) DISCLOSURE.—The Secretary may, for a period of up to 5 years after an award is granted under this section, exempt from mandatory disclosure under section 552 of title 5, United States Code (popularly known as the Freedom of Information Act) information that the Secretary determines would be a privileged or confidential trade secret or commercial or financial information under subsection (b)(4) of such section if the information had been obtained from a non-Government party.

(i) SENSE OF THE CONGRESS.—It is the sense of the Congress that the Secretary should ensure that small businesses engaged in renewable manufacturing be given priority consideration for the assistance awards provided under this section.

(j) AUTHORIZATION OF APPROPRIATIONS.—There is authorized to be appropriated out of funds already authorized to carry out this section $25,000,000 for each of fiscal years 2008 through 2013, to remain available until expended.

TITLE VII — CARBON CAPTURE AND SEQUESTRATION

Subtitle A—Carbon Capture And Sequestration Research, Development, And Demonstration

Sec. 701. Short Title.

This subtitle may be cited as the "Department of Energy Carbon Capture and Sequestration Research, Development, and Demonstration Act of 2007".

Sec. 702. Carbon Capture And Sequestration Research, Development, and Demonstration Program.

(a) AMENDMENT.—Section 963 of the Energy Policy Act of 2005 (42 U.S.C. 16293) is amended—

(1) in the section heading, by striking "RESEARCH AND DEVELOPMENT" and inserting "AND SEQUESTRATION RESEARCH, DEVELOPMENT, AND DEMONSTRATION";

(2) in subsection (a)—

(A) by striking "research and development" and inserting "and sequestration research, development, and demonstration"; and

(B) by striking "capture technologies on combustion- based systems" and inserting "capture and sequestration technologies related to industrial sources of carbon dioxide";

(3) in subsection (b)—

(A) in paragraph (3), by striking "and" at the end;

(B) in paragraph (4), by striking the period at the end and inserting "; and"; and

(C) by adding at the end the following:

"(5) to expedite and carry out large-scale testing of carbon sequestration systems in a range of geologic formations that will provide information on the cost and feasibility of deployment of sequestration technologies."; and

(4) by striking subsection (c) and inserting the following: "(c) PROGRAMMATIC ACTIVITIES.—

"(1) FUNDAMENTAL SCIENCE AND ENGINEERING RESEARCH AND DEVELOPMENT AND DEMONSTRATION SUPPORTING CARBON CAPTURE AND SEQUESTRATION TECHNOLOGIES AND CARBON USE ACTIVITIES.—

"(A) IN GENERAL.—The Secretary shall carry out fundamental science and engineering research (including laboratory-scale experiments, numeric modeling, and simulations) to develop and document the performance of new approaches to capture and sequester, or use carbon dioxide to lead to an overall reduction of carbon dioxide emissions.

"(B) PROGRAM INTEGRATION.—The Secretary shall ensure that fundamental research carried out under this paragraph is appropriately applied to energy technology development activities, the field testing of carbon sequestration, and carbon use activities, including—

"(i) development of new or advanced technologies for the capture and sequestration of carbon dioxide;

"(ii) development of new or advanced technologies that reduce the cost and increase the efficacy of advanced compression of carbon dioxide required for the sequestration of carbon dioxide;

"(iii) modeling and simulation of geologic sequestration field demonstrations;

"(iv) quantitative assessment of risks relating to specific field sites for testing of sequestration technologies;

"(v) research and development of new and advanced technologies for carbon use, including recycling and reuse of carbon dioxide; and

"(vi) research and development of new and advanced technologies for the separation of oxygen from air.

"(2) FIELD VALIDATION TESTING ACTIVITIES.—

"(A) IN GENERAL.—The Secretary shall promote, to the maximum extent practicable, regional carbon sequestration partnerships to conduct geologic sequestration tests involving carbon dioxide injection and monitoring, mitigation, and verification operations in a variety of candidate geologic settings, including—

"(i) operating oil and gas fields;

"(ii) depleted oil and gas fields;

"(iii) unmineable coal seams;

"(iv) deep saline formations;

"(v) deep geologic systems that may be used as engineered reservoirs to extract economical quantities of heat from geothermal resources of low permeability or porosity; and

"(vi) deep geologic systems containing basalt formations.

"(B) OBJECTIVES.—The objectives of tests conducted under this paragraph shall be—

"(i) to develop and validate geophysical tools, analysis, and modeling to monitor, predict, and verify carbon dioxide containment;

"(ii) to validate modeling of geologic formations;

"(iii) to refine sequestration capacity estimated for particular geologic formations;

"(iv) to determine the fate of carbon dioxide concurrent with and following injection into geologic formations;

"(v) to develop and implement best practices for operations relating to, and monitoring of, carbon dioxide injection and sequestration in geologic formations;

"(vi) to assess and ensure the safety of operations related to geologic sequestration of carbon dioxide;

"(vii) to allow the Secretary to promulgate policies, procedures, requirements, and guidance to ensure that the objectives of this subparagraph are met in large- scale testing and deployment activities for carbon capture and sequestration that are funded by the Department of Energy; and

"(viii) to provide information to States, the Environmental Protection Agency, and other appropriate entities to support development of a regulatory framework for commercial-scale sequestration operations that ensure the protection of human health and the environment.

"(3) LARGE-SCALE CARBON DIOXIDE SEQUESTRATION TESTING.—

"(A) IN GENERAL.—The Secretary shall conduct not less than 7 initial large-scale sequestration tests, not including the FutureGen project, for geologic containment of carbon dioxide to collect and validate information on the cost and feasibility of commercial deployment of technologies for geologic containment of carbon dioxide. These 7 tests may include any Regional Partnership projects awarded as of the date of enactment of the Department of Energy Carbon Capture and Sequestration Research, Development, and Demonstration Act of 2007.

"(B) DIVERSITY OF FORMATIONS TO BE STUDIED.—In selecting formations for study under this paragraph, the Secretary shall consider a variety of geologic formations across the United States, and require characterization and modeling of candidate formations, as determined by the Secretary.

"(C) SOURCE OF CARBON DIOXIDE FOR LARGE-SCALE SEQUESTRATION TESTS.—In the process of any acquisition of carbon dioxide for sequestration tests under subparagraph (A), the Secretary shall give preference to sources of carbon dioxide from industrial sources. To the extent feasible, the Secretary shall prefer tests that would facilitate the creation of an integrated system of capture, transportation and sequestration of carbon dioxide. The preference provided for under this subparagraph shall not delay the implementation of the large-scale sequestration tests under this paragraph.

"(D) DEFINITION.—For purposes of this paragraph, the term 'large-scale' means the injection of more than 1,000,000 tons of carbon dioxide from industrial sources annually or a scale that demonstrates the ability to inject and sequester several million metric tons of industrial source carbon dioxide for a large number of years.

"(4) PREFERENCE IN PROJECT SELECTION FROM MERITORIOUS PROPOSALS.—In making competitive awards under this subsection, subject to the requirements of section 989, the Secretary shall—

"(A) give preference to proposals from partnerships among industrial, academic, and government entities; and

"(B) require recipients to provide assurances that all laborers and mechanics employed by contractors and subcontractors in the construction, repair, or alteration of new or existing facilities performed in order to carry out a demonstration or commercial application activity authorized under this subsection shall be paid wages at rates not less than those prevailing on similar construction in the locality, as determined by the Secretary of Labor in accordance with subchapter IV of chapter 31 of title 40, United States Code, and the Secretary of Labor shall, with respect to the labor standards in this paragraph, have the authority and functions set forth in Reorganization Plan Numbered 14 of 1950 (15 Fed. Reg. 3176; 5 U.S.C. Appendix) and section 3145 of title 40, United States Code.

"(5) COST SHARING.—Activities under this subsection shall be considered research and development activities that are subject to the cost sharing requirements of section 988(b).

"(6) PROGRAM REVIEW AND REPORT.—During fiscal year 2011, the Secretary shall—

"(A) conduct a review of programmatic activities carried out under this subsection; and

"(B) make recommendations with respect to continuation of the activities.

"(d) AUTHORIZATION OF APPROPRIATIONS.—There are authorized to be appropriated to carry out this section—

"(1) $240,000,000 for fiscal year 2008;

"(2) $240,000,000 for fiscal year 2009;

"(3) $240,000,000 for fiscal year 2010;

"(4) $240,000,000 for fiscal year 2011; and

"(5) $240,000,000 for fiscal year 2012.".

(b) TABLE OF CONTENTS AMENDMENT.—The item relating to section 963 in the table of contents for the Energy Policy Act of 2005 is amended to read as follows:

"Sec. 963. Carbon capture and sequestration research, development, and demonstration program.".

Sec. 703. Carbon Capture.

(a) PROGRAM ESTABLISHMENT.—

(1) IN GENERAL.—The Secretary shall carry out a program to demonstrate technologies for the large-scale capture of carbon dioxide from industrial sources. In making awards under this program, the Secretary shall select, as appropriate, a

diversity of capture technologies to address the need to capture carbon dioxide from a range of industrial sources.

(2) SCOPE OF AWARD.—Awards under this section shall be only for the portion of the project that—

(A) carries out the large-scale capture (including purification and compression) of carbon dioxide from industrial sources;

(B) provides for the transportation and injection of carbon dioxide; and

(C) incorporates a comprehensive measurement, monitoring, and validation program.

(3) PREFERENCES FOR AWARD.—To ensure reduced carbon dioxide emissions, the Secretary shall take necessary actions to provide for the integration of the program under this paragraph with the large-scale carbon dioxide sequestration tests described in section 963(c)(3) of the Energy Policy Act of 2005 (42 U.S.C. 16293(c)(3)), as added by section 702 of this subtitle. These actions should not delay implementation of these tests. The Secretary shall give priority consideration to projects with the following characteristics:

(A) CAPACITY.—Projects that will capture a high percentage of the carbon dioxide in the treated stream and large volumes of carbon dioxide as determined by the Secretary.

(B) SEQUESTRATION.—Projects that capture carbon dioxide from industrial sources that are near suitable geological reservoirs and could continue sequestration including—

(i) a field testing validation activity under section 963 of the Energy Policy Act of 2005 (42 U.S.C. 16293), as amended by this Act; or

(ii) other geologic sequestration projects approved by the Secretary.

(4) REQUIREMENT.—For projects that generate carbon dioxide that is to be sequestered, the carbon dioxide stream shall be of a sufficient purity level to allow for safe transport and sequestration.

(5) COST-SHARING.—The cost-sharing requirements of section 988 of the Energy Policy Act of 2005 (42 U.S.C. 16352) for research and development projects shall apply to this section.

(b) AUTHORIZATION OF APPROPRIATIONS.—There is authorized to be appropriated to the Secretary to carry out this section $200,000,000 per year for fiscal years 2009 through 2013.

Sec. 704. Review Of Large-Scale Programs.

The Secretary shall enter into an arrangement with the National Academy of Sciences for an independent review and oversight, beginning in 2011, of the programs under section 963(c)(3) of the Energy Policy Act of 2005 (42 U.S.C. 16293(c)(3)), as added by section 702 of this subtitle, and under section 703 of this subtitle, to ensure that the benefits of such programs are maximized. Not later than January 1, 2012, the Secretary shall transmit to the Congress a report on the results of such review and oversight.

Sec. 705. Geologic Sequestration Training And Research.

(a) STUDY.—

(1) IN GENERAL.—The Secretary shall enter into an arrangement with the National Academy of Sciences to undertake a study that—

(A) defines an interdisciplinary program in geology, engineering, hydrology, environmental science, and related disciplines that will support the Nation's capability to capture and sequester carbon dioxide from anthropogenic sources;

(B) addresses undergraduate and graduate education, especially to help develop graduate level programs of research and instruction that lead to advanced degrees with emphasis on geologic sequestration science;

(C) develops guidelines for proposals from colleges and universities with substantial capabilities in the required disciplines that seek to implement geologic sequestration science programs that advance the Nation's capacity to address carbon management through geologic sequestration science; and

(D) outlines a budget and recommendations for how much funding will be necessary to establish and carry out the grant program under subsection (b).

(2) REPORT.—Not later than 1 year after the date of enactment of this Act, the Secretary shall transmit to the Congress a copy of the results of the study provided by the National Academy of Sciences under paragraph (1).

(3) AUTHORIZATION OF APPROPRIATIONS.—There are authorized to be appropriated to the Secretary for carrying out this subsection $1,000,000 for fiscal year 2008.

(b) GRANT PROGRAM.—

(1) ESTABLISHMENT.—The Secretary shall establish a competitive grant program through which colleges and universities may apply for and receive 4-year grants for—

(A) salary and startup costs for newly designated faculty positions in an integrated geologic carbon sequestration science program; and

(B) internships for graduate students in geologic sequestration science.

(2) RENEWAL.—Grants under this subsection shall be renewable for up to 2 additional 3-year terms, based on performance criteria, established by the National Academy of Sciences study conducted under subsection (a), that include the number of graduates of such programs.

(3) INTERFACE WITH REGIONAL GEOLOGIC CARBON SEQUESTRATION PARTNERSHIPS.—To the greatest extent possible, geologic carbon sequestration science programs supported under this subsection shall interface with the research of the Regional Carbon Sequestration Partnerships operated by the Department to provide internships and practical training in carbon capture and geologic sequestration.

(4) AUTHORIZATION OF APPROPRIATIONS.—There are authorized to be appropriated to the Secretary for carrying out this subsection such sums as may be necessary.

Sec. 706. Relation To Safe Drinking Water Act.

The injection and geologic sequestration of carbon dioxide pursuant to this subtitle and the amendments made by this subtitle shall be subject to the requirements of the Safe Drinking Water Act (42 U.S.C. 300f et seq.), including the provisions of part C of such Act (42 U.S.C. 300h et seq.; relating to protection of underground sources of drinking water). Nothing in this subtitle and the amendments made by this subtitle imposes or authorizes the promulgation of any requirement that is inconsistent or in conflict with the requirements of the Safe Drinking Water Act (42 U.S.C. 300f et seq.) or regulations thereunder.

Sec. 707. Safety Research.

(a) PROGRAM.—The Administrator of the Environmental Protection Agency shall conduct a research program to address public health, safety, and environmental impacts that may be associated with capture, injection, and sequestration of greenhouse gases in geologic reservoirs.

(b) AUTHORIZATION OF APPROPRIATIONS.—There are authorized to be appropriated for carrying out this section $5,000,000 for each fiscal year.

Sec. 708. University Based Research And Development Grant Program.

(a) ESTABLISHMENT.—The Secretary, in consultation with other appropriate agencies, shall establish a university based research and development program to study carbon capture and sequestration using the various types of coal.

(b) RURAL AND AGRICULTURAL INSTITUTIONS.—The Secretary shall give special consideration to rural or agricultural based institutions in areas that have regional sources of coal and that offer interdisciplinary programs in the area of environmental science to study carbon capture and sequestration.

(c) AUTHORIZATION OF APPROPRIATIONS.—There are to be authorized to be appropriated $10,000,000 to carry out this section.

Subtitle B—Carbon Capture and Sequestration Assessment and Framework

Sec. 711. Carbon Dioxide Sequestration Capacity Assessment.

(a) DEFINITIONS.—In this section—

(1) ASSESSMENT.—The term "assessment" means the national assessment of onshore capacity for carbon dioxide completed under subsection (f).

(2) CAPACITY.—The term "capacity" means the portion of a sequestration formation that can retain carbon dioxide in accordance with the requirements (including physical, geological, and economic requirements) established under the methodology developed under subsection (b).

(3) ENGINEERED HAZARD.—The term "engineered hazard" includes the location and completion history of any well that could affect potential sequestration.

(4) RISK.—The term "risk" includes any risk posed by geomechanical, geochemical, hydrogeological, structural, and engineered hazards.

(5) SECRETARY.—The term "Secretary" means the Secretary of the Interior, acting through the Director of the United States Geological Survey.

(6) SEQUESTRATION FORMATION.—The term "sequestration formation" means a deep saline formation, unmineable coal seam, or oil or gas reservoir that is capable of accommodating a volume of industrial carbon dioxide.

(b) METHODOLOGY.—Not later than 1 year after the date of Deadline. enactment of this Act, the Secretary shall develop a methodology for conducting an assessment under subsection (f), taking into consideration—

(1) the geographical extent of all potential sequestration formations in all States;
(2) the capacity of the potential sequestration formations;
(3) the injectivity of the potential sequestration formations;
(4) an estimate of potential volumes of oil and gas recoverable by injection and sequestration of industrial carbon dioxide in potential sequestration formations;
(5) the risk associated with the potential sequestration formations; and
(6) the work done to develop the Carbon Sequestration Atlas of the United States and Canada that was completed by the Department.

(c) COORDINATION.—
(1) FEDERAL COORDINATION.—
(A) CONSULTATION.—The Secretary shall consult with the Secretary of Energy and the Administrator of the Environmental Protection Agency on issues of data sharing, format, development of the methodology, and content of the assessment required under this section to ensure the maximum usefulness and success of the assessment.
(B) COOPERATION.—The Secretary of Energy and the Administrator shall cooperate with the Secretary to ensure, to the maximum extent practicable, the usefulness and success of the assessment.
(2) STATE COORDINATION.—The Secretary shall consult with State geological surveys and other relevant entities to ensure, to the maximum extent practicable, the usefulness and success of the assessment.

(d) EXTERNAL REVIEW AND PUBLICATION.—On completion of the methodology under subsection (b), the Secretary shall—
(1) publish the methodology and solicit comments from the public and the heads of affected Federal and State agencies;
(2) establish a panel of individuals with expertise in the Establishment. matters described in paragraphs (1) through (5) of subsection (b) composed, as appropriate, of representatives of Federal agencies, institutions of higher education, nongovernmental organizations, State organizations, industry, and international geoscience organizations to review the methodology and comments received under paragraph (1); and
(3) on completion of the review under paragraph (2), publish in the Federal Register the revised final methodology.

(e) PERIODIC UPDATES.—The methodology developed under this section shall be updated periodically (including at least once every 5 years) to incorporate new data as the data becomes available.

(f) NATIONAL ASSESSMENT.—
(1) IN GENERAL.—Not later than 2 years after the date of publication of the methodology under subsection (d)(1), the Secretary, in consultation with the Secretary of Energy and State geological surveys, shall complete a national assessment of capacity for carbon dioxide in accordance with the methodology.
(2) GEOLOGICAL VERIFICATION.—As part of the assessment under this subsection, the Secretary shall carry out a drilling program to supplement the geological data relevant to determining sequestration capacity of carbon dioxide in geological sequestration formations, including—
(A) well log data;

(B) core data; and
(C) fluid sample data.
(3) PARTNERSHIP WITH OTHER DRILLING PROGRAMS.—As part of the drilling program under paragraph (2), the Secretary shall enter, as appropriate, into partnerships with other entities to collect and integrate data from other drilling programs relevant to the sequestration of carbon dioxide in geological formations.
(4) INCORPORATION INTO NATCARB.—
(A) IN GENERAL.—On completion of the assessment, the Secretary of Energy and the Secretary of the Interior shall incorporate the results of the assessment using—
(i) the NatCarb database, to the maximum extent practicable; or
(ii) a new database developed by the Secretary of Energy, as the Secretary of Energy determines to be necessary.
(B) RANKING.—The database shall include the data necessary to rank potential sequestration sites for capacity and risk, across the United States, within each State, by formation, and within each basin.
(5) REPORT.—Not later than 180 days after the date on which the assessment is completed, the Secretary shall submit to the Committee on Energy and Natural Resources of the Senate and the Committee on Natural Resources of the House of Representatives a report describing the findings under the assessment.
(6) PERIODIC UPDATES.—The national assessment developed under this section shall be updated periodically (including at least once every 5 years) to support public and private sector decisionmaking.
(g) AUTHORIZATION OF APPROPRIATIONS.—There is authorized to be appropriated to carry out this section $30,000,000 for the period of fiscal years 2008 through 2012.

Sec. 712. Assessment Of Carbon Sequestration And Methane And Nitrous Oxide Emissions From Ecosystems.

(a) DEFINITIONS.—In this section:
(1) ADAPTATION STRATEGY.—The term “adaptation strategy” means a land use and management strategy that can be used—
(A) to increase the sequestration capabilities of covered greenhouse gases of any ecosystem; or
(B) to reduce the emissions of covered greenhouse gases from any ecosystem.
(2) ASSESSMENT.—The term “assessment” means the national assessment authorized under subsection (b).
(3) COVERED GREENHOUSE GAS.—The term “covered greenhouse gas” means carbon dioxide, nitrous oxide, and methane gas.
(4) ECOSYSTEM.—The term “ecosystem” means any terrestrial, freshwater aquatic, or coastal ecosystem, including an estuary.
(5) NATIVE PLANT SPECIES.—The term “native plant species” means any noninvasive, naturally occurring plant species within an ecosystem.
(6) SECRETARY.—The term “Secretary” means the Secretary of the Interior.
(b) AUTHORIZATION OF ASSESSMENT.—Not later than 2 years Deadline. after the date on which the final methodology is published under subsection (f)(3)(D), the Secretary shall complete a national assess-

ment of—

(1) the quantity of carbon stored in and released from ecosystems, including from man-caused and natural fires; and

(2) the annual flux of covered greenhouse gases in and out of ecosystems.

(c) COMPONENTS.—In conducting the assessment under subsection (b), the Secretary shall—

(1) determine the processes that control the flux of covered greenhouse gases in and out of each ecosystem;

(2) estimate the potential for increasing carbon sequestration in natural and managed ecosystems through management activities or restoration activities in each ecosystem;

(3) develop near-term and long-term adaptation strategies or mitigation strategies that can be employed—

(A) to enhance the sequestration of carbon in each ecosystem;

(B) to reduce emissions of covered greenhouse gases from ecosystems; and

(C) to adapt to climate change; and

(4) estimate the annual carbon sequestration capacity of ecosystems under a range of policies in support of management activities to optimize sequestration.

(d) USE OF NATIVE PLANT SPECIES.—In developing restoration activities under subsection (c)(2) and management strategies and adaptation strategies under subsection (c)(3), the Secretary shall emphasize the use of native plant species (including mixtures of many native plant species) for sequestering covered greenhouse gas in each ecosystem.

(e) CONSULTATION.—

(1) IN GENERAL.—In conducting the assessment under subsection (b) and developing the methodology under subsection (f), the Secretary shall consult with—

(A) the Secretary of Energy;

(B) the Secretary of Agriculture;

(C) the Administrator of the Environmental Protection Agency;

(D) the Secretary of Commerce, acting through the Under Secretary for Oceans and Atmosphere; and

(E) the heads of other relevant agencies.

(2) OCEAN AND COASTAL ECOSYSTEMS.—In carrying out this section with respect to ocean and coastal ecosystems (including estuaries), the Secretary shall work jointly with the Secretary of Commerce, acting through the Under Secretary for Oceans and Atmosphere.

(f) METHODOLOGY.—

(1) IN GENERAL.—Not later than 1 year after the date of enactment of this Act, the Secretary shall develop a methodology for conducting the assessment.

(2) REQUIREMENTS.—The methodology developed under paragraph (1)—

(A) shall—

(i) determine the method for measuring, monitoring, and quantifying covered greenhouse gas emissions and reductions;

(ii) estimate the total capacity of each ecosystem to sequester carbon; and

(iii) estimate the ability of each ecosystem to reduce emissions of covered greenhouse gases through management practices; and

(B) may employ economic and other systems models, analyses, and estimates, to be developed in consultation with each of the individuals described in subsection (e).

(3) EXTERNAL REVIEW AND PUBLICATION.—On completion of a proposed methodology, the Secretary shall—

(A) publish the proposed methodology;

(B) at least 60 days before the date on which the final methodology is published, solicit comments from—

(i) the public; and

(ii) heads of affected Federal and State agencies;

(C) establish a panel to review the proposed methodology published under subparagraph (A) and any comments received under subparagraph (B), to be composed of members—

(i) with expertise in the matters described in subsections (c) and (d); and

(ii) that are, as appropriate, representatives of Federal agencies, institutions of higher education, nongovernmental organizations, State organizations, industry, and international organizations; and

(D) on completion of the review under subparagraph (C), publish in the Federal Register the revised final methodology.

(g) ESTIMATE; REVIEW.—The Secretary shall—

(1) based on the assessment, prescribe the data, information, and analysis needed to establish a scientifically sound estimate of the carbon sequestration capacity of relevant ecosystems; and

(2) not later than 180 days after the date on which the assessment is completed, submit to the heads of applicable Federal agencies and the appropriate committees of Congress a report that describes the results of the assessment.

(h) DATA AND REPORT AVAILABILITY.—On completion of the assessment, the Secretary shall incorporate the results of the assessment into a web-accessible database for public use.

(i) AUTHORIZATION.—There is authorized to be appropriated to carry out this section \$20,000,000 for the period of fiscal years 2008 through 2012.

Sec. 713. Carbon Dioxide Sequestration Inventory.

Section 354 of the Energy Policy Act of 2005 (42 U.S.C. 15910) is amended—

(1) by redesignating subsection (d) as subsection (e); and

(2) by inserting after subsection (c) the following:

"(d) RECORDS AND INVENTORY.—The Secretary of the Interior, acting through the Bureau of Land Management, shall maintain records on, and an inventory of, the quantity of carbon dioxide stored within Federal mineral leaseholds.".

Sec. 714. Framework For Geological Carbon Sequestration On Public Land.

(a) REPORT.—Not later than 1 year after the date of enactment of this Act, the Secretary of the Interior shall submit to the Committee on Natural Resources of the House of Representatives and the Committee on Energy and Natural Resources of the Senate a report on a recommended framework for managing geological carbon sequestration activities on public land.

(b) CONTENTS.—The report required by subsection (a) shall include the following:

(1) Recommended criteria for identifying candidate geological sequestration sites in each of the following types of geological settings:
(A) Operating oil and gas fields.
(B) Depleted oil and gas fields.
(C) Unmineable coal seams.
(D) Deep saline formations.
(E) Deep geological systems that may be used as engineered reservoirs to extract economical quantities of heat from geothermal resources of low permeability or porosity.
(F) Deep geological systems containing basalt formations.
(G) Coalbeds being used for methane recovery.
(2) A proposed regulatory framework for the leasing of public land or an interest in public land for the long-term geological sequestration of carbon dioxide, which includes an assessment of options to ensure that the United States receives fair market value for the use of public land or an interest in public land for geological sequestration.
(3) A proposed procedure for ensuring that any geological carbon sequestration activities on public land—
(A) provide for public review and comment from all interested persons; and
(B) protect the quality of natural and cultural resources of the public land overlaying a geological sequestration site.
(4) A description of the status of Federal leasehold or Federal mineral estate liability issues related to the geological subsurface trespass of or caused by carbon dioxide stored in public land, including any relevant experience from enhanced oil recovery using carbon dioxide on public land.
(5) Recommendations for additional legislation that may be required to ensure that public land management and leasing laws are adequate to accommodate the long-term geological sequestration of carbon dioxide.
(6) An identification of the legal and regulatory issues specific to carbon dioxide sequestration on land in cases in which title to mineral resources is held by the United States but title to the surface estate is not held by the United States.
(7)(A) An identification of the issues specific to the issuance of pipeline rights-of-way on public land under the Mineral Leasing Act (30 U.S.C. 181 et seq.) or the Federal Land Policy and Management Act of 1976 (43 U.S.C. 1701 et seq.) for natural or anthropogenic carbon dioxide.
(B) Recommendations for additional legislation that may be required to clarify the appropriate framework for issuing rights-of-way for carbon dioxide pipelines on public land.
(c) CONSULTATION WITH OTHER AGENCIES.—In preparing the report under this section, the Secretary of the Interior shall coordinate with—
(1) the Administrator of the Environmental Protection Agency;
(2) the Secretary of Energy; and
(3) the heads of other appropriate agencies.
(d) COMPLIANCE WITH SAFE DRINKING WATER ACT.—The Secretary shall ensure that all recommendations developed under this section are in compliance with all Federal environmental laws, including the Safe Drinking Water Act (42 U.S.C. 300f et seq.) and regulations under that Act.

TITLE VIII—IMPROVED MANAGEMENT OF ENERGY POLICY

Subtitle A—Management Improvements

Sec. 801. National Media Campaign.

(a) IN GENERAL.—The Secretary, acting through the Assistant Secretary for Energy Efficiency and Renewable Energy (referred to in this section as the "Secretary"), shall develop and conduct a national media campaign—

(1) to increase energy efficiency throughout the economy of the United States during the 10-year period beginning on the date of enactment of this Act;

(2) to promote the national security benefits associated with increased energy efficiency; and

(3) to decrease oil consumption in the United States during the 10-year period beginning on the date of enactment of this Act.

(b) CONTRACT WITH ENTITY.—The Secretary shall carry out subsection (a) directly or through—

(1) competitively bid contracts with 1 or more nationally recognized media firms for the development and distribution of monthly television, radio, and newspaper public service announcements; or

(2) collective agreements with 1 or more nationally recognized institutes, businesses, or nonprofit organizations for the funding, development, and distribution of monthly television, radio, and newspaper public service announcements.

(c) USE OF FUNDS.—

(1) IN GENERAL.—Amounts made available to carry out this section shall be used for—

(A) advertising costs, including—

(i) the purchase of media time and space;

(ii) creative and talent costs;

(iii) testing and evaluation of advertising; and

(iv) evaluation of the effectiveness of the media campaign; and

(B) administrative costs, including operational and management expenses.

(2) LIMITATIONS.—In carrying out this section, the Secretary shall allocate not less than 85 percent of funds made available under subsection (e) for each fiscal year for the advertising functions specified under paragraph (1)(A).

(d) REPORTS.—The Secretary shall annually submit to Congress a report that describes—

(1) the strategy of the national media campaign and whether specific objectives of the campaign were accomplished, including—

(A) determinations concerning the rate of change of energy consumption, in both absolute and per capita terms; and

(B) an evaluation that enables consideration of whether the media campaign contributed to reduction of energy consumption;

(2) steps taken to ensure that the national media campaign operates in an effective and efficient manner consistent with the overall strategy and focus of the campaign;

(3) plans to purchase advertising time and space;

(4) policies and practices implemented to ensure that Federal funds are used responsibly to purchase advertising time and space and eliminate the potential for waste, fraud, and abuse; and

(5) all contracts or cooperative agreements entered into with a corporation, partnership, or individual working on behalf of the national media campaign.

(e) AUTHORIZATION OF APPROPRIATIONS.—

(1) IN GENERAL.—There is authorized to be appropriated to carry out this section $5,000,000 for each of fiscal years 2008 through 2012.

(2) DECREASED OIL CONSUMPTION.—The Secretary shall use not less than 50 percent of the amount that is made available under this section for each fiscal year to develop and conduct a national media campaign to decrease oil consumption in the United States over the next decade.

Sec. 802. Alaska Natural Gas Pipeline Administration.

Section 106 of the Alaska Natural Gas Pipeline Act (15 U.S.C. 720d) is amended by adding at the end the following:

"(h) ADMINISTRATION.—

"(1) PERSONNEL APPOINTMENTS.—

"(A) IN GENERAL.—The Federal Coordinator may appoint and terminate such personnel as the Federal Coordinator determines to be appropriate.

"(B) AUTHORITY OF FEDERAL COORDINATOR.—Personnel appointed by the Federal Coordinator under subparagraph (A) shall be appointed without regard to the provisions of title 5, United States Code, governing appointments in the competitive service.

"(2) COMPENSATION.—

"(A) IN GENERAL.—Subject to subparagraph (B), personnel appointed by the Federal Coordinator under paragraph (1)(A) shall be paid without regard to the provisions of chapter 51 and subchapter III of chapter 53 of title 5, United States Code (relating to classification and General Schedule pay rates).

"(B) MAXIMUM LEVEL OF COMPENSATION.—The rate of pay for personnel appointed by the Federal Coordinator under paragraph (1)(A) shall not exceed the maximum level of rate payable for level III of the Executive Schedule (5 U.S.C. 5314).

"(C) ALLOWANCES.—Section 5941 of title 5, United States Code, shall apply to personnel appointed by the Federal Coordinator under paragraph (1)(A).

"(3) TEMPORARY SERVICES.—

"(A) IN GENERAL.—The Federal Coordinator may procure temporary and intermittent services in accordance with section 3109(b) of title 5, United States Code.

"(B) MAXIMUM LEVEL OF COMPENSATION.—The level of compensation of an individual employed on a temporary or intermittent basis under subparagraph (A) shall not exceed the maximum level of rate payable for level III of the Executive Schedule (5 U.S.C. 5314).

"(4) FEES, CHARGES, AND COMMISSIONS.—

"(A) IN GENERAL.—With respect to the duties of the Federal Coordinator, as described in this Act, the Federal Coordinator shall have similar authority to establish, change, and abolish reasonable filing and service fees, charges, and commissions, require deposits of payments, and provide refunds as provided to the Secretary of the Interior in section 304 of the Federal Land Policy and Management Act of 1976 (43 U.S.C. 1734).

"(B) AUTHORITY OF SECRETARY OF THE INTERIOR.—

Subparagraph (A) shall not affect the authority of the Secretary of the Interior to establish, change, and abolish reasonable filing and service fees, charges, and commissions, require deposits of payments, and provide refunds under section 304 of the Federal Land Policy and Management Act of 1976 (43 U.S.C. 1734).

"(C) USE OF FUNDS.—The Federal Coordinator is authorized to use, without further appropriation, amounts collected under subparagraph (A) to carry out this section.".

(i) in the State of Alaska;

(ii) without the use of a dam or impoundment of water; and

(iii) through the use of—

(I) a lake tap (but not a perched alpine lake); or

(II) a run-of-river screened at the point of diversion; and

(B) has a nameplate capacity rating of a wattage that is not more than 15 megawatts.

(2) ELIGIBLE APPLICANT.—The term "eligible applicant" means any—

(A) governmental entity;

(B) private utility;

(C) public utility;

(D) municipal utility;

(E) cooperative utility;

(F) Indian tribes; and

(G) Regional Corporation (as defined in section 3 of the Alaska Native Claims Settlement Act (43 U.S.C. 1602)).

(3) OCEAN ENERGY.—

(A) INCLUSIONS.—The term "ocean energy" includes current, wave, and tidal energy.

(B) EXCLUSION.—The term "ocean energy" excludes thermal energy.

(4) RENEWABLE ENERGY PROJECT.—The term "renewable energy project" means a project—

(A) for the commercial generation of electricity; and

(B) that generates electricity from—

(i) solar, wind, or geothermal energy or ocean energy;

(ii) biomass (as defined in section 203(b) of the Energy Policy Act of 2005 (42 U.S.C. 15852(b)));

(iii) landfill gas; or

(iv) Alaska small hydroelectric power.

(b) RENEWABLE ENERGY CONSTRUCTION GRANTS.—

(1) IN GENERAL.—The Secretary shall use amounts appropriated under this section to make grants for use in carrying out renewable energy projects.

(2) CRITERIA.—Not later than 180 days after the date of Deadline. enactment of this Act, the Secretary shall set forth criteria for use in awarding grants under this section.

(3) APPLICATION.—To receive a grant from the Secretary under paragraph (1), an eligible applicant shall submit to the Secretary an application at such time, in such manner, and containing such information as the Secretary may require, including a written assurance that—

(A) all laborers and mechanics employed by contractors or subcontractors during construction, alteration, or repair that is financed, in whole or in part, by a grant under this section shall be paid wages at rates not less than those prevailing on similar construction in the locality, as determined by the Secretary of Labor in accordance with sections 3141–3144, 3146, and 3147 of title 40, United States Code; and

(B) the Secretary of Labor shall, with respect to the labor standards described in this paragraph, have the authority and functions set forth in Reorganization Plan Numbered 14 of 1950 (5 U.S.C. App.) and section 3145 of title 40, United States Code.

(4) NON-FEDERAL SHARE.—Each eligible applicant that receives a grant under this subsection shall contribute to the total cost of the renewable energy project constructed by the eligible applicant an amount not less than 50 percent of the total cost of the project.

(c) AUTHORIZATION OF APPROPRIATIONS.—There are authorized to be appropriated to the Fund such sums as are necessary to carry out this section.

Sec. 804. Coordination Of Planned Refinery Outages.

(a) DEFINITIONS.—In this section:

(1) ADMINISTRATOR.—The term "Administrator" means the Administrator of the Energy Information Administration.

(2) PLANNED REFINERY OUTAGE.—

(A) IN GENERAL.—The term "planned refinery outage" means a removal, scheduled before the date on which the removal occurs, of a refinery, or any unit of a refinery, from service for maintenance, repair, or modification.

(B) EXCLUSION.—The term "planned refinery outage" does not include any necessary and unplanned removal of a refinery, or any unit of a emergency, or action reasonably anticipated to be necessary to prevent such events.

(3) REFINED PETROLEUM PRODUCT.—The term "refined petroleum product" means any gasoline, diesel fuel, fuel oil, lubricating oil, liquid petroleum gas, or other petroleum distillate that is produced through the refining or processing of crude oil or an oil derived from tar sands, shale, or coal.

(4) REFINERY.—The term "refinery" means a facility used in the production of a refined petroleum product through distillation, cracking, or any other process.

(b) REVIEW AND ANALYSIS OF AVAILABLE INFORMATION.—The Administrator shall, on an ongoing basis—

(1) review information on refinery outages that is available from commercial reporting services;

(2) analyze that information to determine whether the scheduling of a refinery outage may nationally or regionally substantially affect the price or supply of any refined petroleum product by—

(A) decreasing the production of the refined petroleum product; and

(B) causing or contributing to a retail or wholesale supply shortage or disruption;

(3) not less frequently than twice each year, submit to the Secretary a report describing the results of the review and analysis under paragraphs (1) and (2); and

(4) specifically alert the Secretary of any refinery outage that the Administrator determines may nationally or regionally substantially affect the price or supply of a refined petroleum product.

(c) ACTION BY SECRETARY.—On a determination by the Secretary, based on a report or alert under paragraph (3) or (4) of subsection (b), that a refinery outage may affect the price or supply of a refined petroleum product, the Secretary shall make available to refinery operators information on planned refinery outages to encourage reductions of the quantity of refinery capacity that is out of service at any time.

(d) LIMITATION.—Nothing in this section shall alter any existing legal obligation or responsibility of a refinery operator, or create any legal right of action, nor shall this section authorize the Secretary—

(1) to prohibit a refinery operator from conducting a planned refinery outage; or

(2) to require a refinery operator to continue to operate a refinery.

Sec. 805. Assessment Of Resources.

(a) 5-YEAR PLAN.—

(1) ESTABLISHMENT.—The Administrator of the Energy Information Administration (referred to in this section as the "Administrator") shall establish a 5-year plan to enhance the quality and scope of the data collection necessary to ensure the scope, accuracy, and timeliness of the information needed for efficient functioning of energy markets and related financial operations.

(2) REQUIREMENT.—In establishing the plan under paragraph (1), the Administrator shall pay particular attention to—

(A) data series terminated because of budget constraints;

(B) data on demand response;

(C) timely data series of State-level information;

(D) improvements in the area of oil and gas data;

(E) improvements in data on solid byproducts from coal-based energy-producing facilities; and

(F) the ability to meet applicable deadlines under Federal law (including regulations) to provide data required by Congress.

(b) SUBMISSION TO CONGRESS.—The Administrator shall submit to Congress the plan established under subsection (a), including a description of any improvements needed to enhance the ability of the Administrator to collect and process energy information in a manner consistent with the needs of energy markets.

(c) GUIDELINES.—

(1) IN GENERAL.—The Administrator shall—

(A) establish guidelines to ensure the quality, comparability, and scope of State energy data, including data on energy production and consumption by product and sector and renewable and alternative sources, required to provide a comprehensive, accurate energy profile at the State level;

(B) share company-level data collected at the State level with each State involved, in a manner consistent with the legal authorities, confidentiality protections, and stated uses in effect at the time the data were collected, subject to the condition that the State shall agree to reasonable requirements for use of the data, as the Administrator may require;

(C) assess any existing gaps in data obtained and compiled by the Energy Information Administration; and

(D) evaluate the most cost-effective ways to address any data quality and quantity issues in conjunction with State officials.

(2) CONSULTATION.—The Administrator shall consult with State officials and the Federal Energy Regulatory Commission on a regular basis in—

(A) establishing guidelines and determining the scope of State-level data under paragraph (1); and

(B) exploring ways to address data needs and serve data uses.

(d) ASSESSMENT OF STATE DATA NEEDS.—Not later than 1 year after the date of enactment of this Act, the Administrator shall submit to Congress an assessment of State-level data needs, including a plan to address the needs.

(e) AUTHORIZATION OF APPROPRIATIONS.—In addition to any other amounts made available to the Administrator, there are authorized to be appropriated to the Administrator to carry out this section—

(1) $10,000,000 for fiscal year 2008;

(2) $10,000,000 for fiscal year 2009;

(3) $10,000,000 for fiscal year 2010;

(4) $15,000,000 for fiscal year 2011;

(5) $20,000,000 for fiscal year 2012; and

(6) such sums as are necessary for subsequent fiscal years.

Sec. 806. Sense Of Congress Relating To The Use Of Renewable Resources To Generate Energy.

(a) FINDINGS.—Congress finds that—

(1) the United States has a quantity of renewable energy resources that is sufficient to supply a significant portion of the energy needs of the United States;

(2) the agricultural, forestry, and working land of the United States can help ensure a sustainable domestic energy system;

(3) accelerated development and use of renewable energy technologies provide numerous benefits to the United States, including improved national security, improved balance of payments, healthier rural economies, improved environmental quality, and abundant, reliable, and affordable energy for all citizens of the United States;

(4) the production of transportation fuels from renewable energy would help the United States meet rapidly growing domestic and global energy demands, reduce the dependence of the United States on energy imported from volatile regions of the world that are politically unstable, stabilize the cost and availability of energy, and safeguard the economy and security of the United States;

(5) increased energy production from domestic renewable resources would attract substantial new investments in energy infrastructure, create economic growth, develop new jobs for the citizens of the United States, and increase the income for farm, ranch, and forestry jobs in the rural regions of the United States;

(6) increased use of renewable energy is practical and can be cost effective with the implementation of supportive policies and proper incentives to stimulate markets and infrastructure; and

(7) public policies aimed at enhancing renewable energy production and accelerating technological improvements will further reduce energy costs over time and increase market demand.

(b) SENSE OF CONGRESS.—It is the sense of Congress that it is the goal of the United States that, not later than January 1, 2025, the agricultural, forestry, and working land of the United States should—

(1) provide from renewable resources not less than 25 percent of the total energy consumed in the United States; and

(2) continue to produce safe, abundant, and affordable food, feed, and fiber.

Sec. 807. Geothermal Assessment, Exploration Information, And Priority Activities.

(a) IN GENERAL.—Not later than January 1, 2012, the Secretary of the Interior, acting through the Director of the United States Geological Survey, shall—

(1) complete a comprehensive nationwide geothermal resource assessment that examines the full range of geothermal resources in the United States; and

((@)) submit to the the Committee on Natural Resources Reports. of the House of Representatives and the Committee on Energy and Natural Resources of the Senate a report describing the results of the assessment.

(b) PERIODIC UPDATES.—At least once every 10 years, the Secretary shall update the national assessment required under this section to support public and private sector decisionmaking.

(c) AUTHORIZATION OF APPROPRIATIONS.—There are authorized to be appropriated to the Secretary of the Interior to carry out this section—

(1) $15,000,000 for each of fiscal years 2008 through 2012; and

(2) such sums as are necessary for each of fiscal years 2013 through 2022.

Subtitle B—Prohibitions on Market Manipulation and False Information

Sec. 811. Prohibition On Market Manipulation..

It is unlawful for any person, directly or indirectly, to use or employ, in connection with the purchase or sale of crude oil gasoline or petroleum distillates at wholesale, any manipulative or deceptive device or contrivance, in contravention of such rules and regulations as the Federal Trade Commission may prescribe as necessary or appropriate in the public interest or for the protection of United States citizens.

Sec. 812. Prohibition On False Information.

It is unlawful for any person to report information related to the wholesale price of crude oil gasoline or petroleum distillates to a Federal department or agency if—

(1) the person knew, or reasonably should have known, the information to be false or misleading;

(2) the information was required by law to be reported;

(3) the person intended the false or misleading data to affect data compiled by the department or agency for statistical or analytical purposes with respect to the market for crude oil, gasoline, or petroleum distillates.

Sec. 813. Enforcement By The Federal Trade Commission.

(a) ENFORCEMENT.—This subtitle shall be enforced by the Federal Trade Commission in the same manner, by the same means, and with the same jurisdiction as though all applicable terms of the Federal Trade Commission Act (15 U.S.C. 41 et seq.) were incorporated into and made a part of this subtitle.

(b) VIOLATION IS TREATED AS UNFAIR OR DECEPTIVE ACT OR PRACTICE.—The violation of any provision of this subtitle shall be treated as an unfair or deceptive act or practice proscribed under a rule issued under section 18(a)(1)(B) of the Federal Trade Commission Act (15 U.S.C. 57a(a)(1)(B)).

Sec. 814. Penalties.

(a) CIVIL PENALTY.—In addition to any penalty applicable under the Federal Trade Commission Act (15 U.S.C. 41 et seq.), any supplier that violates section 811 or 812 shall be punishable by a civil penalty of not more than $1,000,000.

(b) METHOD.—The penalties provided by subsection (a) shall be obtained in the same manner as civil penalties imposed under section 5 of the Federal Trade Commission Act (15 U.S.C. 45).

(c) MULTIPLE OFFENSES; MITIGATING FACTORS.—In assessing the penalty provided by subsection (a)—

(1) each day of a continuing violation shall be considered a separate violation; and

(2) the court shall take into consideration, among other factors—

(A) the seriousness of the violation; and

(B) the efforts of the person committing the violation to remedy the harm caused by the violation in a timely manner.

Sec. 815. Effect On Other Laws.

(a) OTHER AUTHORITY OF THE COMMISSION.—Nothing in this subtitle limits or affects the authority of the Federal Trade Commission to bring an enforcement action or take any other measure under the Federal Trade Commission Act (15 U.S.C. 41 et seq.) or any other provision of law.

(b) ANTITRUST LAW.—Nothing in this subtitle shall be construed to modify, impair, or supersede the operation of any of the antitrust laws. For purposes of this subsection, the term "antitrust laws" shall have the meaning given it in subsection (a) of the first section of the Clayton Act (15 U.S.C. 12), except that it includes section 5 of the Federal Trade Commission Act (15 U.S.C. 45) to the extent that such section 5 applies to unfair methods of competition.

(c) STATE LAW.—Nothing in this subtitle preempts any State law.

Title IX—International Energy Programs

Sec. 901. Definitions.

In this title:

(1) APPROPRIATE CONGRESSIONAL COMMITTEES.—The term "appropriate congressional committees" means—

(A) the Committee on Foreign Affairs and the Committee on Energy and Commerce of the House of Representatives; and

(B) the Committee on Foreign Relations, the Committee on Energy and Natural Resources, the Committee on Environment and Public Works, and the Committee on Commerce, Science, and Transportation of the Senate.

(2) CLEAN AND EFFICIENT ENERGY TECHNOLOGY.—The term "clean and efficient energy technology" means an energy supply or end-use technology that, compared to a similar technology already in widespread commercial use in a recipient country, will—

(A) reduce emissions of greenhouse gases; or

(B)(i) increase efficiency of energy production; or (ii) decrease intensity of energy usage.

(3) GREENHOUSE GAS.—The term "greenhouse gas" means—

(A) carbon dioxide;

(B) methane;

(C) nitrous oxide;

(D) hydrofluorocarbons;

(E) perfluorocarbons; or

(F) sulfur hexafluoride.

Subtitle A—Assistance to Promote Clean and Efficient Energy Technologies in Foreign Countries

Sec. 911. United States Assistance For Developing Countries.

(a) ASSISTANCE AUTHORIZED.—The Administrator of the United States Agency for International Development shall support policies and programs in developing countries that promote clean and efficient energy technologies—

(1) to produce the necessary market conditions for the private sector delivery of energy and environmental management services;

(2) to create an environment that is conducive to accepting clean and efficient energy technologies that support the overall purpose of reducing greenhouse gas emissions, including—

(A) improving policy, legal, and regulatory frameworks;

(B) increasing institutional abilities to provide energy and environmental management services; and

(C) increasing public awareness and participation in the decision-making of delivering energy and environmental management services; and

(3) to promote the use of American-made clean and efficient energy technologies, products, and energy and environmental management services.

(b) REPORT.—The Administrator of the United States Agency for International Development shall submit to the appropriate congressional committees an annual report on the implementation of this section for each of the fiscal years 2008 through 2012.

(c) AUTHORIZATION OF APPROPRIATIONS.—To carry out this section, there are authorized to be appropriated to the Administrator of the United States Agency for International Development $200,000,000 for each of the fiscal years 2008 through 2012.

Sec. 912. United States Exports And Outreach Programs For India, China, And Other Countries.

(a) ASSISTANCE AUTHORIZED.—The Secretary of Commerce shall direct the United States and Foreign Commercial Service to expand or create a corps of the Foreign Commercial Service officers to promote United States exports in clean and efficient energy technologies and build the capacity of government officials in India, China, and any other country the Secretary of Commerce determines appropriate, to become more familiar with the available technologies—

(1) by assigning or training Foreign Commercial Service attache´s, who have expertise in clean and efficient energy technologies from the United States, to embark on business development and outreach efforts to such countries; and

(2) by deploying the attache´s described in paragraph (1) to educate provincial, state, and local government officials in such countries on the variety of United States-based technologies in clean and efficient energy technologies for the purposes of promoting United States exports and reducing global greenhouse gas emissions.

(b) REPORT.—The Secretary of Commerce shall submit to the appropriate congressional committees an annual report on the implementation of this section for each of the fiscal years 2008 through 2012.

(c) AUTHORIZATION OF APPROPRIATIONS.—To carry out this section, there are authorized to be appropriated to the Secretary of Commerce such sums as may be necessary for each of the fiscal years 2008 through 2012.

Sec. 913. United States Trade Missions To Encourage Private Sector Trade And Investment.

(a) ASSISTANCE AUTHORIZED.—The Secretary of Commerce shall direct the International Trade Administration to expand or create trade missions to and from the United States to encourage private sector trade and investment in clean and efficient energy technologies—

(1) by organizing and facilitating trade missions to foreign countries and by matching United States private sector companies with opportunities in foreign markets so that clean and efficient energy technologies can help to combat increases in global greenhouse gas emissions; and

(2) by creating reverse trade missions in which the Department of Commerce facilitates the meeting of foreign private and public sector organizations with private sector companies in the United States for the purpose of showcasing clean

and efficient energy technologies in use or in development that could be exported to other countries.

(b) REPORT.—The Secretary of Commerce shall submit to the appropriate congressional committees an annual report on the implementation of this section for each of the fiscal years 2008 through 2012.

(c) AUTHORIZATION OF APPROPRIATIONS.—To carry out this section, there are authorized to be appropriated to the Secretary of Commerce such sums as may be necessary for each of the fiscal years 2008 through 2012.

Sec. 914. Actions By Overseas Private Investment Corporation.

(a) SENSE OF CONGRESS.—It is the sense of Congress that the Overseas Private Investment Corporation should promote greater investment in clean and efficient energy technologies by—

(1) proactively reaching out to United States companies that are interested in investing in clean and efficient energy technologies in countries that are significant contributors to global greenhouse gas emissions;

(2) giving preferential treatment to the evaluation and awarding of projects that involve the investment or utilization of clean and efficient energy technologies; and

(3) providing greater flexibility in supporting projects that involve the investment or utilization of clean and efficient energy technologies, including financing, insurance, and other assistance.

(b) REPORT.—The Overseas Private Investment Corporation shall include in its annual report required under section 240A of the Foreign Assistance Act of 1961 (22 U.S.C. 2200a)—

(1) a description of the activities carried out to implement this section; or

(2) if the Corporation did not carry out any activities to implement this section, an explanation of the reasons therefor.

Sec. 915. Actions By United States Trade And Development Agency.

(a) ASSISTANCE AUTHORIZED.—The Director of the Trade and Development Agency shall establish or support policies that—

(1) proactively seek opportunities to fund projects that involve the utilization of clean and efficient energy technologies, including in trade capacity building and capital investment projects;

(2) where appropriate, advance the utilization of clean and efficient energy technologies, particularly to countries that have the potential for significant reduction in greenhouse gas emissions; and

(3) recruit and retain individuals with appropriate expertise or experience in clean, renewable, and efficient energy technologies to identify and evaluate opportunities for projects that involve clean and efficient energy technologies and services.

(b) REPORT.—The President shall include in the annual report President. on the activities of the Trade and Development Agency required under section 661(d) of the Foreign Assistance Act of 1961 (22 U.S.C. 2421(d)) a description of the activities carried out to implement this section.

Sec. 916. Deployment Of International Clean And Efficient Energy Technologies And Investment In Global Energy Markets.

(a) TASK FORCE.—

(1) ESTABLISHMENT.—Not later than 90 days after the date of the enactment of this Act, the President shall establish a Task Force on International Cooperation for Clean and Efficient Energy Technologies (in this section referred to as the "Task Force").

(2) COMPOSITION.—The Task Force shall be composed of representatives, appointed by the head of the respective Federal department or agency, of—

(A) the Council on Environmental Quality;

(B) the Department of Energy;

(C) the Department of Commerce;

(D) the Department of the Treasury;

(E) the Department of State;

(F) the Environmental Protection Agency;

(G) the United States Agency for International Development;

(H) the Export-Import Bank of the United States;

(I) the Overseas Private Investment Corporation:

(J) the Trade and Development Agency;

(K) the Small Business Administration;

(L) the Office of the United States Trade Representative; and

(M) other Federal departments and agencies, as determined by the President.

(3) CHAIRPERSON.—The President shall designate a Chairperson or Co-Chairpersons of the Task Force.

(4) DUTIES.—The Task Force—

(A) shall develop and assist in the implementation of the strategy required under subsection (c); and

(B)(i) shall analyze technology, policy, and market opportunities for the development, demonstration, and deployment of clean and efficient energy technologies on an international basis; and

(ii) shall examine relevant trade, tax, finance, international, and other policy issues to assess which policies, in the United States and in developing countries, would help open markets and improve the export of clean and efficient energy technologies from the United States.

(5) TERMINATION.—The Task Force, including any working group established by the Task Force pursuant to subsection (b), shall terminate 12 years after the date of the enactment of this Act.

(b) WORKING GROUPS.—

(1) ESTABLISHMENT.—The Task Force—

(A) shall establish an Interagency Working Group on the Export of Clean and Efficient Energy Technologies (in this section referred to as the "Interagency Working Group"); and

(B) may establish other working groups as may be necessary to carry out this section.

(2) COMPOSITION.—The Interagency Working Group shall be composed of—

(A) the Secretary of Energy, the Secretary of Commerce, and the Secretary of State, who shall serve as Co-Chairpersons of the Interagency Working Group; and

(B) other members, as determined by the Chairperson or Co-Chairpersons of the Task Force.

(3) DUTIES.—The Interagency Working Group shall coordinate the resources and relevant programs of the Department of Energy, the Department of Commerce, the Department of State, and other relevant Federal departments and agencies to support the export of clean and efficient energy technologies developed or demonstrated in the United States to other countries and the deployment of such clean and efficient energy technologies in such other countries.

(4) INTERAGENCY CENTER.—The Interagency Working Group—

(A) shall establish an Interagency Center on the Export of Clean and Efficient Energy Technologies (in this section referred to as the "Interagency Center") to assist the Interagency Working Group in carrying out its duties required under paragraph (3); and

(B) shall locate the Interagency Center at a site agreed upon by the Co-Chairpersons of the Interagency Working Group, with the approval of the Chairperson or Co-Chairpersons of the Task Force.

(c) STRATEGY.—

(1) IN GENERAL.—Not later than 1 year after the date of the enactment of this Act, the Task Force shall develop and submit to the President and the appropriate congressional committees a strategy to—

(A) support the development and implementation of programs, policies, and initiatives in developing countries to promote the adoption and deployment of clean and efficient energy technologies, with an emphasis on those developing countries that are expected to experience the most significant growth in energy production and use over the next 20 years;

(B) open and expand clean and efficient energy technology markets and facilitate the export of clean and efficient energy technologies to developing countries, in a manner consistent with United States obligations as a member of the World Trade Organization;

(C) integrate into the foreign policy objectives of the United States the promotion of—

(i) the deployment of clean and efficient energy technologies and the reduction of greenhouse gas emissions in developing countries; and

(ii) the export of clean and efficient energy technologies; and

(D) develop financial mechanisms and instruments, including securities that mitigate the political and foreign exchange risks of uses that are consistent with the foreign policy objectives of the United States by combining the private sector market and government enhancements, that—

(i) are cost-effective; and

(ii) facilitate private capital investment in clean and efficient energy technology projects in developing countries.

(2) UPDATES.—Not later than 3 years after the date of submission of the strategy under paragraph (1), and every 3 years thereafter, the Task Force shall update the strategy in accordance with the requirements of paragraph (1).

(d) REPORT.—

(1) IN GENERAL.—Not later than 3 years after the date of submission of the strategy under subsection (c)(1), and every 3 years thereafter, the President shall transmit

to the appropriate congressional committees a report on the implementation of this section for the prior 3-year period.

(2) MATTERS TO BE INCLUDED.—The report required under paragraph (1) shall include the following:

(A) The update of the strategy required under subsection (c)(2) and a description of the actions taken by the Task Force to assist in the implementation of the strategy.

(B) A description of actions taken by the Task Force to carry out the duties required under subsection (a)(4)(B).

(C) A description of assistance provided under this section.

(D) The results of programs, projects, and activities carried out under this section.

(E) A description of priorities for promoting the diffusion and adoption of clean and efficient energy technologies and strategies in developing countries, taking into account economic and security interests of the United States and opportunities for the export of technology of the United States.

(F) Recommendations to the heads of appropriate Federal departments and agencies on methods to streamline Federal programs and policies to improve the role of such Federal departments and agencies in the development, demonstration, and deployment of clean and efficient energy technologies on an international basis.

(G) Strategies to integrate representatives of the private sector and other interested groups on the export and deployment of clean and efficient energy technologies.

(H) A description of programs to disseminate information to the private sector and the public on clean and efficient energy technologies and opportunities to transfer such clean and efficient energy technologies.

(b) AUTHORIZATION OF APPROPRIATIONS.—There are authorized to be appropriated to carry out this section $5,000,000 for each of fiscal years 2008 through 2020.

Sec. 917. United States-Israel Energy Cooperation.

(a) FINDINGS.—Congress finds that—

(1) it is in the highest national security interests of the United States to develop renewable energy sources;

(2) the State of Israel is a steadfast ally of the United States;

(3) the special relationship between the United States and Israel is manifested in a variety of cooperative scientific research and development programs, such as—

(A) the United States-Israel Binational Science Foundation; and

(B) the United States-Israel Binational Industrial Research and Development Foundation;

(4) those programs have made possible many scientific, technological, and commercial breakthroughs in the fields of life sciences, medicine, bioengineering, agriculture, biotechnology, communications, and others;

(5) on February 1, 1996, the Secretary of Energy (referred to in this section as the "Secretary") and the Israeli Minister of Energy and Infrastructure signed an agreement to establish a framework for collaboration between the United States and Israel in energy research and development activities;

(6) Israeli scientists and engineers are at the forefront of research and development in the field of renewable energy sources; and

(7) enhanced cooperation between the United States and Israel for the purpose of research and development of renewable energy sources would be in the national interests of both countries.

(b) GRANT PROGRAM.—

(1) ESTABLISHMENT.—In implementing the agreement entitled the "Agreement between the Department of Energy of the United States of America and the Ministry of Energy and Infrastructure of Israel Concerning Energy Cooperation", dated February 1, 1996, the Secretary shall establish a grant program in accordance with the requirements of sections 988 and 989 of the Energy Policy Act of 2005 (42 U.S.C. 16352, 16353) to support research, development, and commercialization of renewable energy or energy efficiency.

(2) TYPES OF ENERGY.—In carrying out paragraph (1), the Secretary may make grants to promote—

(A)bsolar energy;

(B) biomass energy;

(C) energy efficiency;

(D) wind energy;

(E) geothermal energy;

(F) wave and tidal energy; and

(G) advanced battery technology.

(3) ELIGIBLE APPLICANTS.—An applicant shall be eligible to receive a grant under this subsection if the project of the applicant—

(A) addresses a requirement in the area of improved energy efficiency or renewable energy sources, as determined by the Secretary; and

(B) is a joint venture between—

(i) (I) a for-profit business entity, academic institution, National Laboratory (as defined in section 2 of the Energy Policy Act of 2005 (42 U.S.C. 15801)), or nonprofit entity in the United States; and

(II) a for-profit business entity, academic institution, or nonprofit entity in Israel; or

(ii) (I) the Federal Government; and

(II) the Government of Israel.

(4) APPLICATIONS.—To be eligible to receive a grant under this subsection, an applicant shall submit to the Secretary an application for the grant in accordance with procedures established by the Secretary, in consultation with the advisory board established under paragraph (5).

(5) ADVISORY BOARD.—

(A) ESTABLISHMENT.—The Secretary shall establish an advisory board—

(i) to monitor the method by which grants are awarded under this subsection; and

(ii) to provide to the Secretary periodic performance reviews of actions taken to carry out this subsection.

(B) COMPOSITION.—The advisory board established under subparagraph (A) shall be composed of 3 members, to be appointed by the Secretary, of whom—

(i) 1 shall be a representative of the Federal Government;

(ii) 1 shall be selected from a list of nominees provided by the United States-Israel Binational Science Foundation; and

(iii) 1 shall be selected from a list of nominees provided by the United States-Israel Binational Industrial Research and Development Foundation.

(6) CONTRIBUTED FUNDS.—Notwithstanding section 3302 of title 31, United States Code, the Secretary may accept, retain, and use funds contributed by any person, government entity, or organization for purposes of carrying out this subsection—

(A) without further appropriation; and

(B) without fiscal year limitation.

(7) REPORT.—Not later than 180 days after the date of completion of a project for which a grant is provided under this subsection, the grant recipient shall submit to the Secretary a report that contains—

(A) a description of the method by which the recipient used the grant funds; and

(B) an evaluation of the level of success of each project funded by the grant.

(8) CLASSIFICATION.—Grants shall be awarded under this subsection only for projects that are considered to be unclassified by both the United States and Israel.

(c) TERMINATION.—The grant program and the advisory committee established under this section terminate on the date that is 7 years after the date of enactment of this Act.

(d) AUTHORIZATION OF APPROPRIATIONS.—The Secretary shall use amounts authorized to be appropriated under section 931 of the Energy Policy Act of 2005 (42 U.S.C. 16231) to carry out this section.

Subtitle B—International Clean Energy Foundation

(1) CHIEF EXECUTIVE OFFICER.—The term "Chief Executive Officer" means the chief executive officer of the Foundation appointed pursuant to section 922(b).

(2) FOUNDATION.—The term "Foundation" means the International Clean Energy Foundation established by section 922(a).

Sec. 922. Establishment And Management Of Foundation.

(a) ESTABLISHMENT.—

(1) IN GENERAL.—There is established in the executive branch a foundation to be known as the "International Clean Energy Foundation" that shall be responsible for carrying out the provisions of this subtitle. The Foundation shall be a government corporation, as defined in section 103 of title 5, United States Code.

(2) BOARD OF DIRECTORS.—The Foundation shall be governed by a Board of Directors in accordance with subsection (c).

(3) INTENT OF CONGRESS.—It is the intent of Congress, in establishing the structure of the Foundation set forth in this subsection, to create an entity that serves the long-term foreign policy and energy security goals of reducing global greenhouse gas emissions.

(b) CHIEF EXECUTIVE OFFICER.—

(1) IN GENERAL.—There shall be in the Foundation a Chief Executive Officer who shall be responsible for the management of the Foundation.

(2) APPOINTMENT.—The Chief Executive Officer shall be appointed by the Board, with the advice and consent of the Senate, and shall be a recognized leader in clean and efficient energy technologies and climate change and shall have experience in energy security, business, or foreign policy, chosen on the basis of a rigorous search.

(3) RELATIONSHIP TO BOARD.—The Chief Executive Officer shall report to, and be under the direct authority of, the Board.

(4) COMPENSATION AND RANK.—

(A) IN GENERAL.—The Chief Executive Officer shall be compensated at the rate provided for level III of the Executive Schedule under section 5314 of title 5, United States Code.

(B) AMENDMENT.—Section 5314 of title 5, United States
Code, is amended by adding at the end the following: "Chief Executive Officer, International Clean Energy Foundation.".

(C) AUTHORITIES AND DUTIES.—The Chief Executive Officer shall be responsible for the management of the Foundation and shall exercise the powers and discharge the duties of the Foundation.

(D) AUTHORITY TO APPOINT OFFICERS.—In consultation and with approval of the Board, the Chief Executive Officer shall appoint all officers of the Foundation.

(c) BOARD OF DIRECTORS.—

(1) ESTABLISHMENT.—There shall be in the Foundation a Board of Directors.

(2) DUTIES.—The Board shall perform the functions specified to be carried out by the Board in this subtitle and may prescribe, amend, and repeal bylaws, rules, regulations, and procedures governing the manner in which the business of the Foundation may be conducted and in which the powers granted to it by law may be exercised.

(3) MEMBERSHIP.—The Board shall consist of—

(A) the Secretary of State (or the Secretary's designee), the Secretary of Energy (or the Secretary's designee), and the Administrator of the United States Agency for International Development (or the Administrator's designee); and

(B) four other individuals with relevant experience in matters relating to energy security (such as individuals who represent institutions of energy policy, business organizations, foreign policy organizations, or other relevant organizations) who shall be appointed by the President, by and with the advice and consent of the Senate, of whom—

(i) one individual shall be appointed from among a list of individuals submitted by the Majority Leader of the House of Representatives;

(ii) one individual shall be appointed from among a list of individuals submitted by the Minority Leader of the House of Representatives;

(iii) one individual shall be appointed from among a list of individuals submitted by the Majority Leader of the Senate; and

(iv) one individual shall be appointed from among a list of individuals submitted by the Minority Leader of the Senate.

(4) CHIEF EXECUTIVE OFFICER.—The Chief Executive Officer of the Foundation shall serve as a nonvoting, ex officio member of the Board.

(5) TERMS.—

(A) OFFICERS OF THE FEDERAL GOVERNMENT.—Each member of the Board described in paragraph (3)(A) shall serve for a term that is concurrent with the term of service of the individual's position as an officer within the other Federal department or agency.

(B) OTHER MEMBERS.—Each member of the Board described in paragraph (3)(B) shall be appointed for a term of 3 years and may be reappointed for a term of an additional 3 years.

(C) VACANCIES.—A vacancy in the Board shall be filled in the manner in which the original appointment was made.

(D) ACTING MEMBERS.—A vacancy in the Board may be filled with an appointment of an acting member by the Chairperson of the Board for up to 1 year while a nominee is named and awaits confirmation in accordance with paragraph (3)(B).

(6) CHAIRPERSON.—There shall be a Chairperson of the Board. The Secretary of State (or the Secretary's designee) shall serve as the Chairperson.

(7) QUORUM.—A majority of the members of the Board described in paragraph (3) shall constitute a quorum, which, except with respect to a meeting of the Board during the 135-day period beginning on the date of the enactment of this Act, shall include at least 1 member of the Board described in paragraph (3)(B).

(8) MEETINGS.—The Board shall meet at the call of the Chairperson, who shall call a meeting no less than once a year.

(9) COMPENSATION.—

(A) OFFICERS OF THE FEDERAL GOVERNMENT.—

(i) IN GENERAL.—A member of the Board described in paragraph (3)(A) may not receive additional pay, allowances, or benefits by reason of the member's service on the Board.

(ii) TRAVEL EXPENSES.—Each such member of the Board shall receive travel expenses, including per diem in lieu of subsistence, in accordance with applicable provisions under subchapter I of chapter 57 of title 5, United States Code.

(B) OTHER MEMBERS.—

(i) IN GENERAL.—Except as provided in clause (ii), a member of the Board described in paragraph (3)(B)—

(I) shall be paid compensation out of funds made available for the purposes of this subtitle at the daily equivalent of the highest rate payable under section 5332 of title 5, United States Code, for each day (including travel time) during which the member is engaged in the actual performance of duties as a member of the Board; and

(II) while away from the member's home or regular place of business on necessary travel in the actual performance of duties as a member of the Board, shall be paid per diem, travel, and transportation expenses in the same manner as is provided under subchapter I of chapter 57 of title 5, United States Code.

(i) LIMITATION.—A member of the Board may not be paid compensation under clause (i)(II) for more than 90 days in any calendar year.

Sec. 923. Duties Of Foundation.

The Foundation shall—

(1) use the funds authorized by this subtitle to make grants to promote projects outside of the United States that serve as models of how to significantly reduce the emissions of global greenhouse gases through clean and efficient energy technologies, processes, and services;

(2) seek contributions from foreign governments, especially those rich in energy resources such as member countries of the Organization of the Petroleum Exporting Countries, and private organizations to supplement funds made available under this subtitle;

(3) harness global expertise through collaborative partnerships with foreign governments and domestic and foreign private actors, including nongovernmental organizations and private sector companies, by leveraging public and private capital, technology, expertise, and services towards innovative models that can be instituted to reduce global greenhouse gas emissions;

(4) create a repository of information on best practices and lessons learned on the utilization and implementation of clean and efficient energy technologies and processes to be used for future initiatives to tackle the climate change crisis;

(5) be committed to minimizing administrative costs and to maximizing the availability of funds for grants under this subtitle; and

(6) promote the use of American-made clean and efficient energy technologies, processes, and services by giving preference to entities incorporated in the United States and whose technology will be substantially manufactured in the United States.

Sec. 924. Annual Report.

(a) REPORT REQUIRED.—Not later than March 31, 2008, and each March 31 thereafter, the Foundation shall submit to the appropriate congressional committees a report on the implementation of this subtitle during the prior fiscal year.

(b) CONTENTS.—The report required by subsection (a) shall include—

(1) the total financial resources available to the Foundation during the year, including appropriated funds, the value and source of any gifts or donations accepted pursuant to section 925(a)(6), and any other resources;

(2) a description of the Board's policy priorities for the year and the basis upon which competitive grant proposals were solicited and awarded to nongovernmental institutions and other organizations;

(3) a list of grants made to nongovernmental institutions and other organizations that includes the identity of the institutional recipient, the dollar amount, and the results of the program; and

(4) the total administrative and operating expenses of the Foundation for the year, as well as specific information on—

(A) the number of Foundation employees and the cost of compensation for Board members, Foundation employ- ees, and personal service contractors;

(B) costs associated with securing the use of real property for carrying out the functions of the Foundation;

(C) total travel expenses incurred by Board members and Foundation employees in connection with Foundation activities; and

Sec. 925. Powers Of The Foundation; Related Provisions.

(a) POWERS.—The Foundation—

(1) shall have perpetual succession unless dissolved by a law enacted after the date of the enactment of this Act;

(2) may adopt, alter, and use a seal, which shall be judicially noticed;

(3) may make and perform such contracts, grants, and other agreements with any person or government however designated and wherever situated, as may be necessary for carrying out the functions of the Foundation;

(4) may determine and prescribe the manner in which its obligations shall be incurred and its expenses allowed and paid, including expenses for representation;

(5) may lease, purchase, or otherwise acquire, improve, and use such real property wherever situated, as may be necessary for carrying out the functions of the Foundation;

(6) may accept money, funds, services, or property (real, personal, or mixed), tangible or intangible, made available by gift, bequest grant, or otherwise for the purpose of carrying out the provisions of this title from domestic or foreign private individuals, charities, nongovernmental organizations, corporations, or governments;

(7) may use the United States mails in the same manner and on the same conditions as the executive departments;

(8) may contract with individuals for personal services, who shall not be considered Federal employees for any provision of law administered by the Office of Personnel Management;

(9) may hire or obtain passenger motor vehicles; and

(10) shall have such other powers as may be necessary and incident to carrying out this subtitle.

(b) PRINCIPAL OFFICE.—The Foundation shall maintain its principal office in the metropolitan area of Washington, District of Columbia.

(c) APPLICABILITY OF GOVERNMENT CORPORATION CONTROL ACT.—

(1) IN GENERAL.—The Foundation shall be subject to chapter 91 of subtitle VI of title 31, United States Code, except that the Foundation shall not be authorized to issue obligations or offer obligations to the public.

(2) CONFORMING AMENDMENT.—Section 9101(3) of title 31, United States Code, is amended by adding at the end the following:

"(R) the International Clean Energy Foundation.".

(d) INSPECTOR GENERAL.—

(1) IN GENERAL.—The Inspector General of the Department of State shall serve as Inspector General of the Foundation, and, in acting in such capacity, may conduct reviews, investigations, and inspections of all aspects of the operations and activities of the Foundation.

(2) AUTHORITY OF THE BOARD.—In carrying out the responsibilities under this subsection, the Inspector General shall report to and be under the general supervision of the Board.

(3) REIMBURSEMENT AND AUTHORIZATION OF SERVICES.—

(A) REIMBURSEMENT.—The Foundation shall reimburse the Department of State for all expenses incurred by the Inspector General in connection with the Inspector General's responsibilities under this subsection.

(B) AUTHORIZATION FOR SERVICES.—Of the amount authorized to be appropriated under section 927(a) for a fiscal year, up to $500,000 is authorized to be made available to the Inspector General of the Department of State to conduct reviews, investigations, and inspections of operations and activities of the Foundation.

Sec. 926. General Personnel Authorities.

(a) DETAIL OF PERSONNEL.—Upon request of the Chief Executive Officer, the head of an agency may detail any employee of such agency to the Foundation on a reimbursable basis. Any employee so detailed remains, for the purpose of preserving such employee's allowances, privileges, rights, seniority, and other benefits, an employee of the agency from which detailed.

(b) REEMPLOYMENT RIGHTS.—

(1) IN GENERAL.—An employee of an agency who is serving under a career or career conditional appointment (or the equivalent), and who, with the consent of the head of such agency, transfers to the Foundation, is entitled to be reemployed in such employee's former position or a position of like seniority, status, and pay in such agency, if such employee—

(A) is separated from the Foundation for any reason, other than misconduct, neglect of duty, or malfeasance; and

(B) applies for reemployment not later than 90 days after the date of separation from the Foundation.

(2) SPECIFIC RIGHTS.—An employee who satisfies paragraph (1) is entitled to be reemployed (in accordance with such paragraph) within 30 days after applying for reemployment and, on reemployment, is entitled to at least the rate of basic pay to which such employee would have been entitled had such employee never transferred.

(c) HIRING AUTHORITY.—Of persons employed by the Foundation, no more than 30 persons may be appointed, compensated, or removed without regard to the civil service laws and regulations.

(d) BASIC PAY.—The Chief Executive Officer may fix the rate of basic pay of employees of the Foundation without regard to the provisions of chapter 51 of title 5, United States Code (relating to the classification of positions), subchapter III of chapter 53 of such title (relating to General Schedule pay rates), except that no employee of the Foundation may receive a rate of basic pay that exceeds the rate for level IV of the Executive Schedule under section 5315 of such title.

(e) DEFINITIONS.—In this section—

(1) the term "agency" means an executive agency, as defined by section 105 of title 5, United States Code; and

(2) the term "detail" means the assignment or loan of an employee, without a change of position, from the agency by which such employee is employed to the Foundation.

Sec. 927. Authorization Of Appropriations.

(a) AUTHORIZATION OF APPROPRIATIONS.—To carry out this subtitle, there are authorized to be appropriated $20,000,000 for each of the fiscal years 2009 through 2013.

(b) ALLOCATION OF FUNDS.—

(1) IN GENERAL.—The Foundation may allocate or transfer to any agency of the United States Government any of the funds available for carrying out this subtitle. Such funds shall be available for obligation and expenditure for the purposes for which the funds were authorized, in accordance with authority granted in this subtitle or under authority governing the activities of the United States Government agency to which such funds are allocated or transferred.

(2) NOTIFICATION.—The Foundation shall notify the appropriate congressional committees not less than 15 days prior to an allocation or transfer of funds pursuant to paragraph (1).

Subtitle C—Miscellaneous Provisions

Sec. 931. Energy Diplomacy And Security Within The Department Of State.

(a) STATE DEPARTMENT COORDINATOR FOR INTERNATIONAL ENERGY AFFAIRS.—

(1) IN GENERAL.—The Secretary of State should ensure that energy security is integrated into the core mission of the Department of State.

(2) COORDINATOR FOR INTERNATIONAL ENERGY AFFAIRS.—

There is established within the Office of the Secretary of State a Coordinator for International Energy Affairs, who shall be responsible for—

(A) representing the Secretary of State in interagency efforts to develop the international energy policy of the United States;

(B) ensuring that analyses of the national security implications of global energy and environmental developments are reflected in the decision making process within the Department of State;

(C) incorporating energy security priorities into the activities of the Department of State;

(D) coordinating energy activities of the Department of State with relevant Federal agencies; and

(E) coordinating energy security and other relevant functions within the Department of State currently undertaken by offices within—

(i) the Bureau of Economic, Energy and Business Affairs;

(ii) the Bureau of Oceans and International Environmental and Scientific Affairs; and

(iii) other offices within the Department of State.

(3) AUTHORIZATION OF APPROPRIATIONS.—There are authorized to be appropriated such sums as may be necessary to carry out this subsection.

(b) ENERGY EXPERTS IN KEY EMBASSIES.—Not later than 180 days after the date of the enactment of this Act, the Secretary of State shall submit a report to the Committee on Foreign Relations of the Senate and the Committee on Foreign Affairs of the House of Representatives that includes—

(1) a description of the Department of State personnel who are dedicated to energy matters and are stationed at embassies and consulates in countries that are major energy producers or consumers;

(2) an analysis of the need for Federal energy specialist personnel in United States embassies and other United States diplomatic missions; and

(3) recommendations for increasing energy expertise within United States embassies among foreign service officers and options for assigning to such embassies energy attache´s from the National Laboratories or other agencies within the Department of Energy.

(c) ENERGY ADVISORS.—The Secretary of Energy may make appropriate arrangements with the Secretary of State to assign personnel from the Department of Energy or the National Laboratories of the Department of Energy to serve as dedicated advisors on energy matters in embassies of the United States or other United States diplomatic missions.

(d) REPORT.—Not later than 180 days after the date of the enactment of this Act, and every 2 years thereafter for the following 20 years, the Secretary of State shall submit a report to the Committee on Foreign Relations of the Senate and the Committee on Foreign Affairs of the House of Representatives that describes—

(1) the energy-related activities being conducted by the Department of State, including activities within—

(A) the Bureau of Economic, Energy and Business Affairs;

(B) the Bureau of Oceans and Environmental and Scientific Affairs; and

(C) other offices within the Department of State;

(2) the amount of funds spent on each activity within each office described in paragraph (1); and

(3) the number and qualification of personnel in each embassy (or relevant foreign posting) of the United States whose work is dedicated exclusively to energy matters.

Sec. 932. National Security Council Reorganization.

Section 101(a) of the National Security Act of 1947 (50 U.S.C. 402(a)) is amended—

(1) by redesignating paragraphs (5), (6), and (7) as paragraphs (6), (7), and (8), respectively; and

(2) by inserting after paragraph (4) the following:

"(5) the Secretary of Energy;".

Sec. 933. Annual National Energy Security Strategy Report.

(a) REPORTS.—

(1) IN GENERAL.—Subject to paragraph (2), on the date on which the President submits to Congress the budget for the following fiscal year under section 1105 of title 31, United States Code, the President shall submit to Congress a comprehensive report on the national energy security of the United States.

(2) NEW PRESIDENTS.—In addition to the reports required under paragraph (1), the President shall submit a comprehensive report on the national energy security of the United States by not later than 150 days after the date on which the President assumes the office of President after a presidential election.

(b) CONTENTS.—Each report under this section shall describe the national energy security strategy of the United States, including a comprehensive description of—

(1) the worldwide interests, goals, and objectives of the United States that are vital to the national energy security of the United States;

(2) the foreign policy, worldwide commitments, and national defense capabilities of the United States necessary—

(A) to deter political manipulation of world energy resources; and

(B) to implement the national energy security strategy of the United States;

(3) the proposed short-term and long-term uses of the political, economic, military, and other authorities of the United States—

(A) to protect or promote energy security; and

(B) to achieve the goals and objectives described in paragraph (1);

(4) the adequacy of the capabilities of the United States to protect the national energy security of the United States, including an evaluation of the balance among the capabilities of all elements of the national authority of the United States to support the implementation of the national energy security strategy; and

(5) such other information as the President determines to be necessary to inform Congress on matters relating to the national energy security of the United States.

(c) CLASSIFIED AND UNCLASSIFIED FORM.—Each national energy security strategy report shall be submitted to Congress in—

(1) a classified form; and

(2) an unclassified form.

Sec. 934. Convention On Supplementary Compensation For Nuclear Damage Contingent Cost Allocation.

(a) FINDINGS AND PURPOSE.—

(1) FINDINGS.—Congress finds that—

(A) section 170 of the Atomic Energy Act of 1954 (42 U.S.C. 2210) (commonly known as the "Price-Anderson Act")—

(i) provides a predictable legal framework necessary for nuclear projects; and

(ii) ensures prompt and equitable compensation in the event of a nuclear incident in the United States;

(B) the Price-Anderson Act, in effect, provides operators of nuclear powerplants with insurance for damage arising out of a nuclear incident and funds the insurance primarily through the assessment of a retrospective premium from each operator after the occurrence of a nuclear incident;

(C) the Convention on Supplementary Compensation for Nuclear Damage, done at Vienna on September 12, 1997, will establish a global system—

(i) to provide a predictable legal framework necessary for nuclear energy projects; and

(ii) to ensure prompt and equitable compensation in the event of a nuclear incident;

(D) the Convention benefits United States nuclear suppliers that face potentially unlimited liability for nuclear incidents that are not covered by the Price-Anderson Act by replacing a potentially open-ended liability with a predictable liability regime that, in effect, provides nuclear suppliers with insurance for damage arising out of such an incident;

(E) the Convention also benefits United States nuclear facility operators that may be publicly liable for a Price- Anderson incident by providing an additional early source of funds to compensate damage arising out of the Price-Anderson incident;

(F) the combined operation of the Convention, the Price-Anderson Act, and this section will augment the quantity of assured funds available for victims in a wider variety of nuclear incidents while reducing the potential liability of United States suppliers without increasing potential costs to United States operators;

(G) the cost of those benefits is the obligation of the United States to contribute to the supplementary compensation fund established by the Convention;

(H) any such contribution should be funded in a manner that does not—

(i) upset settled expectations based on the liability regime established under the Price-Anderson Act; or

(ii) shift to Federal taxpayers liability risks for nuclear incidents at foreign installations;

(I) with respect to a Price-Anderson incident, funds already available under the Price-Anderson Act should be used; and

(J) with respect to a nuclear incident outside the United States not covered by the Price-Anderson Act, a retrospective premium should be prorated among nuclear suppliers relieved from potential liability for which insurance is not available.

(2) PURPOSE.—The purpose of this section is to allocate the contingent costs associated with participation by the United States in the international nuclear liability compensation system established by the Convention on Supplementary Compensation for Nuclear Damage, done at Vienna on September 12, 1997—

(A) with respect to a Price-Anderson incident, by using funds made available under section 170 of the Atomic Energy Act of 1954 (42 U.S.C. 2210) to cover the contingent costs in a manner that neither increases the burdens nor decreases the benefits under section 170 of that Act; and

(B) with respect to a covered incident outside the United States that is not a Price-Anderson incident, by allocating the contingent costs equitably, on the basis of risk, among the class of nuclear suppliers relieved by the Convention from the risk of potential liability resulting from any covered incident outside the United States.

(b) DEFINITIONS.—In this section:

(1) COMMISSION.—The term "Commission" means the Nuclear Regulatory Commission.

(2) CONTINGENT COST.—The term "contingent cost" means the cost to the United States in the event of a covered incident the amount of which is equal to the amount of funds the United States is obligated to make available under paragraph 1(b) of Article III of the Convention.

(3) CONVENTION.—The term "Convention" means the Convention on Supplementary Compensation for Nuclear Damage, done at Vienna on September 12, 1997.

(4) COVERED INCIDENT.—The term "covered incident" means a nuclear incident the occurrence of which results in a request for funds pursuant to Article VII of the Convention.

(5) COVERED INSTALLATION.—The term “covered installation” means a nuclear installation at which the occurrence of a nuclear incident could result in a request for funds under Article VII of the Convention.

(6) COVERED PERSON.—

(A) IN GENERAL.—The term “covered person” means—

(i) a United States person; and

(ii) an individual or entity (including an agency or instrumentality of a foreign country) that—

(I) is located in the United States; or

(II) carries out an activity in the United States.

(B) EXCLUSIONS.—The term ‘‘covered person” does not include—

(i) the United States; or

(ii) any agency or instrumentality of the United States.

(7) NUCLEAR SUPPLIER.—The term ‘‘nuclear supplier” means a covered person (or a successor in interest of a covered person) that supplies facilities, equipment, fuel, services, or technology pertaining to the design, construction, operation, or decommissioning of a covered installation; or transports nuclear materials that could result in a covered incident.

(8) PRICE-ANDERSON INCIDENT.—The term ‘‘Price-Anderson incident” means a covered incident for which section 170 of the Atomic Energy Act of 1954 (42 U.S.C. 2210) would make funds available to compensate for public liability (as defined in section 11 of that Act (42 U.S.C. 2014)).

(9) SECRETARY.—The term ‘‘Secretary” means the Secretary of Energy.

(10) UNITED STATES.—

(A) IN GENERAL.—The term ‘‘United States” has the meaning given the term in section 11 of the Atomic Energy Act of 1954 (42 U.S.C. 2014).

(B) INCLUSIONS.—The term ‘‘United States” includes—

(i) the Commonwealth of Puerto Rico;

(ii) any other territory or possession of the United States;

(iii) the Canal Zone; and

(iv) the waters of the United States territorial sea under Presidential Proclamation Number 5928, dated December 27, 1988 (43 U.S.C. 1331 note).

(11) UNITED STATES PERSON.—The term ‘‘United States person” means—

(A) any individual who is a resident, national, or citizen of the United States (other than an individual residing outside of the United States and employed by a person who is not a United States person); and

(B) any corporation, partnership, association, joint stock company, business trust, unincorporated organization, or sole proprietorship that is organized under the laws of the United States.

(c) USE OF PRICE-ANDERSON FUNDS.—

(1) IN GENERAL.—Funds made available under section 170 of the Atomic Energy Act of 1954 (42 U.S.C. 2210) shall be used to cover the contingent cost resulting from any Price- Anderson incident.

(2) EFFECT.—The use of funds pursuant to paragraph (1) shall not reduce the limitation on public liability established under section 170 e. of the Atomic Energy Act of 1954 (42 U.S.C. 2210(e)).

(d) EFFECT ON AMOUNT OF PUBLIC LIABILITY.—

(1) IN GENERAL.—Funds made available to the United States under Article VII of the Convention with respect to a Price-Anderson incident shall be used to satisfy public liability resulting from the Price-Anderson incident.

(2) AMOUNT.—The amount of public liability allowable under section 170 of the Atomic Energy Act of 1954 (42 U.S.C. 2210) relating to a Price-Anderson incident under paragraph (1) shall be increased by an amount equal to the difference between—

(A) the amount of funds made available for the Price- Anderson incident under Article VII of the Convention; and

(B) the amount of funds used under subsection (c) to cover the contingent cost resulting from the Price-Anderson incident.

(e) RETROSPECTIVE RISK POOLING PROGRAM.—

(1) IN GENERAL.—Except as provided under paragraph (2), each nuclear supplier shall participate in a retrospective risk pooling program in accordance with this section to cover the contingent cost resulting from a covered incident outside the United States that is not a Price-Anderson incident.

(2) DEFERRED PAYMENT.—

(A) IN GENERAL.—The obligation of a nuclear supplier to participate in the retrospective risk pooling program shall be deferred until the United States is called on to provide funds pursuant to Article VII of the Convention with respect to a covered incident that is not a Price- Anderson incident.

(B) AMOUNT OF DEFERRED PAYMENT.—The amount of a deferred payment of a nuclear supplier under subparagraph (A) shall be based on the risk-informed assessment formula determined under subparagraph (C).

(C) RISK-INFORMED ASSESSMENT FORMULA.—

(i) IN GENERAL.—Not later than 3 years after the date of the enactment of this Act, and every 5 years thereafter, the Secretary shall, by regulation, determine the risk-informed assessment formula for the allocation among nuclear suppliers of the contingent cost resulting from a covered incident that is not a Price-Anderson incident, taking into account risk factors such as—

(I) the nature and intended purpose of the goods and services supplied by each nuclear supplier to each covered installation outside the United States;

(II) the quantity of the goods and services supplied by each nuclear supplier to each covered installation outside the United States;

(III) the hazards associated with the supplied goods and services if the goods and services fail to achieve the intended purposes;

(IV) the hazards associated with the covered installation outside the United States to which the goods and services are supplied;

(V) the legal, regulatory, and financial infrastructure associated with the covered installation outside the United States to which the goods and services are supplied; and

(VI) the hazards associated with particular forms of transportation.

(ii) FACTORS FOR CONSIDERATION.—In determining the formula, the Secretary may—

(I) exclude—

(aa) goods and services with negligible risk;

(bb) classes of goods and services not intended specifically for use in a nuclear installation;

(cc) a nuclear supplier with a de minimis share of the contingent cost; and

(dd) a nuclear supplier no longer in existence for which there is no identifiable successor; and

(II) establish the period on which the risk assessment is based.

(iii) APPLICATION.—In applying the formula, the Secretary shall not consider any covered installation or transportation for which funds would be available under section 170 of the Atomic Energy Act of 1954 (42 U.S.C. 2210).

(iv) REPORT.—Not later than 5 years after the date of the enactment of this Act, and every 5 years thereafter, the Secretary shall submit to the Committee on Environment and Public Works of the Senate and the Committee on Energy and Commerce of the House of Representatives, a report on whether there is a need for continuation or amendment of this section, taking into account the effects of the implementation of the Convention on the United States nuclear industry and suppliers.

(f) REPORTING.—

(1) COLLECTION OF INFORMATION.—

(A) IN GENERAL.—The Secretary may collect information necessary for developing and implementing the formula for calculating the deferred payment of a nuclear supplier under subsection (e)(2).

(B) PROVISION OF INFORMATION.—Each nuclear supplier and other appropriate persons shall make available to the Secretary such information, reports, records, documents, and other data as the Secretary determines, by regulation, to be necessary or appropriate to develop and implement the formula under subsection (e)(2)(C).

(2) PRIVATE INSURANCE.—The Secretary shall make available to nuclear suppliers, and insurers of nuclear suppliers, information to support the voluntary establishment and maintenance of private insurance against any risk for which nuclear suppliers may be required to pay deferred payments under this section.

(g) EFFECT ON LIABILITY.—Nothing in any other law (including regulations) limits liability for a covered incident to an amount equal to less than the amount prescribed in paragraph 1(a) of Article IV of the Convention, unless the law—

(1) specifically refers to this section; and

(2) explicitly repeals, alters, amends, modifies, impairs, displaces, or supersedes the effect of this subsection.

(h) PAYMENTS TO AND BY THE UNITED STATES.—

(1) ACTION BY NUCLEAR SUPPLIERS.—

(A) NOTIFICATION.—In the case of a request for funds under Article VII of the Convention resulting from a covered incident that is not a Price-Anderson incident, the Secretary shall notify each nuclear supplier of the amount of the deferred payment required to be made by the nuclear supplier.

(B) PAYMENTS.—

(i) IN GENERAL.—Except as provided under clause (ii), not later than 60 days after receipt of a notification under subparagraph (A), a nuclear supplier shall pay to the general fund of the Treasury the deferred payment of the nuclear supplier required under subparagraph (A).

(ii) ANNUAL PAYMENTS.—A nuclear supplier may elect to prorate payment of the deferred payment required under subparagraph (A) in 5 equal annual payments (including interest on the unpaid balance at the prime rate prevailing at the time the first payment is due).

(C) VOUCHERS.—A nuclear supplier shall submit payment certification vouchers to the Secretary of the Treasury in accordance with section 3325 of title 31, United States Code.

(2) USE OF FUNDS.—

(A) IN GENERAL.—Amounts paid into the Treasury under paragraph (1) shall be available to the Secretary of the Treasury, without further appropriation and without fiscal year limitation, for the purpose of making the contributions of public funds required to be made by the United States under the Convention.

(B) ACTION BY SECRETARY OF TREASURY.—The Secretary of the Treasury shall pay the contribution required under the Convention to the court of competent jurisdiction under Article XIII of the Convention with respect to the applicable covered incident.

(3) FAILURE TO PAY.—If a nuclear supplier fails to make a payment required under this subsection, the Secretary may take appropriate action to recover from the nuclear supplier—

(A) the amount of the payment due from the nuclear supplier;

(B) any applicable interest on the payment; and

(C) a penalty of not more than twice the amount of the deferred payment due from the nuclear supplier.

(i) LIMITATION ON JUDICIAL REVIEW; CAUSE OF ACTION.— (1) LIMITATION ON JUDICIAL REVIEW.—

(A) IN GENERAL.—In any civil action arising under the Convention over which Article XIII of the Convention grants jurisdiction to the courts of the United States, any appeal or review by writ of mandamus or otherwise with respect to anuclear incident that is not a Price-Anderson incident shall be in accordance with chapter 83 of title 28, United States Code, except that the appeal or review shall occur in the United States Court of Appeals for the District of Columbia Circuit.

(B) SUPREME COURT JURISDICTION.—Nothing in this paragraph affects the jurisdiction of the Supreme Court of the United States under chapter 81 of title 28, United States Code.

(2) CAUSE OF ACTION.—

(A) IN GENERAL.—Subject to subparagraph (B), in any civil action arising under the Convention over which Article XIII of the Convention grants jurisdiction to the courts of the United States, in addition to any other cause of action that may exist, an individual or entity shall have a cause of action against the operator to recover for nuclear damage suffered by the individual or entity.

(B) REQUIREMENT.—Subparagraph (A) shall apply only if the individual or entity seeks a remedy for nuclear damage (as defined in Article I of the Convention) that was caused by a nuclear incident (as defined in Article I of the Convention) that is not a Price-Anderson incident.

(C) SAVINGS PROVISION.—Nothing in this paragraph may be construed to limit, modify, extinguish, or otherwise affect any cause of action that would have existed in the absence of enactment of this paragraph.

(j) RIGHT OF RECOURSE.—This section does not provide to an operator of a covered installation any right of recourse under the Convention.

(k) PROTECTION OF SENSITIVE UNITED STATES INFORMATION.— Nothing in the Convention or this section requires the disclosure of—

(1) any data that, at any time, was Restricted Data (as defined in section 11 of the Atomic Energy Act of 1954 (42 U.S.C. 2014));

(2) information relating to intelligence sources or methods protected by section 102A(i) of the National Security Act of 1947 (50 U.S.C. 403–1(i)); or

(3) national security information classified under Executive Order 12958 (50 U.S.C. 435 note; relating to classified national security information) (or a successor Executive Order or regulation).

(l) REGULATIONS.—

(1) IN GENERAL.—The Secretary or the Commission, as appropriate, may prescribe regulations to carry out section 170 of the Atomic Energy Act of 1954 (42 U.S.C. 2210) and this section.

(2) REQUIREMENT.—Rules prescribed under this subsection shall ensure, to the maximum extent practicable, that—

(A) the implementation of section 170 of the Atomic Energy Act of 1954 (42 U.S.C. 2210) and this section is consistent and equitable; and

(B) the financial and operational burden on a Commission licensee in complying with section 170 of that Act is not greater as a result of the enactment of this section.

(3) APPLICABILITY OF PROVISION.—Section 553 of title 5, United States Code, shall apply with respect to the promulgation of regulations under this subsection.

(4) EFFECT OF SUBSECTION.—The authority provided under this subsection is in addition to, and does not impair or otherwise affect, any other authority of the Secretary or the Commission to prescribe regulations.

(m) EFFECTIVE DATE.—This section shall take effect on the date of the enactment of this Act.

Sec. 935. Transparency In Extractive Industries Resource Payments.

(a) PURPOSE.—The purpose of this section is to—

(1) ensure greater United States energy security by combating corruption in the governments of foreign countries that receive revenues from the sale of their natural resources; and

(2) enhance the development of democracy and increase political and economic stability in such resource rich foreign countries.

(b) STATEMENT OF POLICY.—It is the policy of the United States—

(1) to increase energy security by promoting anti-corruption initiatives in oil and natural gas rich countries; and

(2) to promote global energy security through promotion of programs such as the Extractive Industries Transparency Initiative (EITI) that seek to instill transparency and accountability into extractive industries resource payments.

(c) SENSE OF CONGRESS.—It is the sense of Congress that the United States should further global energy security and promote democratic development in resource-rich foreign countries by—

(1) encouraging further participation in the EITI by eligible countries and companies; and

(2) promoting the efficacy of the EITI program by ensuring a robust and candid review mechanism.

(d) REPORT.—

(1) REPORT REQUIRED.—Not later than 180 days after the date of the enactment of this Act, and annually thereafter, the Secretary of State, in consultation with the Secretary of Energy, shall submit to the appropriate congressional committees a report on progress made in promoting transparency in extractive industries resource payments.

(2) MATTERS TO BE INCLUDED.—The report required by paragraph (1) shall include a detailed description of United States participation in the EITI, bilateral and multilateral diplomatic efforts to further participation in the EITI, and other United States initiatives to strengthen energy security, deter energy kleptocracy, and promote transparency in the extractive industries.

(e) AUTHORIZATION OF APPROPRIATIONS.—There is authorized to be appropriated $3,000,000 for the purposes of United States contributions to the Multi-Donor Trust Fund of the EITI.

TITLE X—GREEN JOBS

Sec. 1001. Short Title.

This title may be cited as the "Green Jobs Act of 2007".

Sec. 1002. Energy Efficiency And Renewable Energy Worker Training Program.

Section 171 of the Workforce Investment Act of 1998 (29 U.S.C. 2916) is amended by adding at the end the following:

"(e) ENERGY EFFICIENCY AND RENEWABLE ENERGY WORKER TRAINING PROGRAM.—

"(1) GRANT PROGRAM.—

"(A) IN GENERAL.—Not later than 6 months after the Deadline. date of enactment of the Green Jobs Act of 2007, the Secretary, in consultation with the Secretary of Energy, shall establish an energy efficiency and renewable energy worker training program under which the Secretary shall carry out the activities described in paragraph (2) to achieve the purposes of this subsection.

"(B) ELIGIBILITY.—For purposes of providing assistance and services under the program established under this subsection—

"(i) target populations of eligible individuals to be given priority for training and other services shall include—

"(I) workers impacted by national energy and environmental policy;

"(II) individuals in need of updated training related to the energy efficiency and renewable energy industries;

"(III) veterans, or past and present members of reserve components of the Armed Forces;

"(IV) unemployed individuals;

"(V) individuals, including at-risk youth, seeking employment pathways out of poverty and into economic self-sufficiency; and

"(VI) formerly incarcerated, adjudicated, nonviolent offenders; and

"(ii) energy efficiency and renewable energy industries eligible to participate in a program under this subsection include—

"(I) the energy-efficient building, construction, and retrofits industries;

"(II) the renewable electric power industry;

"(III) the energy efficient and advanced drive train vehicle industry;

"(IV) the biofuels industry;

"(V) the deconstruction and materials use industries;

"(VI) the energy efficiency assessment industry serving the residential, commercial, or industrial sectors; and

"(VII) manufacturers that produce sustainable products using environmentally sustainable processes and materials.

"(2) ACTIVITIES.—

"(A) NATIONAL RESEARCH PROGRAM.—Under the program established under paragraph (1), the Secretary, acting through the Bureau of Labor Statistics, where appropriate, shall collect and analyze labor market data to track workforce trends resulting from energy-related initiatives carried out under this subsection. Activities carried out under this paragraph shall include—

"(i) tracking and documentation of academic and occupational competencies as well as future skill needs with respect to renewable energy and energy efficiency technology;

"(ii) tracking and documentation of occupational information and workforce training data with respect to renewable energy and energy efficiency technology;

"(iii) collaborating with State agencies, workforce investments boards, industry, organized labor, and community and nonprofit organizations to disseminate information on successful innovations for labor market services and worker training with respect to renewable energy and energy efficiency technology;

"(iv) serving as a clearinghouse for best practices in workforce development, job placement, and collaborative training partnerships;

"(v) encouraging the establishment of workforce training initiatives with respect to renewable energy and energy efficiency technologies;

"(vi) linking research and development in renewable energy and energy efficiency technology with the development of standards and curricula for current and future jobs;

"(vii) assessing new employment and work practices including career ladder and upgrade training as well as high performance work systems; and

"(viii) providing technical assistance and capacity building to national and State energy partnerships, including industry and labor representatives.

"(B) NATIONAL ENERGY TRAINING PARTNERSHIP GRANTS.—

"(i) IN GENERAL.—Under the program established under paragraph (1), the Secretary shall award National Energy Training Partnerships Grants on a competitive basis to eligible entities to enable such entities to carry out training that leads to economic self-sufficiency and to develop an energy efficiency and renewable energy industries workforce. Grants shall be awarded under this subparagraph so as to ensure geographic diversity with at least 2 grants awarded to entities located in each of the 4 Petroleum Administration for Defense Districts with no subdistricts, and at least 1 grant awarded to an entity located in each of the subdistricts of the Petroleum Administration for Defense District with subdistricts.

"(ii) ELIGIBILITY.—To be eligible to receive a grant under clause (i), an entity shall be a nonprofit partnership that—

"(I) includes the equal participation of industry, including public or private employers, and labor organizations, including joint labor-management training programs, and may include workforce investment boards, community-based organizations, qualified service and conservation corps, educational institutions, small businesses, cooperatives, State and local veterans agencies, and veterans service organizations; and

"(II) demonstrates—

"(aa) experience in implementing and operating worker skills training and education programs;

"(bb) the ability to identify and involve in training programs carried out under this grant, target populations of individuals who would benefit from training and be actively involved in activities related to energy efficiency and renewable energy industries; and

"(cc) the ability to help individuals achieve economic self-sufficiency.

"(iii) PRIORITY.—Priority shall be given to partnerships which leverage additional public and private resources to fund training programs, including cash or in-kind matches from participating employers.

"(C) STATE LABOR MARKET RESEARCH, INFORMATION, AND LABOR EXCHANGE RESEARCH PROGRAM.—

"(i) IN GENERAL.—Under the program established under paragraph (1), the Secretary shall award competitive grants to States to enable such States to administer labor market and labor exchange information programs that include the implementation of the activities described in clause (ii), in coordination with the one-stop delivery system.

"(ii) ACTIVITIES.—A State shall use amounts awarded under a grant under this subparagraph to provide funding to the State agency that administers the Wagner-Peyser Act and State unemployment compensation programs to carry out the following activities using State agency merit staff:

"(I) The identification of job openings in the renewable energy and energy efficiency sector.

"(II) The administration of skill and aptitude testing and assessment for workers.

"(III) The counseling, case management, and referral of qualified job seekers to openings and training programs, including energy efficiency and renewable energy training programs.

"(D) STATE ENERGY TRAINING PARTNERSHIP PROGRAM.—

"(i) IN GENERAL.—Under the program established under paragraph (1), the Secretary shall award competitive grants to States to enable such States to administer renewable energy and energy efficiency workforce development programs that include the implementation of the activities described in clause (ii).

"(ii) PARTNERSHIPS.—A State shall use amounts awarded under a grant under this subparagraph to award competitive grants to eligible State Energy Sector Partnerships to enable such Partnerships to coordinate with existing apprenticeship and labor management training programs and implement training programs that lead to the economic self-sufficiency of trainees.

"(iii) ELIGIBILITY.—To be eligible to receive a grant under this subparagraph, a State Energy Sector Partnership shall—

"(I) consist of nonprofit organizations that include equal participation from industry, including public or private nonprofit employers, and labor organizations, including joint labor- management training programs, and may include representatives from local governments, the workforce investment system, including one-stop career centers, community based organizations, qualified service and conservation corps, community colleges, and other post-secondary institutions, small businesses, cooperatives, State and local veterans agencies, and veterans service organizations;

"(II) demonstrate experience in implementing and operating worker skills training and education programs; and

"(III) demonstrate the ability to identify and involve in training programs, target populations of workers who would benefit from training and be actively involved in activities related to energy efficiency and renewable energy industries.

"(iv) PRIORITY.—In awarding grants under this subparagraph, the Secretary shall give priority to States that demonstrate that activities under the grant—

"(I) meet national energy policies associated with energy efficiency, renewable energy, and the reduction of emissions of greenhouse gases;

"(II) meet State energy policies associated with energy efficiency, renewable energy, and the reduction of emissions of greenhouse gases; and

"(III) leverage additional public and private resources to fund training programs, including cash or in-kind matches from participating employers.

"(v) COORDINATION.—A grantee under this subparagraph shall coordinate activities carried out under the grant with existing other appropriate training programs, including apprenticeship and labor management training programs, including such activities referenced in paragraph

(3)(A), and implement training programs that lead to the economic self-sufficiency of trainees.

"(E) PATHWAYS OUT OF POVERTY DEMONSTRATION PROGRAM.—

"(i) IN GENERAL.—Under the program established under paragraph (1), the Secretary shall award competitive grants of sufficient size to eligible entities to enable such entities to carry out training that leads to economic self-sufficiency. The Secretary shall give priority to entities that serve individuals in families with income of less than 200 percent of the sufficiency standard for the local areas where the training is conducted that specifies, as defined by the State, or where such standard is not established, the income needs of families, by family size, the number and ages of children in the family, and sub-State geographical considerations. Grants shall be awarded to ensure geographic diversity.

"(ii) ELIGIBLE ENTITIES.—To be eligible to receive a grant an entity shall be a partnership that—

"(I) includes community-based nonprofit organizations, educational institutions with expertise in serving low-income adults or youth, public or private employers from the industry sectors described in paragraph (1)(B)(ii), and labor organizations representing workers in such industry sectors;

"(II) demonstrates a record of successful experience in implementing and operating worker skills training and education programs;

"(III) coordinates activities, where appropriate, with the workforce investment system; and

"(IV) demonstrates the ability to recruit individuals for training and to support such individuals to successful completion in training programs carried out under this grant, targeting populations of workers who are or will be engaged in activities related to energy efficiency and renewable energy industries.

"(iii) PRIORITIES.—In awarding grants under this paragraph, the Secretary shall give priority to applicants that—

"(I) target programs to benefit low-income workers, unemployed youth and adults, high school dropouts, or other underserved sectors of the workforce within areas of high poverty;

"(II) ensure that supportive services are integrated with education and training, and delivered by organizations with direct access to and experience with targeted populations;

"(III) leverage additional public and private resources to fund training programs, including cash or in-kind matches from participating employers;

"(IV) involve employers and labor organizations in the determination of relevant skills and competencies and ensure that the certificates or credentials that result from the training are employer-recognized;

"(V) deliver courses at alternative times (such as evening and weekend programs) and locations most convenient and accessible to participants and link adult remedial education with occupational skills training; and

"(VI) demonstrate substantial experience in administering local, municipal, State, Federal, foundation, or private entity grants.

"(iv) DATA COLLECTION.—Grantees shall collect and Reports. report the following information:

"(I) The number of participants.

"(II) The demographic characteristics of participants, including race, gender, age, parenting status, participation in other Federal programs, education and literacy level at entry, significant barriers to employment (such as limited English proficiency, criminal record, addiction or mental health problem requiring treatment, or mental disability).

"(III) The services received by participants, including training, education, and supportive services.

"(IV) The amount of program spending per participant.

"(V) Program completion rates.

"(VI) Factors determined as significantly interfering with program participation or completion.

"(VII) The rate of job placement and the rate of employment retention after 1 year.

"(VIII) The average wage at placement, including any benefits, and the rate of average wage increase after 1 year.

"(IX) Any post-employment supportive services provided.

"(X) The Secretary shall assist grantees in the collection of data under this clause by making available, where practicable, low-cost means of tracking the labor market outcomes of participants, and by providing standardized reporting forms, where appropriate.

"(3) ACTIVITIES.—

"(A) IN GENERAL.—Activities to be carried out under a program authorized by subparagraph (B), (D), or (E) of paragraph (2) shall be coordinated with existing systems or providers, as appropriate. Such activities may include—

"(i) occupational skills training, including curriculum development, on-the-job training, and classroom training;

"(ii) safety and health training;

"(iii) the provision of basic skills, literacy, GED, English as a second language, and job readiness training;

"(iv) individual referral and tuition assistance for a community college training program, or any training program leading to an industry-recognized certificate;

"(v) internship programs in fields related to energy efficiency and renewable energy;

"(vi) customized training in conjunction with an existing registered apprenticeship program or labor- management partnership;

"(vii) incumbent worker and career ladder training and skill upgrading and retraining;

"(viii) the implementation of transitional jobs strategies; and

"(ix)the provision of supportive services.

"(B) OUTREACH ACTIVITIES.—In addition to the activities authorized under subparagraph (A), activities authorized for programs under subparagraph (E) of paragraph (2) may include the provision of outreach, recruitment, career guidance, and case management services.

"(4) WORKER PROTECTIONS AND NONDISCRIMINATION REQUIREMENTS.—

"(A) APPLICATION OF WIA.—The provisions of sections 181 and 188 of the Workforce Investment Act of 1998 (29 U.S.C. 2931 and 2938) shall apply to all programs carried out with assistance under this subsection.

"(B) CONSULTATION WITH LABOR ORGANIZATIONS.—If a labor organization represents a substantial number of workers who are engaged in similar work or training in an area that is the same as the area that is proposed to be funded under this Act, the labor organization shall be provided an opportunity to be consulted and to submit comments in regard to such a proposal.

"(5) PERFORMANCE MEASURES.—

"(A) IN GENERAL.—The Secretary shall negotiate and reach agreement with the eligible entities that receive grants and assistance under this section on performance measures for the indicators of performance referred to in subparagraphs (A) and (B) of section 136(b)(2) that will be used to evaluate the performance of the eligible entity in carrying out the activities described in subsection (e)(2). Each performance measure shall consist of such an indicator of performance, and a performance level referred to in subparagraph (B).

"(B) PERFORMANCE LEVELS.—The Secretary shall negotiate and reach agreement with the eligible entity regarding the levels of performance expected to be achieved by the eligible entity on the indicators of performance.

"(6) REPORT.—

"(A) STATUS REPORT.—Not later than 18 months after the date of enactment of the Green Jobs Act of 2007, the Secretary shall transmit a report to the Senate Committee on Energy and Natural Resources, the Senate Committee on Health, Education, Labor, and Pensions, the House Committee on Education and Labor, and the House Committee on Energy and Commerce on the training program established by this subsection. The report shall include a description of the entities receiving funding and the activities carried out by such entities.

"(B) EVALUATION.—Not later than 3 years after the Deadline. date of enactment of such Act, the Secretary shall transmit to the Senate Committee on Energy and Natural Resources, the Senate Committee on Health, Education, Labor, and Pensions, the House Committee on Education and Labor, and the House Committee on Energy and Commerce an assessment of such program and an evaluation of the activities carried out by entities receiving funding from such program.

"(7) DEFINITION.—As used in this subsection, the term 'renewable energy' has the meaning given such term in section 203(b)(2) of the Energy Policy Act of 2005 (Public Law 109– 58).

"(8) AUTHORIZATION OF APPROPRIATIONS.—There is authorized to be appropriated to carry out this subsection, $125,000,000 for each fiscal year, of which—

"(A) not to exceed 20 percent of the amount appropriated in each such fiscal year shall be made available for, and shall be equally divided between, national labor market research and information under paragraph (2)(A) and State labor market information and labor exchange research under paragraph (2)(C), and not more than 2 percent of such amount shall be for the evaluation and report required under paragraph (4);

"(B) 20 percent shall be dedicated to Pathways Out of Poverty Demonstration Programs under paragraph (2)(E); and

"(C) the remainder shall be divided equally between National Energy Partnership Training Grants under paragraph (2)(B) and State energy training partnership grants under paragraph (2)(D).".

TITLE XI—ENERGY TRANSPORTATION AND INFRASTRUCTURE

Subtitle A—Department of Transportation

Sec. 1101. Office Of Climate Change And Environment.

(a) IN GENERAL.—Section 102 of title 49, United States Code, is amended—

by redesignating subsection (g) as subsection (h); and

by inserting after subsection (f) the following:

"(g) OFFICE OF CLIMATE CHANGE AND ENVIRONMENT.—

"(1) ESTABLISHMENT.—There is established in the Department an Office of Climate Change and Environment to plan, coordinate, and implement—

"(A) department-wide research, strategies, and actions under the Department's statutory authority to reduce transportation-related energy use and mitigate the effects of climate change; and

"(B) department-wide research strategies and actions to address the impacts of climate change on transportation systems and infrastructure.

"(2) CLEARINGHOUSE.—The Office shall establish a clearing-house of solutions, including cost-effective congestion reduction approaches, to reduce air pollution and transportation-related energy use and mitigate the effects of climate change.".

(b) COORDINATION.—The Office of Climate Change and Environment of the Department of Transportation shall coordinate its activities with the United States Global Change Research Program.

(c) TRANSPORTATION SYSTEM'S IMPACT ON CLIMATE CHANGE AND FUEL EFFICIENCY.—

(1) STUDY.—The Office of Climate Change and Environment, in coordination with the Environmental Protection Agency and in consultation with the United States Global Change Research Program, shall conduct a study to examine the impact of the Nation's transportation system on climate change and the fuel efficiency savings and clean air impacts of major transportation projects, to identify solutions to reduce air pollution and transportation-related energy use and mitigate the effects of climate change, and to examine the potential fuel savings that could result from changes in the current transportation system and through the use of intelligent transportation systems that help businesses and consumers to plan their travel and avoid delays, including Web-based real-time transit information systems, congestion information systems, carpool information systems, parking information systems, freight route management systems, and traffic management systems.

(2) REPORT.—Not later than 1 year after the date of enactment of this Act, the Secretary of Transportation, in coordination with the Administrator of the Environmental Protection Agency, shall transmit to the Committee on Transportation and Infrastructure and the Committee on Energy and Commerce of the House of Representatives and the Committee on Commerce, Science, and Transportation and the Committee on Environment and Public Works of the Senate a report that contains the results of the study required under this section. (d) AUTHORIZATION OF APPROPRIATIONS.—There are authorized to be appropriated to the Secretary of Transportation for the Office of Climate Change and Environment to carry out its duties under section 102(g) of title 49, United States Code (as amended by this Act), such sums as may be necessary for fiscal years 2008 through 2011.

Subtitle B—Railroads

Sec. 1111. Advanced Technology Locomotive Grant Pilot Program.

(a) IN GENERAL.—The Secretary of Transportation, in consultation with the Administrator of the Environmental Protection Agency, shall establish and carry out a pilot program for making grants to railroad carriers (as defined in section 20102 of title 49, United States Code) and State and local governments—

(1) for assistance in purchasing hybrid or other energy- efficient locomotives, including hybrid switch and generator- set locomotives; and

(2) to demonstrate the extent to which such locomotives increase fuel economy, reduce emissions, and lower costs of operation.

(b) LIMITATION.—Notwithstanding subsection (a), no grant under this section may be used to fund the costs of emissions reductions that are mandated under Federal law.

(c) GRANT CRITERIA.—In selecting applicants for grants under this section, the Secretary of Transportation shall consider—

(1) the level of energy efficiency that would be achieved by the proposed project;

(2) the extent to which the proposed project would assist in commercial deployment of hybrid or other energy-efficient locomotive technologies;

(3) the extent to which the proposed project complements other private or governmental partnership efforts to improve air quality or fuel efficiency in a particular area; and

(4) the extent to which the applicant demonstrates innovative strategies and a financial commitment to increasing energy efficiency and reducing greenhouse gas emissions of its railroad operations.

(d) COMPETITIVE GRANT SELECTION PROCESS.—

(1) APPLICATIONS.—A railroad carrier or State or local government seeking a grant under this section shall submit for approval by the Secretary of Transportation an application for the grant containing such information as the Secretary of Transportation may require.

(2) COMPETITIVE SELECTION.—The Secretary of Transportation shall conduct a national solicitation for applications for grants under this section and shall select grantees on a competitive basis.

(e) FEDERAL SHARE.—The Federal share of the cost of a project under this section shall not exceed 80 percent of the project cost.

(f) REPORT.—Not later than 3 years after the date of enactment of this Act, the Secretary of Transportation shall submit to Congress a report on the results of the pilot program carried out under this section.

(g) AUTHORIZATION OF APPROPRIATIONS.—There is authorized to be appropriated to the Secretary of Transportation $10,000,000 for each of the fiscal years 2008 through 2011 to carry out this section. Such funds shall remain available until expended.

Sec. 1112. Capital Grants For Class II And Class III Railroads.

(a) AMENDMENT.—Chapter 223 of title 49, United States Code, is amended to read as follows:

"CHAPTER 223—CAPITAL GRANTS FOR CLASS II AND CLASS III RAILROADS

"Sec. "22301. Capital grants for class II and class III railroads.

§ 22301. Capital grants for class II and class III railroads

"(a) ESTABLISHMENT OF PROGRAM.—

"(1) ESTABLISHMENT.—The Secretary of Transportation shall establish a program for making capital grants to class II and class III railroads. Such grants shall be for projects in the public interest that—

"(A)(i) rehabilitate, preserve, or improve railroad track (including roadbed, bridges, and related track structures) used primarily for freight transportation;

(ii) facilitate the continued or greater use of railroad transportation for freight shipments; and

(iii) reduce the use of less fuel efficient modes of transportation in the transportation of such shipments; and

"(B) demonstrate innovative technologies and advanced research and development that increase fuel economy, reduce greenhouse gas emissions, and lower the costs of operation.

"(2) PROVISION OF GRANTS.—Grants may be provided under this chapter—

"(A) directly to the class II or class III railroad; or

"(B) with the concurrence of the class II or class III railroad, to a State or local government.

"(3) STATE COOPERATION.—Class II and class III railroad applicants for a grant under this chapter are encouraged to utilize the expertise and assistance of State transportation agencies in applying for and administering such grants. State transportation agencies are encouraged to provide such expertise and assistance to such railroads.

"(4) REGULATIONS.—Not later than October 1, 2008, the Deadline. Secretary shall issue final regulations to implement the pro-
gram under this section.

"(b) MAXIMUM FEDERAL SHARE.—The maximum Federal share for carrying out a project under this section shall be 80 percent of the project cost. The non-Federal share may be provided by any non-Federal source in cash, equipment, or supplies. Other in-kind contributions may be approved by the Secretary on a case- by-case basis consistent with this chapter.

"(c) USE OF FUNDS.—Grants provided under this section shall be used to implement track capital projects as soon as possible. In no event shall grant funds be contractually obligated for a project later than the end of the third Federal fiscal year following the year in which the grant was awarded. Any funds not so obligated by the end of such fiscal year shall be returned to the Secretary for reallocation.

"(d) EMPLOYEE PROTECTION.—The Secretary shall require as a condition of any grant made under this section that the recipient railroad provide a fair arrangement at least as protective of the interests of employees who are affected by the project to be funded with the grant as the terms imposed under section 11326(a), as in effect on the date of the enactment of this chapter.

"(e) LABOR STANDARDS.—

"(1) PREVAILING WAGES.—The Secretary shall ensure that laborers and mechanics employed by contractors and subcontractors in construction work financed by a grant made under this section will be paid wages not less than those prevailing on similar construction in the locality, as determined by the Secretary of Labor under subchapter IV of chapter 31 of title 40 (commonly known as the 'Davis-Bacon Act'). The Secretary shall make a grant under this section only after being assured that required labor standards will be maintained on the construction work.

"(2) WAGE RATES.—Wage rates in a collective bargaining agreement negotiated under the Railway Labor Act (45 U.S.C. 151 et seq.) are deemed for purposes of this subsection to comply with the subchapter IV of chapter 31 of title 40.

"(f) STUDY.—The Secretary shall conduct a study of the projects carried out with grant assistance under this section to determine the extent to which the program helps promote a reduction in fuel use associated with the transportation of freight and demonstrates innovative technologies that increase fuel economy, reduce greenhouse gas emissions, and lower the costs of operation. Not Deadline. later than March 31, 2009, the Secretary shall submit a report Reports. to the Committee on Transportation and Infrastructure of the House of Representatives and the Committee on Commerce, Science, and Transportation of the Senate on the study, including any recommendations the Secretary considers appropriate regarding the program.

"(g) AUTHORIZATION OF APPROPRIATIONS.—There is authorized to be appropriated to the Secretary $50,000,000 for each of fiscal years 2008 through 2011 for carrying out this section.".

(b) CLERICAL AMENDMENT.—The item relating to chapter 223 in the table of chapters of subtitle V of title 49, United States Code, is amended to read as follows:

"223. CAPITAL GRANTS FOR CLASS II AND CLASS III RAILROADS. 22301".

Subtitle C—Marine Transportation

Sec. 1121. Short Sea Transportation Initiative.

(a) IN GENERAL.—Title 46, United States Code, is amended by adding after chapter 555 the following:

"CHAPTER 556—SHORT SEA TRANSPORTATION

"Sec. 55601. Short sea transportation program. "Sec. 55602. Cargo and shippers.

"Sec. 55603. Interagency coordination.

"Sec. 55604. Research on short sea transportation. "Sec. 55605. Short sea transportation defined.

"§ 55601. Short sea transportation program

"(a) ESTABLISHMENT.—The Secretary of Transportation shall establish a short sea transportation program and designate short sea transportation projects to be conducted under the program to mitigate landside congestion.

"(b) PROGRAM ELEMENTS.—The program shall encourage the use of short sea transportation through the development and expansion of—

"(1) documented vessels;

"(2) shipper utilization;

"(3) port and landside infrastructure; and

"(4) marine transportation strategies by State and local governments.

"(c) SHORT SEA TRANSPORTATION ROUTES.—The Secretary shall designate short sea transportation routes as extensions of the surface transportation system to focus public and private efforts to use the waterways to relieve landside congestion along coastal corridors. The Secretary may collect and disseminate data for the designation and delineation of short sea transportation routes.

"(d) PROJECT DESIGNATION.—The Secretary may designate a project to be a short sea transportation project if the Secretary determines that the project may—

"(1) offer a waterborne alternative to available landside transportation services using documented vessels; and

"(2) provide transportation services for passengers or freight (or both) that may reduce congestion on landside infrastructure using documented vessels.

"(e) ELEMENTS OF PROGRAM.—For a short sea transportation project designated under this section, the Secretary may—

"(1) promote the development of short sea transportation services;

"(2) coordinate, with ports, State departments of transportation, localities, other public agencies, and the private sector and on the development of landside facilities and infrastructure to support short sea transportation services; and

"(3) develop performance measures for the short sea transportation program.

"(f) MULTISTATE, STATE AND REGIONAL TRANSPORTATION PLANNING.—The Secretary, in consultation with Federal entities and State and local governments, shall develop strategies to encourage the use of short sea transportation for transportation of passengers and cargo. The Secretary shall—

"(1) assess the extent to which States and local governments include short sea transportation and other marine transportation solutions in their transportation planning;

"(2) encourage State departments of transportation to develop strategies, where appropriate, to incorporate short sea transportation, ferries, and other marine transportation solutions for regional and interstate transport of freight and passengers in their transportation planning; and

"(3) encourage groups of States and multi-State transportation entities to determine how short sea transportation can address congestion, bottlenecks, and other interstate transportation challenges.

"§ 55602. Cargo and shippers

"(a) MEMORANDUMS OF AGREEMENT.—The Secretary of Transportation shall enter into memorandums of understanding with the heads of other Federal entities to transport federally owned or generated cargo using a short sea transportation project designated under section 55601 when practical or available.

"(b) SHORT-TERM INCENTIVES.—The Secretary shall consult shippers and other participants in transportation logistics and develop proposals for short-term incentives to encourage the use of short sea transportation.

"§ 55603. Interagency coordination

"The Secretary of Transportation shall establish a board to identify and seek solutions to impediments hindering effective use of short sea transportation. The board shall include representatives of the Environmental Protection Agency and other Federal, State, and local governmental entities and private sector entities.

"§ 55604. Research on short sea transportation

"The Secretary of Transportation, in consultation with the Administrator of the Environmental Protection Agency, may conduct research on short sea transportation, regarding—

"(1) the environmental and transportation benefits to be derived from short sea transportation alternatives for other forms of transportation;

"(2) technology, vessel design, and other improvements that would reduce emissions, increase fuel economy, and lower costs of short sea transportation and increase the efficiency of inter- modal transfers; and

"(3) solutions to impediments to short sea transportation projects designated under section 55601.

"§ 55605. Short sea transportation defined

"In this chapter, the term 'short sea transportation' means the carriage by vessel of cargo—

"(1) that is—

"(A) contained in intermodal cargo containers and loaded by crane on the vessel; or

"(B) loaded on the vessel by means of wheeled technology; and

"(2) that is—

"(A) loaded at a port in the United States and unloaded either at another port in the United States or at a port in Canada located in the Great Lakes Saint Lawrence Seaway System; or

"(B) loaded at a port in Canada located in the Great Lakes Saint Lawrence Seaway System and unloaded at a port in the United States.".

(b) CLERICAL AMENDMENT.—The table of chapters at the beginning of subtitle V of such title is amended by inserting after the item relating to chapter 555 the following:

"556. Short Sea Transportation ....55601".

(c) REGULATIONS.—

(1) INTERIM REGULATIONS.—Not later than 90 days after the date of enactment of this Act, the Secretary of Transportation shall issue temporary regulations to implement the program under this section. Subchapter II of chapter 5 of title 5,

United States Code, does not apply to a temporary regulation issued under this paragraph or to an amendment to such a temporary regulation.

(2) FINAL REGULATIONS.—Not later than October 1, 2008, the Secretary of Transportation shall issue final regulations to implement the program under this section.

Sec. 1122. Short Sea Shipping Eligibility For Capital Construction Fund.

(a) DEFINITION OF QUALIFIED VESSEL.—Section 53501 of title 46, United States Code, is amended—

(1) in paragraph (5)(A)(iii) by striking "or noncontiguous domestic" and inserting "noncontiguous domestic, or short sea transportation trade"; and

(2) by inserting after paragraph (6) the following:

"(7) SHORT SEA TRANSPORTATION TRADE.—The term 'short sea transportation trade' means the carriage by vessel of cargo—

"(A) that is—

"(i) contained in intermodal cargo containers and loaded by crane on the vessel; or

"(ii) loaded on the vessel by means of wheeled technology; and

"(B) that is—

"(i) loaded at a port in the United States and unloaded either at another port in the United States or at a port in Canada located in the Great Lakes Saint Lawrence Seaway System; or

"(ii) loaded at a port in Canada located in the Great Lakes Saint Lawrence Seaway System and unloaded at a port in the United States.".

(b) ALLOWABLE PURPOSE.—Section 53503(b) of such title is amended by striking "or noncontiguous domestic trade" and inserting "noncontiguous domestic, or short sea transportation trade".

Sec. 1123. Short Sea Transportation Report.

Not later than 1 year after the date of enactment of this Act, the Secretary of Transportation, in consultation with the Administrator of the Environmental Protection Agency, shall submit to the Committee on Transportation and Infrastructure of the House of Representatives and the Committee on Commerce, Science, and Transportation of the Senate a report on the short sea transportation program established under the amendments made by section 1121. The report shall include a description of the activities conducted under the program, and any recommendations for further legislative or administrative action that the Secretary of Transportation considers appropriate.

Subtitle D—Highways

Sec. 1131. Increased Federal Share For Cmaq Projects.

Section 120(c) of title 23, United States Code, is amended—

(1) in the subsection heading by striking "FOR CERTAIN SAFETY PROJECTS";

(2) by striking "The Federal share" and inserting the following:

"(1) CERTAIN SAFETY PROJECTS.—The Federal share"; and

(3) by adding at the end the following:

"(2) CMAQ PROJECTS.—The Federal share payable on account of a project or program carried out under section 149 with funds obligated in fiscal year 2008 or 2009, or both, shall be not less than 80 percent and, at the discretion of the State, may be up to 100 percent of the cost thereof.".

Sec. 1132. Distribution Of Rescissions.

(a) IN GENERAL.—Any unobligated balances of amounts that are appropriated from the Highway Trust Fund for a fiscal year, and apportioned under chapter 1 of title 23, United States Code, before, on, or after the date of enactment of this Act and that are rescinded in fiscal year 2008 or fiscal year 2009 shall be distributed by the Secretary of Transportation within each State (as defined in section 101 of such title) among all programs for which funds are apportioned under such chapter for such fiscal year, to the extent sufficient funds remain available for obligation, in the ratio that the amount of funds apportioned for each program under such chapter for such fiscal year, bears to the amount of funds apportioned for all such programs under such chapter for such fiscal year.

(b) ADJUSTMENTS.—A State may make adjustments to the distribution of a rescission within the State for a fiscal year under subsection (a) by transferring the amounts to be rescinded among the programs for which funds are apportioned under chapter 1 of title 23, United States Code, for such fiscal year, except that in making such adjustments the State may not rescind from any such program more than 110 percent of the funds to be rescinded from the program for the fiscal year as determined by the Secretary of Transportation under subsection (a).

(c) TREATMENT OF TRANSPORTATION ENHANCEMENT SET-ASIDE AND FUNDS SUBALLOCATED TO SUBSTATE AREAS.—Funds set aside under sections 133(d)(2) and 133(d)(3) of title 23, United States Code, shall be treated as being apportioned under chapter 1 of such title for purposes of subsection (a).

Sec. 1133. Sense Of Congress Regarding Use Of Complete Streets Design Techniques.

It is the sense of Congress that in constructing new roadways or rehabilitating existing facilities, State and local governments should consider policies accommodate all users, including motorists, pedestrians, cyclists, transit riders, and people of all ages and abilities, in order to—

(1) serve all surface transportation users by creating a more interconnected and intermodal system;

(2) create more viable transportation options; and

(3) facilitate the use of environmentally friendly options, such as public transportation, walking, and bicycling.

Title XII—Small Business Energy Programs

Sec. 1201. Express Loans For Renewable Energy And Energy Efficiency.

Section 7(a)(31) of the Small Business Act (15 U.S.C. 636(a)(31)) is amended by adding at the end the following:

"(F) EXPRESS LOANS FOR RENEWABLE ENERGY AND ENERGY EFFICIENCY.—

"(i) DEFINITIONS.—In this subparagraph—

"(I) the term 'biomass'—

"(aa) means any organic material that is available on a renewable or recurring basis, including—

"(AA) agricultural crops;

"(BB) trees grown for energy production;

"(CC) wood waste and wood residues; "(DD) plants (including aquatic plants and grasses);

"(EE) residues;

"(FF) fibers;

"(GG) animal wastes and other waste materials; and

"(HH) fats, oils, and greases (including recycled fats, oils, and greases); and

"(bb) does not include—

"(AA) paper that is commonly recycled; or

"(BB) unsegregated solid waste;

"(I) the term 'energy efficiency project' means the installation or upgrading of equipment that results in a significant reduction in energy usage; and

"(II) the term 'renewable energy system' means a system of energy derived from—

"(aa) a wind, solar, biomass (including biodiesel), or geothermal source; or

"(bb) hydrogen derived from biomass or water using an energy source described in item (aa).

"(ii) LOANS.—The Administrator may make a loan under the Express Loan Program for the purpose of—

"(I) purchasing a renewable energy system;

or

Sec. 1202. Pilot Program For Reduced 7(A) Fees For Purchase Of Energy Efficient Technologies.

Section 7(a) of the Small Business Act (15 U.S.C. 636(a)) is amended by adding at the end the following:

"(32) LOANS FOR ENERGY EFFICIENT TECHNOLOGIES.—

"(A) DEFINITIONS.—In this paragraph—

"(i) the term 'cost' has the meaning given that term in section 502 of the Federal Credit Reform Act of 1990 (2 U.S.C. 661a);

"(ii) the term 'covered energy efficiency loan' means a loan—

"(I) made under this subsection; and

"(II) the proceeds of which are used to purchase energy efficient designs, equipment, or fixtures, or to reduce the energy consumption of the borrower by 10 percent or more; and

"(iii) the term 'pilot program' means the pilot program established under subparagraph (B)

"(B) ESTABLISHMENT.—The Administrator shall establish and carry out a pilot program under which the Administrator shall reduce the fees for covered energy efficiency loans.

"(C) DURATION.—The pilot program shall terminate at the end of the second full fiscal year after the date that the Administrator establishes the pilot program.

"(D) MAXIMUM PARTICIPATION.—A covered energy efficiency loan shall include the maximum participation levels by the Administrator permitted for loans made under this subsection.

"(E) FEES.—

"(i) IN GENERAL.—The fee on a covered energy efficiency loan shall be equal to 50 percent of the fee otherwise applicable to that loan under paragraph (18).

"(ii) WAIVER.—The Administrator may waive clause (i) for a fiscal year if—

"(I) for the fiscal year before that fiscal year, the annual rate of default of covered energy efficiency loans exceeds that of loans made under this subsection that are not covered energy efficiency loans;

"(II) the cost to the Administration of making loans under this subsection is greater than zero and such cost is directly attributable to the cost of making covered energy efficiency loans; and

"(III) no additional sources of revenue authority are available to reduce the cost of making loans under this subsection to zero.

"(iii) EFFECT OF WAIVER.—If the Administrator waives the reduction of fees under clause (ii), the Administrator—

"(I) shall not assess or collect fees in an amount greater than necessary to ensure that the cost of the program under this subsection is not greater than zero; and

"(II) shall reinstate the fee reductions under clause (i) when the conditions in clause (ii) no longer apply.

"(iv) NO INCREASE OF FEES.—The Administrator shall not increase the fees under paragraph (18) on loans made under this subsection that are not covered energy efficiency loans as a direct result of the pilot program.

"(F) GAO REPORT.—

"(i) IN GENERAL.—Not later than 1 year after the date that the pilot program terminates, the Comptroller General of the United States shall submit to the Committee on Small Business of the House of Representatives and the Committee on Small Business and Entrepreneurship of the Senate a report on the pilot program.

"(ii) CONTENTS.—The report submitted under clause (i) shall include—

"(I) the number of covered energy efficiency loans for which fees were reduced under the pilot program;

"(II) a description of the energy efficiency savings with the pilot program;

"(III) a description of the impact of the pilot program on the program under this subsection;

"(IV) an evaluation of the efficacy and potential fraud and abuse of the pilot program; and

"(V) recommendations for improving the pilot program.".

Sec. 1203. Small Business Energy Efficiency.

(a) DEFINITIONS.—In this section—

(1) the terms "Administration" and "Administrator" mean the Small Business Administration and the Administrator thereof, respectively;

(2) the term "association" means the association of small business development centers established under section 21(a)(3)(A) of the Small Business Act (15 U.S.C.). The term "disability" has the meaning given that term in section 3 of the Americans with Disabilities Act of 1990 (42 U.S.C. 12102);

(3) the term "Efficiency Program" means the Small Business Energy Efficiency Program established under subsection (c)(1);

(4) the term "electric utility" has the meaning given that term in section 3 of the Public Utility Regulatory Policies Act of 1978 (16 U.S.C. 2602);

(5) the term "high performance green building" has the meaning given that term in section 401;

(6) the term "on-bill financing" means a low interest or no interest financing agreement between a small business concern and an electric utility for the purchase or installation of equipment, under which the regularly scheduled payment of that small business concern to that electric utility is not reduced by the amount of the reduction in cost attributable to the new equipment and that amount is credited to the electric utility, until the cost of the purchase or installation is repaid;

(7) the term "small business concern" has the same meaning as in section 3 of the Small Business Act (15 U.S.C. 632);

(8) the term "small business development center" means a small business development center described in section 21 of the Small Business Act (15 U.S.C. 648);

(9) the term "telecommuting" means the use of telecommunications to perform work functions under circumstances which reduce or eliminate the need to commute;

(10) the term "Telecommuting Pilot Program" means the pilot program established under subsection (d)(1)(A); and

(11) the term "veteran" has the meaning given that term in section 101 of title 38, United States Code.

(b) IMPLEMENTATION OF SMALL BUSINESS ENERGY EFFICIENCY PROGRAM.—

(1) IN GENERAL.—Not later than 90 days after the date of enactment of this Act, the Administrator shall promulgate final rules establishing the Government-wide program authorized under subsection (d) of section 337 of the Energy Policy and Conservation Act (42 U.S.C. 6307) that ensure compliance with that subsection by not later than 6 months after such date of enactment.

(2) PROGRAM REQUIRED.—The Administrator shall develop and coordinate a Government-wide program, building on the Energy Star for Small Business program, to assist small business concerns in—

(A) becoming more energy efficient;

(B) understanding the cost savings from improved energy efficiency; and

(C) identifying financing options for energy efficiency upgrades.

(3) CONSULTATION AND COOPERATION.—The program required by paragraph (2) shall be developed and coordinated—

(A) in consultation with the Secretary of Energy and the Administrator of the Environmental Protection Agency; and

(B) in cooperation with any entities the Administrator considers appropriate, such as industry trade associations, industry members, and energy efficiency organizations.

(4) AVAILABILITY OF INFORMATION.—The Administrator shall make available the information and materials developed under the program required by paragraph (2) to—

(A) small business concerns, including smaller design, engineering, and construction firms; and

(B) other Federal programs for energy efficiency, such as the Energy Star for Small Business program.

(5) STRATEGY AND REPORT.—

(A) STRATEGY REQUIRED.—The Administrator shall develop a strategy to educate, encourage, and assist small business concerns in adopting energy efficient building fixtures and equipment.

(B) REPORT.—Not later than December 31, 2008, the Administrator shall submit to Congress a report containing a plan to implement the strategy developed under subparagraph (A).

(c) SMALL BUSINESS SUSTAINABILITY INITIATIVE.—

(1) AUTHORITY.—The Administrator shall establish a Small Business Energy Efficiency Program to provide energy efficiency assistance to small business concerns through small business development centers.

(2) SMALL BUSINESS DEVELOPMENT CENTERS.—

(A) IN GENERAL.—In carrying out the Efficiency Program, the Administrator shall enter into agreements with small business development centers under which such centers shall—

(i) provide access to information and resources on energy efficiency practices, including on-bill financing options;

(ii) conduct training and educational activities;

(iii) offer confidential, free, one-on-one, in-depth energy audits to the owners and operators of small business concerns regarding energy efficiency practices;

(iv) give referrals to certified professionals and other providers of energy efficiency assistance who meet such standards for educational, technical, and professional competency as the Administrator shall establish;

(v) to the extent not inconsistent with controlling State public utility regulations, act as a facilitator between small business concerns, electric utilities, lenders, and the Administration to facilitate on-bill financing arrangements;

(vi) provide necessary support to small business concerns to—

(I) evaluate energy efficiency opportunities and opportunities to design or construct high performance green buildings;

(II) evaluate renewable energy sources, such as the use of solar and small wind to supplement power consumption;

(III) secure financing to achieve energy efficiency or to design or construct high performance green buildings; and

implement energy efficiency projects;

(vii) assist owners of small business concerns with the development and commercialization of clean technology products, goods, services, and processes that use renewable energy sources, dramatically reduce the use of natural resources, and cut or eliminate greenhouse gas emissions through—

(I) technology assessment;

(II) intellectual property;

(III) Small Business Innovation Research submissions under section 9 of the Small Business Act (15 U.S.C. 638);

(IV) strategic alliances;

(V) business model development; and

(VI) preparation for investors; and

(viii) help small business concerns improve environmental performance by shifting to less hazardous materials and reducing waste and emissions, including by providing assistance for small business

concerns to adapt the materials they use, the processes they operate, and the products and services they produce.

(B) REPORTS.—Each small business development center participating in the Efficiency Program shall submit to the Administrator and the Administrator of the Environmental Protection Agency an annual report that includes—

(i) a summary of the energy efficiency assistance provided by that center under the Efficiency Program;

(ii) the number of small business concerns assisted by that center under the Efficiency Program;

(iii) statistics on the total amount of energy saved as a result of assistance provided by that center under the Efficiency Program; and

(iv) any additional information determined necessary by the Administrator, in consultation with the association.

(C) REPORTS TO CONGRESS.—Not later than 60 days after the date on which all reports under subparagraph (B) relating to a year are submitted, the Administrator shall submit to the Committee on Small Business and Entrepreneurship of the Senate and the Committee on Small Business of the House of Representatives a report summarizing the information regarding the Efficiency Program submitted by small business development centers participating in that program.

(3) ELIGIBILITY.—A small business development center shall be eligible to participate in the Efficiency Program only if that center is certified under section 21(k)(2) of the Small Business Act (15 U.S.C. 648(k)(2)).

(4) SELECTION OF PARTICIPATING STATE PROGRAMS.—From among small business development centers submitting applications to participate in the Efficiency Program, the Administrator—

(A) shall, to the maximum extent practicable, select small business development centers in such a manner so as to promote a nationwide distribution of centers participating in the Efficiency Program; and

(B) may not select more than 1 small business development center in a State to participate in the Efficiency Program.

(5) MATCHING REQUIREMENT.—Subparagraphs (A) and (B) of section 21(a)(4) of the Small Business Act (15 U.S.C. 648(a)(4)) shall apply to assistance made available under the Efficiency Program.

(6) GRANT AMOUNTS.—Each small business development center selected to participate in the Efficiency Program under paragraph (4) shall be eligible to receive a grant in an amount equal to—

(A) not less than $100,000 in each fiscal year; and

(B) not more than $300,000 in each fiscal year.

(7) EVALUATION AND REPORT.—The Comptroller General of the United States shall—

(A) not later than 30 months after the date of disbursement of the first grant under the Efficiency Program, initiate an evaluation of that program; and

(B) not later than 6 months after the date of the initiation of the evaluation under subparagraph (A), submit to the Administrator, the Committee on Small Business and Entrepreneurship of the Senate, and the Committee on Small Business of the House of Representatives, a report containing—

(i) the results of the evaluation; and

(ii) any recommendations regarding whether the Efficiency Program, with or without modification, should be extended to include the participation of all small business development centers.

(8) GUARANTEE.—To the extent not inconsistent with State law, the Administrator may guarantee the timely payment of a loan made to a small business concern through an on- bill financing agreement on such terms and conditions as the Administrator shall establish through a formal rulemaking, after providing notice and an opportunity for comment.

(9) IMPLEMENTATION.—Subject to amounts approved in advance in appropriations Acts and separate from amounts approved to carry out section 21(a)(1) of the Small Business Act (15 U.S.C. 648(a)(1)), the Administrator may make grants or enter into cooperative agreements to carry out this subsection.

(10) AUTHORIZATION OF APPROPRIATIONS.—There are authorized to be appropriated such sums as are necessary to make grants and enter into cooperative agreements to carry out this subsection.

(11) TERMINATION.—The authority under this subsection shall terminate 4 years after the date of disbursement of the first grant under the Efficiency Program.

(d) SMALL BUSINESS TELECOMMUTING.—

(1) PILOT PROGRAM.—

(A) IN GENERAL.—The Administrator shall conduct, in not more than 5 of the regions of the Administration, a pilot program to provide information regarding telecom- muting to employers that are small business concerns and to encourage such employers to offer telecommuting options to employees.

(B) SPECIAL OUTREACH TO INDIVIDUALS WITH DISABILITIES.—In carrying out the Telecommuting Pilot Program, the Administrator shall make a concerted effort to provide information to—

(i) small business concerns owned by or employing individuals with disabilities, particularly veterans who are individuals with disabilities;

(ii) Federal, State, and local agencies having knowledge and expertise in assisting individuals with disabilities, including veterans who are individuals with disabilities; and

(iii) any group or organization, the primary purpose of which is to aid individuals with disabilities or veterans who are individuals with disabilities.

(C) PERMISSIBLE ACTIVITIES.—In carrying out the Telecommuting Pilot Program, the Administrator may—

(i) produce educational materials and conduct presentations designed to raise awareness in the small business community of the benefits and the ease of telecommuting;

(ii) conduct outreach—

(I) to small business concerns that are considering offering telecommuting options; and

(II) as provided in subparagraph (B); and

(iii) acquire telecommuting technologies and equipment to be used for demonstration purposes.

(D) SELECTION OF REGIONS.—In determining which regions will participate in the Telecommuting Pilot Program, the Administrator shall give priority consideration to regions in which Federal agencies and private-sector employers have demonstrated a strong regional commitment to telecommuting.

(2) REPORT TO CONGRESS.—Not later than 2 years after the date on which funds are first appropriated to carry out this subsection, the Administrator shall transmit to the Committee on Small Business and Entrepreneurship of the Senate and the Committee on Small Business of the House of Representatives a report containing the results of an evaluation of the Telecommuting Pilot Program and any recommendations regarding whether the pilot program, with or without modification, should be extended to include the participation of all regions of the Administration.

(3) TERMINATION.—The Telecommuting Pilot Program shall terminate 4 years after the date on which funds are first appropriated to carry out this subsection.

(4) AUTHORIZATION OF APPROPRIATIONS.—There is authorized to be appropriated to the Administration $5,000,000 to carry out this subsection.

(e) ENCOURAGING INNOVATION IN ENERGY EFFICIENCY.—Section 9 of the Small Business Act (15 U.S.C. 638) is amended by adding at the end the following:

"(z) ENCOURAGING INNOVATION IN ENERGY EFFICIENCY.—

"(1) FEDERAL AGENCY ENERGY-RELATED PRIORITY.—In carrying out its duties under this section relating to SBIR and STTR solicitations by Federal departments and agencies, the Administrator shall—

"(A) ensure that such departments and agencies give high priority to small business concerns that participate in or conduct energy efficiency or renewable energy system research and development projects; and

"(B) include in the annual report to Congress under subsection (b)(7) a determination of whether the priority described in subparagraph (A) is being carried out.

"(2) CONSULTATION REQUIRED.—The Administrator shall consult with the heads of other Federal departments and agencies in determining whether priority has been given to small business concerns that participate in or conduct energy efficiency or renewable energy system research and development projects, as required by this subsection.

"(3) GUIDELINES.—The Administrator shall, as soon as is practicable after the date of enactment of this subsection, issue guidelines and directives to assist Federal agencies in meeting the requirements of this subsection.

"(4) DEFINITIONS.—In this subsection—

"(A) the term 'biomass'—

"(i) means any organic material that is available on a renewable or recurring basis, including—

"(I) agricultural crops;

"(II) trees grown for energy production;

"(III) wood waste and wood residues;

"(IV) plants (including aquatic plants and grasses);

"(V) residues;

"(VI) fibers;

"(VII) animal wastes and other waste materials; and

"(VIII) fats, oils, and greases (including recycled fats, oils, and greases); and

"(ii) does not include—

"(I) paper that is commonly recycled; or

"(II) unsegregated solid waste;

"(B) the term 'energy efficiency project' means the installation or upgrading of equipment that results in a significant reduction in energy usage; and

"(C) the term 'renewable energy system' means a system of energy derived from—

"(i) a wind, solar, biomass (including biodiesel), or geothermal source; or

"(ii) hydrogen derived from biomass or water using an energy source described in clause (i).".

Sec. 1204. Larger 504 Loan Limits To Help Business Develop Energy Efficient Technologies And Purchases.

(a) ELIGIBILITY FOR ENERGY EFFICIENCY PROJECTS.—Section 501(d)(3) of the Small Business Investment Act of 1958 (15 U.S.C. 695(d)(3)) is amended—

(1) in subparagraph (G) by striking "or" at the end;

(2) in subparagraph (H) by striking the period at the end and inserting a comma;

(3) by inserting after subparagraph (H) the following:

"(I) reduction of energy consumption by at least 10 percent,

"(J) increased use of sustainable design, including designs that reduce the use of greenhouse gas emitting fossil fuels, or low-impact design to produce buildings that reduce the use of non-renewable resources and minimize environmental impact, or

"(K) plant, equipment and process upgrades of renewable energy sources such as the small-scale production of energy for individual buildings or communities consumption, commonly known as micropower, or renewable fuels producers including biodiesel and ethanol producers."; and

(4) by adding at the end the following: "In subparagraphs (J) and (K), terms have the meanings given those terms under the Leadership in Energy and Environmental Design (LEED) standard for green building certification, as determined by the Administrator.".

(b) LOANS FOR PLANT PROJECTS USED FOR ENERGY-EFFICIENT PURPOSES.—Section 502(2)(A) of the Small Business Investment Act of 1958 (15 U.S.C. 696(2)(A)) is amended—

(1) in clause (ii) by striking "and" at the end;

(2) in clause (iii) by striking the period at the end and inserting a semicolon; and

(3) by adding at the end the following:

"(iv) $4,000,000 for each project that reduces the borrower's energy consumption by at least 10 percent; and

"(v) $4,000,000 for each project that generates renewable energy or renewable fuels, such as biodiesel or ethanol production.".

Sec. 1205. Energy Saving Debentures.

(a) IN GENERAL.—Section 303 of the Small Business Investment Act of 1958 (15 U.S.C. 683) is amended by adding at the end the following:

"(k) ENERGY SAVING DEBENTURES.—In addition to any other authority under this Act, a small business investment company licensed in the first fiscal year after the date of enactment of this subsection or any fiscal year thereafter may issue Energy Saving debentures.".

(b) DEFINITIONS.—Section 103 of the Small Business Investment Act of 1958 (15 U.S.C. 662) is amended—

(1) in paragraph (16), by striking "and" at the end;

(2) in paragraph (17), by striking the period at the end and inserting a semicolon; and

(3) by adding at the end the following:

"(18) the term 'Energy Saving debenture' means a deferred interest debenture that—

"(A) is issued at a discount;

"(B) has a 5-year maturity or a 10-year maturity;

"(C) requires no interest payment or annual charge for the first 5 years;

"(D) is restricted to Energy Saving qualified investments; and

"(E) is issued at no cost (as defined in section 502 of the Credit Reform Act of 1990) with respect to purchasing and guaranteeing the debenture; and

"(19) the term 'Energy Saving qualified investment' means investment in a small business concern that is primarily engaged in researching, manufacturing, developing, or providing products, goods, or services that reduce the use or consumption of non-renewable energy resources.".

Sec. 1206. Investments In Energy Saving Small Businesses.

(a) MAXIMUM LEVERAGE.—Section 303(b)(2) of the Small Business Investment Act of 1958 (15 U.S.C. 303(b)(2)) is amended by 15 USC 683. adding at the end the following:

"(D) INVESTMENTS IN ENERGY SAVING SMALL BUSINESSES.—

"(i) N GENERAL.—Subject to clause (ii), in calculating the outstanding leverage of a company for purposes of subparagraph (A), the

Administrator shall exclude the amount of the cost basis of any Energy Saving qualified investment in a smaller enterprise made in the first fiscal year after the date of enactment of this subparagraph or any fiscal year thereafter by a company licensed in the applicable fiscal year.

"(ii) LIMITATIONS.—

"(I) AMOUNT OF EXCLUSION.—The amount excluded under clause (i) for a company shall not exceed 33 percent of the private capital of that company.

"(II) MAXIMUM INVESTMENT.—A company shall not make an Energy Saving qualified investment in any one entity in an amount equal to more than 20 percent of the private capital of that company.

"(III) OTHER TERMS.—The exclusion of amounts under clause (i) shall be subject to such terms as the Administrator may impose to ensure that there is no cost (as that term is defined in section 502 of the Federal Credit Reform Act of 1990 (2 U.S.C. 661a)) with respect to purchasing or guaranteeing any debenture involved.".

(b) MAXIMUM AGGREGATE AMOUNT OF LEVERAGE.—Section 303(b)(4) of the Small Business Investment Act of 1958 (15 U.S.C. 303(b)(4)) is amended by adding at the end the following:

"(E) INVESTMENTS IN ENERGY SAVING SMALL BUSINESSES.—

"(i) IN GENERAL.—Subject to clause (ii), in calculating the aggregate outstanding leverage of a company for purposes of subparagraph (A), the Administrator shall exclude the amount of the cost basis of any Energy Saving qualified investment in a smaller enterprise made in the first fiscal year after the date of enactment of this subparagraph or any fiscal year thereafter by a company licensed in the applicable fiscal year.

"(ii) LIMITATIONS.—

"(I) AMOUNT OF EXCLUSION.—The amount excluded under clause (i) for a company shall not exceed 33 percent of the private capital of that company.

"(II) MAXIMUM INVESTMENT.—A company shall not make an Energy Saving qualified investment in any one entity in an amount equal to more than 20 percent of the private capital of that company.

"(III) OTHER TERMS.—The exclusion of amounts under clause (i) shall be subject to such terms as the Administrator may impose to ensure that there is no cost (as that term is defined in section 502 of the Federal Credit Reform Act of 1990 (2 U.S.C. 661a)) with respect to purchasing or guaranteeing any debenture involved.".

Sec. 1207. Renewable Fuel Capital Investment Company.

Title III of the Small Business Investment Act of 1958 (15 U.S.C. 681 et seq.) is amended by adding at the end the following:

"PART C—RENEWABLE FUEL CAPITAL INVESTMENT PILOT PROGRAM

"(1) OPERATIONAL ASSISTANCE.—The term 'operational assistance' means management, marketing, and other technical assistance that assists a small business concern with business development.

"(2) PARTICIPATION AGREEMENT.—The term 'participation agreement' means an agreement, between the Administrator and a company granted final approval under section 384(e), that—

"(A) details the operating plan and investment criteria of the company; and

"(B) requires the company to make investments in smaller enterprises primarily engaged in researching, manufacturing, developing, producing, or bringing to market goods, products, or services that generate or support the production of renewable energy.

"(3) RENEWABLE ENERGY.—The term 'renewable energy' means energy derived from resources that are regenerative or that cannot be depleted, including solar, wind, ethanol, and biodiesel fuels.

"(4) RENEWABLE FUEL CAPITAL INVESTMENT COMPANY.—The term 'Renewable Fuel Capital Investment company' means a company—

"(A) that—

"(i) has been granted final approval by the Administrator under section 384(e); and

"(ii) has entered into a participation agreement with the Administrator; or

"(B) that has received conditional approval under section 384(c).

"(5) STATE.—The term 'State' means each of the several States, the District of Columbia, the Commonwealth of Puerto Rico, the Virgin Islands, Guam, American Samoa, the Commonwealth of the Northern Mariana Islands, and any other commonwealth, territory, or possession of the United States.

"(6) VENTURE CAPITAL.—The term 'venture capital' means capital in the form of equity capital investments, as that term is defined in section 303(g)(4).

"Sec. 382. Purposes.

"The purposes of the Renewable Fuel Capital Investment Program established under this part are—

"(1) to promote the research, development, manufacture, production, and bringing to market of goods, products, or services that generate or support the production of renewable energy by encouraging venture capital investments in smaller enterprises primarily engaged such activities; and

"(2) to establish a venture capital program, with the mission of addressing the unmet equity investment needs of smaller enterprises engaged in researching, developing, manufacturing, producing, and bringing to market goods, products, or services that generate or support the production of renewable energy, to be administered by the Administrator—

"(A) to enter into participation agreements with Renewable Fuel Capital Investment companies;

"(B) to guarantee debentures of Renewable Fuel Capital Investment companies to enable each such company to make venture capital investments in smaller enterprises engaged in the research, development, manufacture, production, and bringing to market of goods, products, or services that generate or support the production of renewable energy; and

"(C) to make grants to Renewable Fuel Investment Capital companies, and to other entities, for the purpose of providing operational assistance to smaller enterprises financed, or expected to be financed, by such companies.

"Sec. 383. Establishment.

"The Administrator shall establish a Renewable Fuel Capital Investment Program, under which the Administrator may—

"(1) enter into participation agreements for the purposes described in section 382; and

"(2) guarantee the debentures issued by Renewable Fuel Capital Investment companies as provided in section 385.

"Sec. 384. Selection Of Renewable Fuel Capital Investment Companies.

"(a) ELIGIBILITY.—A company is eligible to apply to be designated as a Renewable Fuel Capital Investment company if the company—

"(1) is a newly formed for-profit entity or a newly formed for-profit subsidiary of an existing entity;

"(2) has a management team with experience in alternative energy financing or relevant venture capital financing; and

"(3) has a primary objective of investment in smaller enterprises that research, manufacture, develop, produce, or bring to market goods, products, or services that generate or support the production of renewable energy.

"(b) APPLICATION.—A company desiring to be designated as a Renewable Fuel Capital Investment company shall submit an application to the Administrator that includes—

"(1) a business plan describing how the company intends to make successful venture capital investments in smaller enterprises primarily engaged in the research, manufacture, development, production, or bringing to market of goods, products, or services that generate or support the production of renewable energy;

"(2) information regarding the relevant venture capital qualifications and general reputation of the management of the company;

"(3) a description of how the company intends to seek to address the unmet capital needs of the smaller enterprises served;

"(4) a proposal describing how the company intends to use the grant funds provided under this part to provide operational assistance to smaller enterprises financed by the company, including information regarding whether the company has employees with appropriate professional licenses or will contract with another entity when the services of such an individual are necessary;

"(5) with respect to binding commitments to be made to the company under this part, an estimate of the ratio of cash to in-kind contributions;

"(6) a description of whether and to what extent the company meets the criteria under subsection (c)(2) and the objectives of the program established under this part;

"(7) information regarding the management and financial strength of any parent firm, affiliated firm, or any other firm essential to the success of the business plan of the company; and such other information as the Administrator may require.

"(c) CONDITIONAL APPROVAL.—

"(1) IN GENERAL.—From among companies submitting applications under subsection (b), the Administrator shall conditionally approve companies to operate as Renewable Fuel Capital Investment companies.

"(2) SELECTION CRITERIA.—In conditionally approving companies under paragraph (1), the Administrator shall consider—

"(A) the likelihood that the company will meet the goal of its business plan;

"(B) the experience and background of the management team of the company;

"(C) the need for venture capital investments in the geographic areas in which the company intends to invest;

"(D) the extent to which the company will concentrate its activities on serving the geographic areas in which it intends to invest;

"(E) the likelihood that the company will be able to satisfy the conditions under subsection (d);

"(F) the extent to which the activities proposed by the company will expand economic opportunities in the geographic areas in which the company intends to invest;

"(G) the strength of the proposal by the company to provide operational assistance under this part as the proposal relates to the ability of the company to meet applicable cash requirements and properly use in-kind contributions, including the use of resources for the services of licensed professionals, when necessary, whether provided by employees or contractors; and

"(H) any other factor determined appropriate by the Administrator.

"(3) NATIONWIDE DISTRIBUTION.—From among companies submitting applications under subsection (b), the Administrator shall consider the selection criteria under paragraph (2) and shall, to the maximum extent practicable, approve at least one company from each geographic region of the Administration.

"(d) REQUIREMENTS TO BE MET FOR FINAL APPROVAL.—

"(1) IN GENERAL.—The Administrator shall grant each conditionally approved company 2 years to satisfy the requirements of this subsection.

"(2) CAPITAL REQUIREMENT.—Each conditionally approved company shall raise not less than $3,000,000 of private capital or binding capital commitments from 1 or more investors (which shall not be departments or agencies of the Federal Government) who meet criteria established by the Administrator.

"(3) NONADMINISTRATION RESOURCES FOR OPERATIONAL ASSISTANCE.—

"(A) IN GENERAL.—In order to provide operational assistance to smaller enterprises expected to be financed by the company, each conditionally approved company shall have binding commitments (for contribution in cash or in-kind) from sources other than the Administration that meet criteria established by the Administrator; and payable or available over a multiyear period by the Administrator (not to exceed 10 years).

"(B) EXCEPTION.—The Administrator may, in the discretion of the Administrator and based upon a showing of special circumstances and good cause, consider an applicant to have satisfied the requirements of subparagraph (A) if the applicant has a viable plan that reasonably projects the capacity of the applicant to raise the amount (in cash or in-kind) required under subparagraph (A); and binding commitments in an amount equal to not less than 20 percent of the total amount required under paragraph (A).

"(C) LIMITATION.—The total amount of a in-kind contributions by a company shall be not more than 50 percent of the total contributions by a company.

"(e) FINAL APPROVAL; DESIGNATION.—The Administrator shall, with respect to each applicant conditionally approved under subsection (c)—

"(1) grant final approval to the applicant to operate as a Renewable Fuel Capital Investment company under this part and designate the applicant as such a company, if the applicant satisfies the requirements of subsection (d) on or before the expiration of the time period described in that subsection; and enters into a participation agreement with the Administrator; or

"(2) if the applicant fails to satisfy the requirements of subsection (d) on or before the expiration of the time period described in paragraph (1) of that subsection, revoke the conditional approval granted under that subsection.

Sec. 385. Debentures.

"(a) IN GENERAL.—The Administrator may guarantee the timely payment of principal and interest, as scheduled, on debentures issued by any Renewable Fuel Capital Investment company.

"(b) TERMS AND CONDITIONS.—The Administrator may make guarantees under this section on such terms and conditions as it determines appropriate, except that—

"(1) the term of any debenture guaranteed under this section shall not exceed 15 years; and

"(2) a debenture guaranteed under this section—

"(A) shall carry no front-end or annual fees;

"(B) shall be issued at a discount;

"(C) shall require no interest payments during the 5-year period beginning on the date the debenture is issued;

"(D) shall be prepayable without penalty after the end of the 1-year period beginning on the date the debenture is issued; and

"(E) shall require semiannual interest payments after the period described in subparagraph (C).

"(c) FULL FAITH AND CREDIT OF THE UNITED STATES.—The full faith and credit of the United States is pledged to pay all amounts that may be required to be paid under any guarantee under this part.

"(d) MAXIMUM GUARANTEE.—

"(1) IN GENERAL.—Under this section, the Administrator may guarantee the debentures issued by a Renewable Fuel Capital Investment company only to the extent that the total face amount of outstanding guaranteed debentures of such company does not exceed 150 percent of the private capital of the company, as determined by the Administrator.

"(2) TREATMENT OF CERTAIN FEDERAL FUNDS.—For purposes of paragraph (1), private capital shall include capital Federal funds, if capital is contributed by an investor other than a department or agency of the Federal Government.

"Sec. 386. Issuance And Guarantee Of Trust Certificates.

"(a) ISSUANCE.—The Administrator may issue trust certificates representing ownership of all or a fractional part of debentures issued by a Renewable Fuel Capital Investment company and guaranteed by the Administrator under this part, if such certificates are based on and backed by a trust or pool approved by the Administrator and composed solely of guaranteed debentures.

"(b) GUARANTEE.—

"(1) IN GENERAL.—The Administrator may, under such terms and conditions as it determines appropriate, guarantee the timely payment of the principal of and interest on trust certificates issued by the Administrator or its agents for purposes of this section.

"(2) LIMITATION.—Each guarantee under this subsection shall be limited to the extent of principal and interest on the guaranteed debentures that compose the trust or pool.

"(3) PREPAYMENT OR DEFAULT.—If a debenture in a trust or pool is prepaid, or in the event of default of such a debenture, the guarantee of timely payment of principal and interest on the trust certificates shall be reduced in proportion to the amount of principal and interest such prepaid debenture represents in the trust or pool. Interest on prepaid or defaulted debentures shall accrue and be guaranteed by the Administrator only through the date of payment of the guarantee. At any time during its term, a trust certificate may be called for redemption due to prepayment or default of all debentures.

"(c) FULL FAITH AND CREDIT OF THE UNITED STATES.—The full faith and credit of the United States is pledged to pay all amounts that may be required to be paid under any guarantee of a trust certificate issued by the Administrator or its agents under this section.

"(d) FEES.—The Administrator shall not collect a fee for any guarantee of a trust certificate under this section, but any agent of the Administrator may collect a fee approved by the Administrator for the functions described in subsection (f)(2).

"(e) SUBROGATION AND OWNERSHIP RIGHTS.—

"(1) SUBROGATION.—If the Administrator pays a claim under a guarantee issued under this section, it shall be subrogated fully to the rights satisfied by such payment.

"(2) OWNERSHIP RIGHTS.—No Federal, State, or local law shall preclude or limit the exercise by the Administrator of its ownership rights in the debentures residing in a trust or pool against which trust certificates are issued under this section.

"(f) MANAGEMENT AND ADMINISTRATION.—

"(1) REGISTRATION.—The Administrator may provide for a central registration of all trust certificates issued under this section.

"(2) CONTRACTING OF FUNCTIONS.—

"(A) IN GENERAL.—The Administrator may contract with an agent or agents to carry out on behalf of the Administrator the pooling and the central registration functions provided for in this section, including, not withstanding any other provision of law—

"(i) maintenance, on behalf of and under the direction of the Administrator, of such commercial bank accounts or investments in obligations of the United States as may be necessary to facilitate the creation of trusts or pools backed by debentures guaranteed under this part; and

"(ii) the issuance of trust certificates to facilitate the creation of such trusts or pools.

"(B) FIDELITY BOND OR INSURANCE REQUIREMENT.—Any agent performing functions on behalf of the Administrator under this paragraph shall provide a fidelity bond or insurance in such amounts as the Administrator determines to be necessary to fully protect the interests of the United States.

"(3) REGULATION OF BROKERS AND DEALERS.—The Administrator may regulate brokers and dealers in trust certificates issued under this section.

"(4) ELECTRONIC REGISTRATION.—Nothing in this subsection may be construed to prohibit the use of a book-entry or other electronic form of registration for trust certificates issued under this section.

Sec. 387. Fees.

"(a) IN GENERAL.—Except as provided in section 386(d), the Administrator may charge such fees as it determines appropriate with respect to any guarantee or grant issued under this part, in an amount established annually by the Administrator, as necessary to reduce to zero the cost (as defined in section 502 of the Federal Credit Reform Act of 1990) to the Administration of purchasing and guaranteeing debentures under this part, which amounts shall be paid to and retained by the Administration.

"(b) OFFSET.—The Administrator may, as provided by section 388, offset fees charged and collected under subsection (a).

"(c) ANNUAL ADJUSTMENT.—Each fee contribution under subsection (a) shall be effective for 1 fiscal year and shall be adjusted as necessary for each fiscal year thereafter to ensure that amounts under subsection (a) are fully used. The fee contribution for a fiscal year shall be based on the outstanding commitments made and the guarantees and grants that the Administrator projects will be made during that fiscal year, given the program level authorized by law for that fiscal year and any other factors that the Administrator determines appropriate.

"Sec. 389. Operational Assistance Grants.

"(a) IN GENERAL.—

"(1) AUTHORITY.—The Administrator may make grants to Renewable Fuel Capital Investment companies to provide operational assistance to smaller enterprises financed, or expected to be financed, by such companies or other entities.

"(2) TERMS.—A grant under this subsection shall be made over a multiyear period not to exceed 10 years, under such other terms as the Administrator may require.

"(3) GRANT AMOUNT.—The amount of a grant made under this subsection to a Renewable Fuel Capital Investment company shall be equal to the lesser of—

"(A) 10 percent of the resources (in cash or in-kind) raised by the company under section 384(d)(2); or

"(B) $1,000,000.

"(4) PRO RATA REDUCTIONS.—If the amount made available to carry out this section is insufficient for the Administrator to provide grants in the amounts provided for in paragraph (3), the Administrator shall make pro rata reductions in the amounts otherwise payable to each company and entity under such paragraph.

"(5) GRANTS TO CONDITIONALLY APPROVED COMPANIES.—

"(A) IN GENERAL.—Subject to subparagraphs (B) and (C), upon the request of a company conditionally approved under section 384(c), the Administrator shall make a grant to the company under this subsection.

"(B) REPAYMENT BY COMPANIES NOT APPROVED.—If a company receives a grant under this paragraph and does not enter into a participation agreement for final approval, the company shall, subject to controlling Federal law, repay the amount of the grant to the Administrator.

"(C) DEDUCTION OF GRANT TO APPROVED COMPANY.—

If a company receives a grant under this paragraph and receives final approval under section 384(e), the Administrator shall deduct the amount of the grant from the total grant amount the company receives for operational assistance.

"(D) AMOUNT OF GRANT.—No company may receive a grant of more than $100,000 under this paragraph.

"(b) SUPPLEMENTAL GRANTS.—

"(1) IN GENERAL.—The Administrator may make supplemental grants to Renewable Fuel Capital Investment companies and to other entities, as authorized by this part, under such terms as the Administrator may require, to provide additional operational assistance to smaller enterprises financed, or expected to be financed, by the companies.

"(2) MATCHING REQUIREMENT.—The Administrator may require, as a condition of any supplemental grant made under this subsection, that the company or entity receiving the grant provide from resources (in a cash or in kind), other then those provided by the Administrator, a matching contribution equal to the amount of the supplemental grant.

"(c) LIMITATION.—None of the assistance made available under this section may be used for any overhead or general and administrative expense of a Renewable Fuel Capital Investment company.

"Sec. 390. Bank Participation.

"(a) IN GENERAL.—Except as provided in subsection (b), any national bank, any member bank of the Federal Reserve System, and (to the extent permitted under applicable State law) any insured bank that is not a member of such system, may invest in any Renewable Fuel Capital Investment company, or in any entity established to invest solely in Renewable Fuel Capital Investment companies.

"(b) LIMITATION.—No bank described in subsection (a) may make investments described in such subsection that are greater than 5 percent of the capital and surplus of the bank.

"Sec. 391. Federal Financing Bank.

"Notwithstanding section 318, the Federal Financing Bank may acquire a debenture issued by a Renewable Fuel Capital Investment company under this part.

"Sec. 392. Reporting Requirement.

"Each Renewable Fuel Capital Investment company that participates in the program established under this part shall provide to the Administrator such information as the Administrator may require, including—

"(1) information related to the measurement criteria that the company proposed in its program application; and

"(2) in each case in which the company makes, under this part, an investment in, or a loan or a grant to, a business that is not primarily engaged in the research, development, manufacture, or bringing to market or renewable energy sources, a report on the nature, origin, and revenues of the business in which investments are made.

"Sec. 393. Examinations.

"(a) IN GENERAL.—Each Renewable Fuel Capital Investment company that participates in the program established under this part shall be subject to examinations made at the direction of the Investment Division of the Administration in accordance with this section.

"(b) ASSISTANCE OF PRIVATE SECTOR ENTITIES.—Examinations under this section may be conducted with the assistance of a private sector entity that has both the qualifications and the expertise necessary to conduct such examinations.

"(c) COSTS.—

"(1) ASSESSMENT.—

"(A) IN GENERAL.—The Administrator may assess the cost of examinations under this section, including compensation of the examiners, against the company examined.

"(B) PAYMENT.—Any company against which the Administrator assesses costs under this paragraph shall pay such costs.

"(2) DEPOSIT OF FUNDS.—Funds collected under this section shall be deposited in the account for salaries and expenses of the Administration.

"Sec. 394. Miscellaneous.

"To the extent such procedures are not inconsistent with the requirements of this part, the Administrator may take such action as set forth in sections 309, 311, 312, and 314 and an officer, director, employee, agent, or other participant in the management or conduct of the affairs of a Renewable Fuel Capital Investment company shall be subject to the requirements of such sections.

"Sec. 395. Removal Or Suspension Of Directors Or Officers.

"Using the procedures for removing or suspending a director or an officer of a licensee set forth in section 313 (to the extent such procedures are not inconsistent with the requirements of this part), the Administrator may remove or suspend any director or officer of any Renewable Fuel Capital Investment company.

"Sec. 396. Regulations.

"The Administrator may issue such regulations as the Administrator determines necessary to carry out the provisions of this part in accordance with its purposes.

"Sec. 397. Authorizations Of Appropriations.

"(a) IN GENERAL.—Subject to the availability of appropriations, the Administrator is authorized to make $15,000,000 in operational assistance grants under section 389 for each of fiscal years 2008 and 2009.

"(b) FUNDS COLLECTED FOR EXAMINATIONS.—Funds deposited under section 393(c)(2) are authorized to be appropriated only for the costs of examinations under section 393 and for the costs of other oversight activities with respect to the program established under this part.

"Sec. 398. Termination.

"The program under this part shall terminate at the end of the second full fiscal year after the date that the Administrator establishes the program under this part.".

Sec. 1208. Study And Report.

The Administrator of the Small Business Administration shall conduct a study of the Renewable Fuel Capital Investment Program under part C of title III of the Small Business Investment Act of 1958, as added by this Act. Not later than 3 years after the date of enactment of this Act, the Administrator shall complete the study under this section and submit to Congress a report regarding the results of the study.

TITLE XIII — SMART GRID

Sec. 1301. Statement Of Policy On Modernization Of Electricity Grid.

It is the policy of the United States to support the modernization of the Nation's electricity transmission and distribution system to maintain a reliable and secure electricity infrastructure that can meet future demand growth and to achieve each of the following, which together characterize a Smart Grid:

(1) Increased use of digital information and controls technology to improve reliability, security, and efficiency of the electric grid.

(2) Dynamic optimization of grid operations and resources, with full cyber-security.

(3) Deployment and integration of distributed resources and generation, including renewable resources.

(4) Development and incorporation of demand response, demand-side resources, and energy-efficiency resources.

(5) Deployment of "smart" technologies (real-time, automated, interactive technologies that optimize the physical operation of appliances and consumer devices) for metering, communications concerning grid operations and status, and distribution automation.

(6) Integration of "smart" appliances and consumer devices.

(7) Deployment and integration of advanced electricity storage and peak-shaving technologies, including plug-in electric and hybrid electric vehicles, and thermal-storage air conditioning.

(8) Provision to consumers of timely information and control options.

(9) Development of standards for communication and inter- operability of appliances and equipment connected to the electric grid, including the infrastructure serving the grid.

(10) Identification and lowering of unreasonable or unnecessary barriers to adoption of smart grid technologies, practices, and services.

Sec. 1302. Smart Grid System Report.

The Secretary, acting through the Assistant Secretary of the Office of Electricity Delivery and Energy Reliability (referred to in this section as the "OEDER") and through the Smart Grid Task Force established in section 1303, shall, after consulting with any interested individual or entity as appropriate, no later than 1 year after enactment, and every 2 years thereafter, report to Congress concerning the status of smart grid deployments nationwide and any regulatory or government barriers to continued deployment. The report shall provide the current status and prospects of smart grid development, including information on technology penetration, communications network capabilities, costs, and obstacles. It may include recommendations for State and Federal policies or actions helpful to facilitate the transition to a smart grid. To the extent appropriate, it should take a regional perspective. In preparing this report, the Secretary shall solicit advice and contributions from the Smart Grid Advisory Committee created in section 1303; from other involved Federal agencies including but not limited to the Federal Energy Regulatory Commission ("Commission"), the National Institute of Standards and Technology ("Institute"), and the Department of Homeland Security; and from other stakeholder groups not already represented on the Smart Grid Advisory Committee.

Sec. 1303. Smart Grid Advisory Committee And Smart Grid Task Force.

(a) SMART GRID ADVISORY COMMITTEE.—

(1) ESTABLISHMENT.—The Secretary shall establish, within Deadline. 90 days of enactment of this Part, a Smart Grid Advisory Committee (either as an independent entity or as a designated sub-part of a larger advisory committee on electricity matters). The Smart Grid Advisory Committee shall include eight or more members appointed by the Secretary who have sufficient experience and expertise to represent the full range of smart grid technologies and services, to represent both private and non-Federal public sector stakeholders. One member shall be appointed by the Secretary to Chair the Smart Grid Advisory Committee.

(2) MISSION.—The mission of the Smart Grid Advisory Committee shall be to advise the Secretary, the Assistant Secretary, and other relevant Federal officials concerning the development of smart grid technologies, the progress of a national transition to the use of smart-grid technologies and services, the evolution of widely-accepted technical and practical standards and protocols to allow interoperability and inter-communication among smart-grid capable devices, and the optimum means of using Federal incentive authority to encourage such progress.

(3) APPLICABILITY OF FEDERAL ADVISORY COMMITTEE ACT.—

The Federal Advisory Committee Act (5 U.S.C. App.) shall apply to the Smart Grid Advisory Committee.

(b) SMART GRID TASK FORCE.—

(1) ESTABLISHMENT.—The Assistant Secretary of the Office Deadline. of Electricity Delivery and Energy Reliability shall establish, within 90 days of enactment of this Part, a Smart Grid Task Force composed of designated employees from the various divisions of that office who have responsibilities related to the transition to smart-grid technologies and practices. The Assistant Secretary or his designee shall be identified as the Director of the Smart Grid Task Force. The Chairman of the Federal Energy Regulatory Commission and the Director of the National Institute of Standards and Technology shall each designate at least one employee to participate on the Smart Grid Task Force. Other members may come from other agencies at the invitation of the Assistant Secretary or the nomination of the head of such other agency. The Smart Grid Task Force shall, without disrupting the work of the Divisions or Offices from which its members are drawn, provide an identifiable Federal entity to embody the Federal role in the national transition toward development and use of smart grid technologies.

(2) MISSION.—The mission of the Smart Grid Task Force shall be to insure awareness, coordination and integration of the diverse activities of the Office and elsewhere in the Federal Government related to smart-grid technologies and practices, including but not limited to: smart grid research and development; development of widely accepted smart-grid standards and protocols; the relationship of smart-grid technologies and practices to electric utility regulation; the relationship of smart- grid technologies and practices to infrastructure development, system reliability and security; and the relationship of smart- grid technologies and practices to other facets of electricity supply, demand, transmission, distribution, and policy. The Smart Grid Task Force shall collaborate with the Smart Grid Advisory Committee and other Federal agencies and offices. The Smart Grid Task Force shall meet at the call of its Director as necessary to accomplish its mission.

(c) AUTHORIZATION.—There are authorized to be appropriated for the purposes of this section such sums as are necessary to the Secretary to support the operations of the Smart Grid Advisory Committee and Smart Grid Task Force for each of fiscal years 2008 through 2020.

Sec. 1304. Smart Grid Technology Research, Development, And Demonstration.

(a) POWER GRID DIGITAL INFORMATION TECHNOLOGY.—The Secretary, in consultation with the Federal Energy Regulatory Commission and other appropriate agencies, electric utilities, the States, and other stakeholders, shall carry out a program—

(1) to develop advanced techniques for measuring peak load reductions and energy-efficiency savings from smart metering, demand response, distributed generation, and electricity storage systems;

(2) to investigate means for demand response, distributed generation, and storage to provide ancillary services;

(3) to conduct research to advance the use of wide-area measurement and control networks, including data mining, visualization, advanced computing, and secure and dependable communications in a highly-distributed environment;

(4) to test new reliability technologies, including those concerning communications network capabilities, in a grid control room environment against a representative set of local outage and wide area blackout scenarios;

(5) to identify communications network capacity needed to implement advanced technologies.

(6) to investigate the feasibility of a transition to time- of-use and real-time electricity pricing;

(7) to develop algorithms for use in electric transmission system software applications;

(8) to promote the use of underutilized electricity generation capacity in any substitution of electricity for liquid fuels in the transportation system of the United States; and

(9) in consultation with the Federal Energy Regulatory Commission, to propose interconnection protocols to enable electric utilities to access electricity stored in vehicles to help meet peak demand loads.

(b) SMART GRID REGIONAL DEMONSTRATION INITIATIVE.—

(1) IN GENERAL.—The Secretary shall establish a smart grid regional demonstration initiative (referred to in this subsection as the "Initiative") composed of demonstration projects specifically focused on advanced technologies for use in power grid sensing, communications, analysis, and power flow control. The Secretary shall seek to leverage existing smart grid deployments.

(2) GOALS.—The goals of the Initiative shall be—

(A) to demonstrate the potential benefits of concentrated investments in advanced grid technologies on a regional grid;

(B) to facilitate the commercial transition from the current power transmission and distribution system technologies to advanced technologies;

(C) to facilitate the integration of advanced technologies in existing electric networks to improve system performance, power flow control, and reliability;

(D) to demonstrate protocols and standards that allow for the measurement and validation of the energy savings and fossil fuel emission reductions associated with the installation and use of energy efficiency and demand response technologies and practices; and

(E) to investigate differences in each region and regulatory environment regarding best practices in implementing smart grid technologies.

(3) DEMONSTRATION PROJECTS.—

(A) IN GENERAL.—In carrying out the initiative, the Secretary shall carry out smart grid demonstration projects in up to 5 electricity control areas, including rural areas and at least 1 area in which the majority of generation and transmission assets are controlled by a tax-exempt entity.

(B) COOPERATION.—A demonstration project under subparagraph (A) shall be carried out in cooperation with the electric utility that owns the grid facilities in the electricity control area in which the demonstration project is carried out.

(C) FEDERAL SHARE OF COST OF TECHNOLOGY INVESTMENTS.—The Secretary shall provide to an electric utility described in subparagraph (B) financial assistance for use in paying an amount equal to not more than 50 percent of the cost of qualifying advanced grid technology investments made by the electric utility to carry out a demonstration project.

(D) INELIGIBILITY FOR GRANTS.—No person or entity participating in any demonstration project conducted under this subsection shall be eligible for grants under section 1306 for otherwise qualifying investments made as part of that demonstration project.

(c) AUTHORIZATION OF APPROPRIATIONS.—There are authorized to be appropriated—

(1) to carry out subsection (a), such sums as are necessary for each of fiscal years 2008 through 2012; and

(2) to carry out subsection (b), $100,000,000 for each of fiscal years 2008 through 2012.

Sec. 1305. Smart Grid Interoperability Framework.

(a) INTEROPERABILITY FRAMEWORK.—The Director of the National Institute of Standards and Technology shall have primary responsibility to coordinate the development of a framework that includes protocols and model standards for information management to achieve interoperability of smart grid devices and systems. Such protocols and standards shall further align policy, business, and technology approaches in a manner that would enable all electric resources, including demand-side resources, to contribute to an efficient, reliable electricity network. In developing such protocols and standards—

(1) the Director shall seek input and cooperation from the Commission, OEDER and its Smart Grid Task Force, the Smart Grid Advisory Committee, other relevant Federal and State agencies; and

(2) the Director shall also solicit input and cooperation from private entities interested in such protocols and standards, including but not limited to the Gridwise Architecture Council, the International Electrical and Electronics Engineers, the National Electric Reliability Organization recognized by the Federal Energy Regulatory Commission, and National Electrical Manufacturer's Association.

(b) SCOPE OF FRAMEWORK.—The framework developed under subsection (a) shall be flexible, uniform and technology neutral, including but not limited to technologies for managing smart grid information, and designed—

(1) to accommodate traditional, centralized generation and transmission resources and consumer distributed resources, including distributed generation, renewable generation, energy storage, energy efficiency, and demand response and enabling devices and systems;

(2) to be flexible to incorporate—

(A) regional and organizational differences; and

(B) technological innovations;

(3) to consider the use of voluntary uniform standards for certain classes of mass-produced electric appliances and equipment for homes and businesses that enable customers, at their election and consistent with applicable State and Federal laws, and are manufactured with the ability to respond to electric grid emergencies and demand response signals by curtailing all, or a portion of, the electrical power consumed by the appliances or equipment in response to an emergency or demand response signal, including through—

(A) load reduction to reduce total electrical demand;

(B) adjustment of load to provide grid ancillary services; and

(C) in the event of a reliability crisis that threatens an outage, short-term load shedding to help preserve the stability of the grid; and

(4) such voluntary standards should incorporate appropriate manufacturer lead time.

(c) TIMING OF FRAMEWORK DEVELOPMENT.—The Institute shall begin work pursuant to this section within 60 days of enactment. The Institute shall provide and publish an initial report on progress toward recommended or consensus standards and protocols within 1 year after enactment, further reports at such times as developments warrant in the judgment of the Institute, and a final report when the Institute determines that the work is completed or that a Federal role is no longer necessary.

(d) STANDARDS FOR INTEROPERABILITY IN FEDERAL JURISDICTION.—At any time after the Institute's work has led to sufficient consensus in the Commission's judgment, the Commission shall institute a rulemaking proceeding to adopt such standards and protocols as may be necessary to insure smart-grid functionality and interoperability in interstate transmission of electric power, and regional and wholesale electricity markets.

(e) AUTHORIZATION.—There are authorized to be appropriated for the purposes of this section $5,000,000 to the Institute to support the activities required by this subsection for each of fiscal years 2008 through 2012.

Sec. 1306. Federal Matching Fund For Smart Grid Investment Costs.

(a) MATCHING FUND.—The Secretary shall establish a Smart Grid Investment Matching Grant Program to provide reimbursement of one-fifth (20 percent) of qualifying Smart Grid investments.

(b) QUALIFYING INVESTMENTS.—Qualifying Smart Grid investments may include any of the following made on or after the date of enactment of this Act:

(1) In the case of appliances covered for purposes of establishing energy conservation standards under part B of title III of the Energy Policy and Conservation Act of 1975 (42 U.S.C. 6291 et seq.), the documented expenditures incurred by a manufacturer of such appliances associated with purchasing or designing, creating the ability to manufacture, and manufacturing and installing for one calendar year, internal devices that allow the appliance to engage in Smart Grid functions.

(2) In the case of specialized electricity-using equipment, including motors and drivers, installed in industrial or commercial applications, the documented expenditures incurred by its owner or its manufacturer of installing devices or modifying that equipment to engage in Smart Grid functions.

(3) In the case of transmission and distribution equipment fitted with monitoring and communications devices to enable smart grid functions, the documented expenditures incurred by the electric utility to purchase and install such monitoring and communications devices.

(4) In the case of metering devices, sensors, control devices, and other devices integrated with and attached to an electric utility system or retail distributor or marketer of electricity that are capable of engaging in Smart Grid functions, the documented expenditures incurred by the electric utility, distributor, or marketer and its customers to purchase and install such devices.

(5) In the case of software that enables devices or computers to engage in Smart Grid functions, the documented purchase costs of the software.

(6) In the case of entities that operate or coordinate operations of regional electric grids, the documented expenditures for purchasing and installing such equipment that allows Smart Grid functions to operate and be combined or coordinated among multiple electric utilities and between that region and other regions.

(7) In the case of persons or entities other than electric utilities owning and operating a distributed electricity generator, the documented expenditures of enabling that generator to be monitored, controlled, or otherwise integrated into grid operations and electricity flows on the grid utilizing Smart Grid functions.

(8) In the case of electric or hybrid-electric vehicles, the documented expenses for devices that allow the vehicle to engage in Smart Grid functions (but not the costs of electricity storage for the vehicle).

(9) The documented expenditures related to purchasing and implementing Smart Grid functions in such other cases as the Secretary shall identify. In making such

grants, the Secretary shall seek to reward innovation and early adaptation, even if success is not complete, rather than deployment of proven and commercially viable technologies.

(c) INVESTMENTS NOT INCLUDED.—Qualifying Smart Grid investments do not include any of the following:

(1) Investments or expenditures for Smart Grid technologies, devices, or equipment that are eligible for specific tax credits or deductions under the Internal Revenue Code, as amended.

(2) Expenditures for electricity generation, transmission, or distribution infrastructure or equipment not directly related to enabling Smart Grid functions.

(3) After the final date for State consideration of the Smart Grid Information Standard under section 1307 (paragraph (17) of section 111(d) of the Public Utility Regulatory Policies Act of 1978), an investment that is not in compliance with such standard.

(4) After the development and publication by the Institute of protocols and model standards for interoperability of smart grid devices and technologies, an investment that fails to incorporate any of such protocols or model standards.

(5) Expenditures for physical interconnection of generators or other devices to the grid except those that are directly related to enabling Smart Grid functions.

(6) Expenditures for ongoing salaries, benefits, or personnel costs not incurred in the initial installation, training, or start up of smart grid functions.

(7) Expenditures for travel, lodging, meals or other personal costs.

(8) Ongoing or routine operation, billing, customer relations, security, and maintenance expenditures.

(9) Such other expenditures that the Secretary determines not to be Qualifying Smart Grid Investments by reason of the lack of the ability to perform Smart Grid functions or lack of direct relationship to Smart Grid functions.

(d) SMART GRID FUNCTIONS.—The term "smart grid functions" means any of the following:

(1) The ability to develop, store, send and receive digital information concerning electricity use, costs, prices, time of use, nature of use, storage, or other information relevant to device, grid, or utility operations, to or from or by means of the electric utility system, through one or a combination of devices and technologies.

(2) The ability to develop, store, send and receive digital information concerning electricity use, costs, prices, time of use, nature of use, storage, or other information relevant to device, grid, or utility operations to or from a computer or other control device.

(3) The ability to measure or monitor electricity use as a function of time of day, power quality characteristics such as voltage level, current, cycles per second, or source or type of generation and to store, synthesize or report that information by digital means.

(4) The ability to sense and localize disruptions or changes in power flows on the grid and communicate such information instantaneously and automatically for purposes of enabling automatic protective responses to sustain reliability and security of grid operations.

(5) The ability to detect, prevent, communicate with regard to, respond to, or recover from system security threats, including cyber-security threats and terrorism, using digital information, media, and devices.

(6) The ability of any appliance or machine to respond to such signals, measurements, or communications automatically or in a manner programmed by its owner or operator without independent human intervention.

(7) The ability to use digital information to operate functionalities on the electric utility grid that were previously electro-mechanical or manual.

(8) The ability to use digital controls to manage and modify electricity demand, enable congestion management, assist in voltage control, provide operating reserves, and provide frequency regulation.

(9) Such other functions as the Secretary may identify as being necessary or useful to the operation of a Smart Grid.

(e) The Secretary shall—

(1) establish and publish in the Federal Register, within 1 year after the enactment of this Act procedures by which applicants who have made qualifying Smart Grid investments can seek and obtain reimbursement of one-fifth of their documented expenditures;

(2) establish procedures to ensure that there is no duplication or multiple reimbursement for the same investment or costs, that the reimbursement goes to the party making the actual expenditures for Qualifying Smart Grid Investments, and that the grants made have significant effect in encouraging and facilitating the development of a smart grid;

(3) maintain public records of reimbursements made, recipients, and qualifying Smart Grid investments which have received reimbursements;

(4) establish procedures to provide, in cases deemed by the Secretary to be warranted, advance payment of moneys up to the full amount of the projected eventual reimbursement, to creditworthy applicants whose ability to make Qualifying Smart Grid Investments may be hindered by lack of initial capital, in lieu of any later reimbursement for which that applicant qualifies, and subject to full return of the advance payment in the event that the Qualifying Smart Grid investment is not made; and

(5) have and exercise the discretion to deny grants for investments that do not qualify in the reasonable judgment of the Secretary.

(f) AUTHORIZATION OF APPROPRIATIONS.—There are authorized to be appropriated to the Secretary such sums as are necessary for the administration of this section and the grants to be made pursuant to this section for fiscal years 2008 through 2012.

Sec. 1307. State Consideration Of Smart Grid.

(a) Section 111(d) of the Public Utility Regulatory Policies Act of 1978 (16 U.S.C. 2621(d)) is amended by adding at the end the following:

"(16) CONSIDERATION OF SMART GRID INVESTMENTS.—

"(A) IN GENERAL.—Each State shall consider requiring that, prior to undertaking investments in nonadvanced grid technologies, an electric utility of the State demonstrate to the State that the electric utility considered an investment in a qualified smart grid system based on appropriate factors, including—

"(i) total costs;

"(ii) cost-effectiveness;

"(iii) improved reliability;

"(iv) security;

"(v) system performance; and

"(vi) societal benefit.

"(B) RATE RECOVERY.—Each State shall consider authorizing each electric utility of the State to recover from ratepayers any capital, operating expenditure, or other costs of the electric utility relating to the deployment of a qualified smart grid system, including a reasonable rate of return on the capital expenditures of the electric utility for the deployment of the qualified smart grid system.

"(C) OBSOLETE EQUIPMENT.—Each State shall consider authorizing any electric utility or other party of the State to deploy a qualified smart grid system to recover in a timely manner the remaining book-value costs of any equipment rendered obsolete by the deployment of the qualified smart grid system, based on the remaining depreciable life of the obsolete equipment.

"(17) SMART GRID INFORMATION.—

"(A) STANDARD.—All electricity purchasers shall be provided direct access, in written or electronic machine-readable form as appropriate, to information from their electricity provider as provided in subparagraph (B).

"(B) INFORMATION.—Information provided under this section, to the extent practicable, shall include:

"(i) PRICES.—Purchasers and other interested persons shall be provided with information on—

"(I) time-based electricity prices in the wholesale electricity market; and

"(II) time-based electricity retail prices or rates that are available to the purchasers.

"(ii) USAGE.—Purchasers shall be provided with the number of electricity units, expressed in kwh, purchased by them.

"(iii) INTERVALS AND PROJECTIONS.—Updates of information on prices and usage shall be offered on not less than a daily basis, shall include hourly price and use information, where available, and shall include a day-ahead projection of such price information to the extent available.

"(iv) SOURCES.—Purchasers and other interested persons shall be provided annually with written information on the sources of the power provided by the utility, to the extent it can be determined, by type of generation, including greenhouse gas emissions associated with each type of generation, for intervals during which such information is available on a cost- effective basis.

"(C) ACCESS.—Purchasers shall be able to access their own information at any time through the Internet and on other means of communication elected by that utility for Smart Grid applications. Other interested persons shall be able to access information not specific to any purchaser through the Internet. Information specific to any purchaser shall be provided solely to that purchaser.".

(b) COMPLIANCE.—

(1) TIME LIMITATIONS.—Section 112(b) of the Public Utility Regulatory Policies Act of 1978 (16 U.S.C. 2622(b)) is amended by adding the following at the end thereof:

"(6)(A) Not later than 1 year after the enactment of this Deadlines. paragraph, each State regulatory authority (with respect to each electric utility for which it has ratemaking authority) and each nonregulated utility shall commence the consideration referred to in section 111, or set a hearing date for consideration, with respect to the standards established by paragraphs (17) through (18) of section 111(d).

"(B) Not later than 2 years after the date of the enactment of this paragraph, each State regulatory authority (with respect to each electric utility for which it has ratemaking authority), and each nonregulated electric utility, shall complete the consideration, and shall make the determination, referred to in section 111 with respect to each standard established by paragraphs (17) through (18) of section 111(d).".

(2) FAILURE TO COMPLY.—Section 112(c) of the Public Utility Regulatory Policies Act of 1978 (16 U.S.C. 2622(c)) is amended by adding the following at the end: "In the case of the standards established by paragraphs (16) through (19) of section 111(d), the reference contained in this subsection to the date of enactment of this Act shall be deemed to be a reference to the date of enactment of such paragraphs.".

(3) PRIOR STATE ACTIONS.—Section 112(d) of the Public Utility Regulatory Policies Act of 1978 (16 U.S.C. 2622(d)) is amended by inserting "and paragraphs (17) through (18)" before "of section 111(d)".

Sec. 1308. Study Of The Effect Of Private Wire Laws On The Development Of Combined Heat And Power Facilities.

(a) STUDY.—

(1) IN GENERAL.—The Secretary, in consultation with the States and other appropriate entities, shall conduct a study of the laws (including regulations) affecting the siting of privately owned electric distribution wires on and across public rights-of-way.

(2) REQUIREMENTS.—The study under paragraph (1) shall include—

(A) an evaluation of—

(i) the purposes of the laws; and

(ii) the effect the laws have on the development of combined heat and power facilities;

(B) a determination of whether a change in the laws would have any operating, reliability, cost, or other impacts on electric utilities and the customers of the electric utilities; and

(C) an assessment of—

(i) whether privately owned electric distribution wires would result in duplicative facilities; and

(ii) whether duplicative facilities are necessary or desirable.

(b) REPORT.—Not later than 1 year after the date of enactment of this Act, the Secretary shall submit to Congress a report that describes the results of the study conducted under subsection (a).

Sec. 1309. Doe Study Of Security Attributes Of Smart Grid Systems.

(a) DOE STUDY.—The Secretary shall, within 18 months after the date of enactment of this Act, submit a report to Congress that provides a quantitative assessment and determination of the existing and potential impacts of the deployment of Smart Grid systems on improving the security of the Nation's electricity infrastructure and operating capability. The report shall include but not be limited to specific recommendations on each of the following:

(1) How smart grid systems can help in making the Nation's electricity system less vulnerable to disruptions due to intentional acts against the system.

(2) How smart grid systems can help in restoring the integrity of the Nation's electricity system subsequent to disruptions.

(3) How smart grid systems can facilitate nationwide, inter- operable emergency communications and control of the Nation's electricity system during times of localized, regional, or nationwide emergency.

(4) What risks must be taken into account that smart grid systems may, if not carefully created and managed, create vulnerability to security threats of any sort, and how such risks may be mitigated.

(b) CONSULTATION.—The Secretary shall consult with other Federal agencies in the development of the report under this section, including but not limited to the Secretary of Homeland Security, the Federal Energy Regulatory Commission, and the Electric Reliability Organization certified by the Commission under section 215(c) of the Federal Power Act (16 U.S.C. 824o) as added by section 1211 of the Energy Policy Act of 2005 (Public Law 109– 58; 119 Stat. 941).

TITLE XIV—POOL AND SPA SAFETY

Sec. 1401. Short Title.

This title may be cited as the "Virginia Graeme Baker Pool and Spa Safety Act".

Sec. 1402. Findings.

Congress finds the following:

(1) Of injury-related deaths, drowning is the second leading cause of death in children aged 1 to 14 in the United States.

(2) In 2004, 761 children aged 14 and under died as a result of unintentional drowning.

(3) Adult supervision at all aquatic venues is a critical safety factor in preventing children from drowning.

(4) Research studies show that the installation and proper use of barriers or fencing, as well as additional layers of protection, could substantially reduce the number of childhood residential swimming pool drownings and near drownings.

Sec. 1403. Definitions.

In this title:

(1) ASME/ANSI.—The term "ASME/ANSI" as applied to a safety standard means such a standard that is accredited by the American National Standards Institute and published by the American Society of Mechanical Engineers.

(2) BARRIER.—The term "barrier" includes a natural or constructed topographical feature that prevents unpermitted access by children to a swimming pool, and, with respect to a hot tub, a lockable cover.

(3) COMMISSION.—The term "Commission" means the Consumer Product Safety Commission.

(4) MAIN DRAIN.—The term "main drain" means a submerged suction outlet typically located at the bottom of a pool or spa to conduct water to a recirculating pump.

(5) SAFETY VACUUM RELEASE SYSTEM.—The term "safety vacuum release system" means a vacuum release system capable of providing vacuum release at a suction outlet caused by a high vacuum occurrence due to a suction outlet flow blockage.

(6) SWIMMING POOL; SPA.—The term "swimming pool" or "spa" means any outdoor or indoor structure intended for swimming or recreational bathing, including in-ground and above- ground structures, and includes hot tubs, spas, portable spas, and non-portable wading pools.

(7) UNBLOCKABLE DRAIN.—The term "unblockable drain" means a drain of any size and shape that a human body cannot sufficiently block to create a suction entrapment hazard.

Sec 1404. Federal Swimming Pool And Spa Drain Cover Standard.

(a) CONSUMER PRODUCT SAFETY RULE.—The requirements described in subsection (b) shall be treated as a consumer product safety rule issued by the Consumer Product Safety Commission under the Consumer Product Safety Act (15 U.S.C. 2051 et seq.).

(b) DRAIN COVER STANDARD.—Effective 1 year after the date Effective date. of enactment of this title, each swimming pool or spa drain cover manufactured, distributed, or entered into commerce in the United States shall conform to the entrapment protection standards of the ASME/ANSI A112.19.8 performance standard, or any successor standard regulating such swimming pool or drain cover.

(c) PUBLIC POOLS.—

(1) REQUIRED EQUIPMENT.—

(A) IN GENERAL.—Beginning 1 year after the date of Effective date. enactment of this title—

(i) each public pool and spa in the United States shall be equipped with anti-entrapment devices or systems that comply with the ASME/ANSI A112.19.8 performance standard, or any successor standard; and

(ii) each public pool and spa in the United States with a single main drain other than an unblockable drain shall be equipped, at a minimum, with 1 or more of the following devices or systems designed to prevent entrapment by pool or spa drains that meets the requirements of subparagraph (B):

(I) SAFETY VACUUM RELEASE SYSTEM.—A safety vacuum release system which ceases operation of the pump, reverses the circulation flow, or otherwise provides a vacuum release at a suction outlet when a blockage is detected, that has been tested by an independent third party and found to conform to ASME/ANSI standard A112.19.17 or ASTM standard F2387.

(II) SUCTION-LIMITING VENT SYSTEM.—A suction-limiting vent system with a tamper-resistant atmospheric opening.

(III) GRAVITY DRAINAGE SYSTEM.—A gravity drainage system that utilizes a collector tank.

(IV) AUTOMATIC PUMP SHUT-OFF SYSTEM.—An automatic pump shut-off system.

(V) DRAIN DISABLEMENT.—A device or system that disables the drain.

(VI) OTHER SYSTEMS.—Any other system determined by the Commission to be equally effective as, or better than, the systems described in subclauses (I) through (V) of this clause at preventing or eliminating the risk of injury or death associated with pool drainage systems.

(A) APPLICABLE STANDARDS.—Any device or system described in subparagraph (ii) shall meet the requirements of any ASME/ANSI or ASTM performance standard if there is such a standard for such a device or system, or any applicable consumer product safety standard.

(2) PUBLIC POOL AND SPA DEFINED.—In this subsection, the term "public pool and spa" means a swimming pool or spa that is—

(A) open to the public generally, whether for a fee or free of charge;

(B) open exclusively to—

(i) members of an organization and their guests;

(ii) residents of a multi-unit apartment building, apartment complex, residential real estate development, or other multi-family residential area (other than a municipality, township, or other local government jurisdiction); or

(iii) patrons of a hotel or other public accommodations facility; or

(C) operated by the Federal Government (or by a concessionaire on behalf of the Federal Government) for the benefit of members of the Armed Forces and their dependents or employees of any department or agency and their dependents.

(3) ENFORCEMENT.—Violation of paragraph (1) shall be considered to be a violation of section 19(a)(1) of the Consumer Product Safety Act (15 U.S.C. 2068(a)(1)) and may also be enforced under section 17 of that Act (15 U.S.C. 2066).

Sec. 1405. State Swimming Pool Safety Grant Program.

(a) IN GENERAL.—Subject to the availability of appropriations authorized by subsection (e), the Commission shall establish a grant program to provide assistance to eligible States.

(b) ELIGIBILITY.—To be eligible for a grant under the program, a State shall—

(1) demonstrate to the satisfaction of the Commission that it has a State statute, or that, after the date of enactment of this title, it has enacted a statute, or amended an existing statute, and provides for the enforcement of, a law that—

(A) except as provided in section 1406(a)(1)(A)(i), applies to all swimming pools in the State; and

(B) meets the minimum State law requirements of section 1406; and

(2) submit an application to the Commission at such time, in such form, and containing such additional information as the Commission may require.

(c) AMOUNT OF GRANT.—The Commission shall determine the amount of a grant awarded under this title, and shall consider—

(1) the population and relative enforcement needs of each qualifying State; and

(2) allocation of grant funds in a manner designed to provide the maximum benefit from the program in terms of protecting children from drowning or entrapment, and, in making that allocation, shall give priority to States that have not received a grant under this title in a preceding fiscal year.

(d) USE OF GRANT FUNDS.—A State receiving a grant under this section shall use—

(1) at least 50 percent of amounts made available to hire and train enforcement personnel for implementation and enforcement of standards under the State swimming pool and spa safety law; and

(2) the remainder—

(A) to educate pool construction and installation companies and pool service companies about the standards;

(B) to educate pool owners, pool operators, and other members of the public about the standards under the swimming pool and spa safety law and about the prevention of drowning or entrapment of children using swimming pools and spas; and

(C) to defray administrative costs associated with such training and education programs.

(d) AUTHORIZATION OF APPROPRIATIONS.—There are authorized to be appropriated to the Commission for each of fiscal years 2009 and 2010 $2,000,000 to carry out this section, such sums to remain available until expended. Any amounts appropriated pursuant to this subsection that remain unexpended and unobligated at the end of fiscal year 2010 shall be retained by the Commission and credited to the appropriations account that funds enforcement of the Consumer Product Safety Act.

Sec. 1406. Minimum State Law Requirements.

(a) IN GENERAL.—

(1) SAFETY STANDARDS.—A State meets the minimum State law requirements of this section if—

(A) the State requires by statute—

(i) the enclosure of all outdoor residential pools and spas by barriers to entry that will effectively prevent small children from gaining unsupervised and unfettered access to the pool or spa;

(ii) that all pools and spas be equipped with devices and systems designed to prevent entrapment by pool or spa drains;

(iii) that pools and spas built more than 1 year after the date of the enactment of such statute have—

(I) more than 1 drain;

(II) 1 or more unblockable drains; or

(III) no main drain;

(iv) every swimming pool and spa that has a main drain, other than an unblockable drain, be equipped with a drain cover that meets the consumer product safety standard established by section 1404; and

(v) that periodic notification is provided to owners of residential swimming pools or spas about compliance with the entrapment protection standards of the ASME/ANSI A112.19.8 performance standard, or any successor standard; and

(B) the State meets such additional State law requirements for pools and spas as the Commission may establish after public notice and a 30-day public comment period.

(2) NO LIABILITY INFERENCE ASSOCIATED WITH STATE NOTIFICATION REQUIREMENT.—The minimum State law notification requirement under paragraph (1)(A)(v) shall not be construed to imply any liability on the part of a State related to that requirement.

(3) USE OF MINIMUM STATE LAW REQUIREMENTS.—The Commission—

(A) shall use the minimum State law requirements under paragraph (1) solely for the purpose of determining the eligibility of a State for a grant under section 1405 of this Act; and

(B) may not enforce any requirement under paragraph (1) except for the purpose of determining the eligibility of a State for a grant under section 1405 of this Act.

(4) REQUIREMENTS TO REFLECT NATIONAL PERFORMANCE STANDARDS AND COMMISSION GUIDELINES.—In establishing minimum State law requirements under paragraph (1), the Commission shall—

(A) consider current or revised national performance standards on pool and spa barrier protection and entrapment prevention; and

(B) ensure that any such requirements are consistent with the guidelines contained in the Commission's publication 362, entitled "Safety Barrier Guidelines for Home Pools", the Commission's publication entitled "Guidelines for Entrapment Hazards: Making Pools and Spas Safer", and any other pool safety guidelines established by the Commission.

(b) STANDARDS.—Nothing in this section prevents the Commission from promulgating standards regulating pool and spa safety or from relying on an applicable national performance standard.

(c) BASIC ACCESS-RELATED SAFETY DEVICES AND EQUIPMENT REQUIREMENTS TO BE CONSIDERED.—In establishing minimum State law requirements for swimming pools and spas under subsection (a)(1), the Commission shall consider the following requirements:

(1) COVERS.—A safety pool cover.

(2) GATES.—A gate with direct access to the swimming pool or spa that is equipped with a self-closing, self-latching device.

(3) DOORS.—Any door with direct access to the swimming pool or spa that is equipped with an audible alert device or alarm which sounds when the door is opened.

(4) POOL ALARM.—A device designed to provide rapid detection of an entry into the water of a swimming pool or spa.

(d) ENTRAPMENT, ENTANGLEMENT, AND EVISCERATION PREVENTION STANDARDS TO BE REQUIRED.—

(1) IN GENERAL.—In establishing additional minimum State law requirements for swimming pools and spas under subsection (a)(1), the Commission shall require, at a minimum, 1 or more of the following (except for pools constructed without a single main drain):

(A) SAFETY VACUUM RELEASE SYSTEM.—A safety vacuum release system which ceases operation of the pump, reverses the circulation flow, or otherwise provides a vacuum release at a suction outlet when a blockage is detected, that has been tested by an independent third party and found to conform to ASME/ANSI standard A112.19.17 or ASTM standard F2387, or any successor standard.

(B) SUCTION-LIMITING VENT SYSTEM.—A suction-limiting vent system with a tamper-resistant atmospheric opening.

(C) GRAVITY DRAINAGE SYSTEM.—A gravity drainage system that utilizes a collector tank.

(D) AUTOMATIC PUMP SHUT-OFF SYSTEM.—An automatic pump shut-off system.

(E) DRAIN DISABLEMENT.—A device or system that disables the drain.

(F) OTHER SYSTEMS.—Any other system determined by the Commission to be equally effective as, or better than, the systems described in subparagraphs (A) through (E) of this paragraph at preventing or eliminating the risk of injury or death associated with pool drainage systems.

(2) APPLICABLE STANDARDS.—Any device or system described in subparagraphs (B) through (E) of paragraph (1) shall meet the requirements of any ASME/ANSI or ASTM performance standard if there is such a standard for such a device or system, or any applicable consumer product safety standard.

Sec. 1407. Education Program

(a) IN GENERAL.—The Commission shall establish and carry out an education program to inform the public of methods to prevent drowning and entrapment in swimming pools and spas. In carrying out the program, the Commission shall develop—

(1) educational materials designed for pool manufacturers, pool service companies, and pool supply retail outlets;

(2) educational materials designed for pool owners and operators; and

(3) a national media campaign to promote awareness of pool and spa safety.

(b) AUTHORIZATION OF APPROPRIATIONS.—There are authorized to be appropriated to the Commission for each of the fiscal years 2008 through 2012 $5,000,000 to carry out the education program authorized by subsection (a).

Sec. 1408. Cpsc Report.

Not later than 1 year after the last day of each fiscal year for which grants are made under section 1405, the Commission shall submit to Congress a report evaluating the implementation of the grant program authorized by that section.

TITLE XV — REVENUE PROVISIONS

Sec. 1500. Amendment Of 1986 Code.

Except as otherwise expressly provided, whenever in this title an amendment or repeal is expressed in terms of an amendment to, or repeal of, a section or other provision, the reference shall be considered to be made to a section or other provision of the Internal Revenue Code of 1986.

Sec. 1501. Extension Of Additional 0.2 Percent Futa Surtax.

(a) IN GENERAL.—Section 3301 (relating to rate of tax) is amended—

by striking "2007" in paragraph (1) and inserting "2008", and

by striking "2008" in paragraph (2) and inserting "2009".

(b) EFFECTIVE DATE.—The amendments made by this section shall apply to wages paid after December 31, 2007.

Sec. 1502. 7-Year Amortization of Geological and Geophysical Expenditures for Certain Major Integrated Oil Companies.

(a) IN GENERAL.—Subparagraph (A) of section 167(h)(5) (relating to special rule for major integrated oil companies) is amended by striking "5-year" and inserting "7-year".

(b) EFFECTIVE DATE.—The amendment made by this section shall apply to amounts paid or incurred after the date of the enactment of this Act.

TITLE XVI — EFFECTIVE DATE

Sec. 1601. Effective Date.

This Act and the amendments made by this Act take effect on the date that is 1 day after the date of enactment of this Act.

In: Global Energy Security
Editor: Roger T. Attenberg

ISBN: 978-1-60692-087-9
©2009 Nova Science Publishers, Inc.

Chapter 3

THE EUROPEAN UNION'S ENERGY SECURITY CHALLENGES*

Paul Belkin

ABSTRACT

Recent increases in energy prices and a steady escalation in global energy demand — expected to rise by nearly 60% over the next 20 years — have led U.S. policy-makers to engage in a wide ranging debate over how best to address the country's future energy requirements. Similarly, energy security has become a policy priority for the European Union (EU) and its 27 member states. The EU imports about 50% of its energy needs. Barring significant changes, the European Commission expects this figure to rise to 65% by 2030. About half of the EU's natural gas imports and 30% of its imported oil come from Russia. Europe's growing dependence on Russian energy, and long-term energy agreements between Russian firms and some European governments have fueled speculation that Moscow is using the "energy weapon" to try to influence European foreign and economic policy.

The EU has traditionally exerted little if any influence over individual member state energy policy. However, in March 2007, in the face of increasing concern about Europe's reliance on Russian energy, and growing public pressure to address global climate change, EU member states agreed to forge an "Energy Policy for Europe." They have agreed on a set of EU-wide targets — some legally binding — to increase the use of renewable energy, and reduce carbon emissions. However, member states continue to pursue divergent external energy policies, particularly toward Russia, and some European countries remain reluctant to cede national control over energy markets.

The United States and Europe have steadily broadened the transatlantic energy dialogue to include joint promotion of collective energy security, energy efficiency and alternative energy sources. At the April 2007 U.S.-EU summit, leaders on both sides of the Atlantic agreed to advance cooperation to develop alternative and renewable energy technologies. However, U.S. officials have expressed concern at some European member states' unwillingness to exert more pressure on Russia to comply with EU market principles. On the other hand, European leaders appear increasingly frustrated with U.S.

---

* Excerpted from Congressional Research Service Report for Congress updated January 30, 2008; Order Code RL33636.

reluctance to pursue binding multilateral regulatory frameworks to reduce carbon emissions and promote energy efficiency.

Members of Congress have expressed an interest in efforts to increase European energy security, particularly vis-a-vis Russia. In the first session of the 110th Congress, committees held hearings that touched on the issue of European energy dependence on Russia, and on European efforts to increase energy efficiency and renewable energy use. The second session of the 1 10th Congress may also hold hearings and pursue legislation on various aspects of EU energy policy.

This report examines some of Europe's critical energy security challenges and EU efforts to coordinate a common European energy strategy. It also includes an overview of broader transatlantic energy security cooperation and will be updated as needed. For additional information, see CRS Report RL34261, *Russian Energy Policy Toward Neighboring Countries*, by Steven Woehrel, and CRS Report RS22409, *NATO and Energy Security,* by Paul Gallis.

INTRODUCTION[1]

Although the European Union's (EU) 27 member states have ceded some national sovereignty (or competency) to EU institutions in a variety of areas, including economic and trade policy, energy policy remains primarily the responsibility of the member states. Decisions regarding long-term oil or gas purchases, the development and improvement of energy-related infrastructure, and the use of particular fuels continue to be made at the national level by individual member states.

At their March 2007 summit, EU member states moved to boost European coordination to help secure and diversify energy supplies, increase the development and use of renewable and alternative energy resources within the EU, and reduce energy demand and consumption. Although member state governments remain reluctant to cede national sovereignty over energy-security aspects of their foreign policies, they have set binding EU-wide targets for the use of renewable energies and biofuels, and have agreed to ambitious but non-binding energy efficiency and carbon emission reduction targets for the year 2020. In addition, member states are considering potentially significant reforms to further liberalize energy markets. Nonetheless, most observers expect member states to continue to retain significant national control over national energy markets and external relations with energy- producing countries. EU heads of State are set to discuss specific policy proposals in these areas at their spring 2008 summit.

Europe's renewed interest in energy security has been influenced by both internal and external factors. Internally, steadily rising energy prices, declining European energy production and a fragmented internal energy market have contributed to anxieties over Europe's ability to meet future energy demand. The strain on global demand exerted by the emerging economies of countries such as China and India, persistent instability in energy producing regions, the threat of terrorist strikes against energy infrastructure, and Russia's apparent willingness to use its energy power for political ends, are all raising concerns in Europe over how to address external influences that could affect future energy requirements.[2] Recent calls for EU-wide energy coordination have been driven by rising European concern

[1] For additional information see CRS Report RS22409, *NATO and Energy Security*, by Paul Gallis.

about the effects of energy production and consumption on global climate change. To this end, EU member states' recent energy policy decisions center largely on promoting energy efficiency, developing renewable energy and clean fuel sources, and reducing overall greenhouse gas emissions.

As uncertainties surrounding global energy supply and demand persist, issues pertaining to U.S., European, and global energy security are may gain importance in the second session of the 110th Congress. Members of both parties have introduced legislation aimed at increasing energy independence and energy security and reducing carbon emissions. Several of these proposals envision carbon trading schemes similar to those in use in Europe.[3] In several hearings during the first session of the 110th Congress, Members of Congress voiced concern about Europe's dependence on Russian energy resources, and the potential for Russian manipulation of European energy markets. To this end, Senator Richard Lugar has raised the possibility of a more proactive role for NATO in guaranteeing energy security for Alliance members.

The Context of Europe's Energy Security Debate

Background

Collectively, EU member states import half of their energy needs. Barring significant policy changes, this figure is expected to rise to 65% by 2030.[4] Today, oil, natural gas, and coal account for 80% of the energy consumed in the EU.

Europe's energy imports come primarily from Russia and the Middle East, where approximately 70% of global oil and gas supplies originate. Yet, the Middle East region is fraught with war, terrorism and politically unstable regimes. Iraq's oil production has not reached pre-war levels, and there is fear that terrorist groups could target pipelines and production facilities throughout the region. Iran has threatened to cut back oil production if forced to abandon its nuclear power program. With regard to Russia, recent political and economic behavior exhibited by Moscow has raised the dual specter of reliability and "energy politics."

High demand has also raised questions regarding the future availability of global oil and gas reserves. Although significant shortages are not projected for the next several decades, uncertainties over future exploration and production in areas such as Russia and the Middle East have also raised concerns about long-term supply availability. The International Energy Agency (IEA) estimates that close to $16 trillion in new investments may be needed over the next 30 years to meet future global energy demand.[5]

European concern regarding the security of its energy supply was first prompted by the Arab oil embargo of the early 1970s. Specifically, the embargo highlighted three main issues.

[2] Jos Van Gennip, Energy Security, NATO Parliamentary Assembly paper, 2006.

[3] For more information on the EU's carbon emission trading system see CRS Report RL33581, Climate Change: The European Union's Emissions Trading System (EU-ETS) by Larry Parker.

[4] See Energy Overview, Council of the European Commission, June 2006; An Energy Policy for Europe, Communication From the Commission to the European Council and the European Parliament. COM(2007) 1. January 10, 2007.

[5] See International Energy Agency (IEA), World Energy Investment Outlook, 2005.

First, it exposed a need for increased energy policy collaboration among European countries and between Europe and the energy producing world. Second, it became clear that institutional mechanisms for increased coordination in the event of future supply disruptions were essential. Third, consensus emerged that Europe should prepare strategies to prevent it from becoming the victim of future attempts by exporting nations to use energy as a political or economic weapon.[6] The 1974 creation of the International Energy Agency (IEA), which has become Europe's primary instrument for monitoring and analyzing world energy markets, was one response to the embargo. In addition, European countries sought to develop strategies to diversify energy supply.

After the embargo, European countries began to identify Russia and other Eurasian countries as potential energy suppliers. At the time, Soviet Russia was beginning to realize its energy producing potential but required major investments in its energy sector. The prospect of future cooperation in the energy field began to play a key role in European perspectives on developing relations with the Soviet Union.

In 1991, the European Union launched the Energy Charter Declaration, an initiative intended to promote energy cooperation and diversify Europe's energy supply.[7] The Declaration gave way to the 1994 Energy Charter Treaty that entered into legal force in 1998 and established a framework of rules and agreements to promote international energy cooperation. To date, 51 countries and the EU have signed or acceded to the Treaty.[8] The Treaty seeks to create a level playing field of rules regarding the promotion of foreign energy investments; free trade in energy materials, products and equipment; freedom of energy transit through pipelines and grids; promoting energy efficiency; and providing mechanisms for addressing disputes.[9]

Since the signing of the Energy Charter Treaty, the European Commission has used its existing competency in competition and environment and consumer protection policy to attempt to shape a European energy policy in a variety of ways.[10] These include promoting an internal gas and electricity market, encouraging the development of alternative energy supplies, and, in cooperation with the office of the High Representative for Common Foreign and Security Policy, pursuing a more cooperative approach to external relations with current and future energy suppliers.

Turning Point

A 2005 German-Russian gas pipeline agreement and more recent Russian manipulation of gas and oil flows to the European market have sparked a newfound sense of urgency among European leaders regarding the need for a more coordinated strategy. These events

---

[6] Daniel Yergin, "Ensuring Energy Security," Foreign Affairs, March/April 2006.

[7] For more detailed information on the Treaty see An Introduction to the Energy Charter Treaty. The Energy Charter Organization. At [http://www.encharter.org].

[8] Although the United States signed the 1991 Energy Charter Declaration, it has not signed the Energy Charter Treaty, so it retains the status of observer to the Charter process. U.S. officials have cited a preference to pursue energy-related matters on a bilateral basis.

[9] An Introduction to the Energy Charter Treaty, Op. Cit.

[10] The European Commission is the EU's executive and holds the sole right of legislative initiative. However, in many policy-making areas, the Commission remains primarily an administrative body serving the representatives of national governments, which make up the EU's main decision-making body, the Council of Ministers,. For more information see CRS Report RS21372, The European Union: Questions and Answers, by Kristin Archick.

correspond with growing concern among the European public and political classes regarding the link between energy production and consumption and global climate change.

In 2005, Germany and Russia agreed to build a gas pipeline connecting the countries under the Baltic Sea, the so-called North European Gas Pipeline (NEGP). While Germany maintains that the pipeline will significantly enhance German and therefore European energy supply and security, a number of EU member states, including Poland and Lithuania, have protested the decision. They counter that by running the pipeline under sea so it bypasses both countries, and that by failing to coordinate with EU neighbors when negotiating with Russia, Germany's actions pose a threat to their and broader European energy security.[11] Furthermore, prominent Swedish officials have voiced concerns that the pipeline will provide Russia with a platform to increase both military surveillance and its military presence in the strategically important Baltic Sea.[12] The German-Russian agreement and subsequent responses from Poland, Lithuania, and more recently, Sweden, have reignited calls for a more coordinated European energy strategy.

As internal strife over the German pipeline decision continues, disputes between Russia and Ukraine and Russia and Belarus have exposed some undesirable consequences of European dependence on Russian energy resources. In late December 2005, Russia's gas monopoly, Gazprom, temporarily suspended gas flows to Ukraine as part of a dispute over gas price increases. Within hours of the shut off, several European countries, including Austria, Italy, Poland, and Germany, reported drops in their own pipeline pressure by as much as 30%.[13] The gas crisis lasted only a few days, and after Russia and Ukraine reached an agreement on gas prices, gas was flowing again.

An almost identical dispute between Russia and Belarus with similar consequences for European countries, particularly Germany, occurred in early January 2007. This time, Russian oil pipeline operator Transneft shut down the Druzhba oil pipeline through which Germany receives 20% of its oil imports. Germany and the EU sharply rebuked Russia's decision, and Russia resumed oil delivery after three days of price negotiations with Belarus.

Many European observers have characterized the Russia-Ukraine and RussiaBelarus gas and oil crises as "wake up" calls exposing Europe's energy security vulnerability even to unintended supply disruptions. More importantly, however, the crises raised the dual questions of Russia's reliability as an energy partner and Moscow's willingness to use its energy power as a political weapon. In response, European leaders have advocated coordinating decisions on energy supply so as to present a unified front to producer nations like Russia, and have promoted energy efficiency and more efficient energy use. Despite these calls, some European countries and energy companies have continued to pursue long-term energy deals bilaterally. In the past year, companies from Italy, Austria, Bulgaria, and Serbia have signed contracts with Russian energy companies which some observers contend could pose a threat to collective European energy security.

---

[11] "Polish Press Slams Germany's Schroeder over Gas Pipeline Deal," Agence FrancePresse, December 12, 2005.

[12] "Sweden Afraid of Russian Spooks," Spiegel Online. November 15, 2006.

[13] See "Q&A: Ukraine Gas Row," BBC News, January 4, 2006; and CRS Report RS22378, Russia's Cutoff of Natural Gas to Ukraine: Context and Implications, by Jim Nichol, Steven Woehrel, and Bernard A. Gelb.

European Energy Consumption: By the Numbers

The EU's 27 member states account for approximately 17% of the world's total energy consumption.[14] In 2005, about 80% of the energy consumed within the EU was from fossil fuels. Figure 1 provides an overview of the EU's energy consumption by fuel source.

Europe imports about 50% of its total energy supply — slightly over 80% of its oil and close to 57% of its natural gas. Its dependence on imported energy sources, particularly natural gas, is expected to grow substantially in the coming decades. Commission estimates suggest that if current trends continue, Europe will import 65% of its total energy requirements by 2030.[15] Russia, Norway, the Middle East, and North Africa are the largest suppliers of EU energy. In 2005, Russia accounted for 45% of the EU's natural gas imports and about 29% of its oil imports.[16]

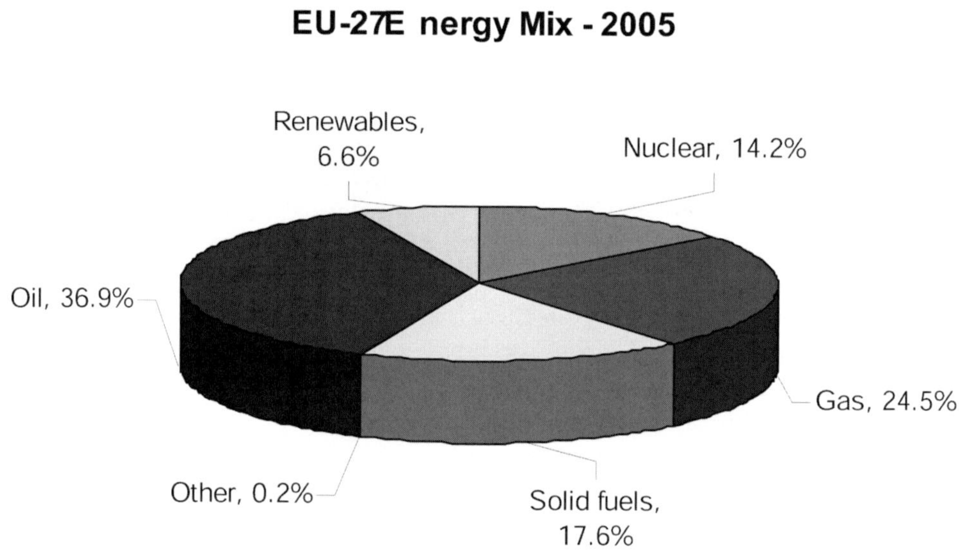

Source: European Commission, "Energy and Transport in Figures, *Statistical Pocket Book 2007.*"

Figure 1. EU Energy Mix

Forecasters predict that natural gas consumption in the EU will double over the next 25 years, and gas has rapidly become Europe's fuel of choice for power generation. European natural gas consumption currently represents 18% of world consumption. European gas imports are expected to reach slightly over 80% of total consumption by 2030.[17] In 2004, EU

---

[14] Country Analysis Briefs: European Union, Energy Information Administration, January 2006.

[15] "An Energy Policy for Europe," Op. Cit.

[16] "EU Energy Policy Data," European Commission Document SEC(2007) 12, January 10, 2007; The European Commission's Energy and Transport in Figures, Statistical Pocket Book 2007, available at [http://ec.europa.eu/dgs/energy_transport /figures/pocketbook/2006_en.htm].

[17] *Ibid.*

member states and Norway accounted for just over half of the EU's natural gas supply. The other half was imported primarily from Russia (29%) and Algeria (13%). Several EU member states are totally dependent on Russian natural gas for their domestic energy consumption. **Table 1** illustrates the levels of dependency on Russian natural gas in selected nations of the EU.

Table 1. Imported Gas and Gas from Russia

Country	Dependence on Imported Gas, 2005	Total Gas Consumed, Imported from Russia
Austria	88%	74%
Czech Republic	98%	70%
Estonia	100%	100%
France	98%	26%
Finland	100%	100%
Germany	81%	39%
Italy	85%	30%
Poland	70%	50%

Source: International Energy Agency; Eurostat; British Petroleum.

AN ENERGY POLICY FOR EUROPE?

Concern over Europe's dependence on Russian energy resources has corresponded with growing public calls for EU-wide action on climate change. European leaders have responded by renewing efforts to establish a more cohesive European energy policy. During their March 2007 summit, EU heads of state adopted a series of European Commission proposals that they expect will form the foundation of an "Energy Policy for Europe." The adopted measures are among a larger group of recommendations the Commission laid out in a March 2006 "Green Paper" and more detailed action plans unveiled in January 2007 and 2008.[18] The Commission proposals focus on three broad interconnected goals: increasing European-wide energy security; enhancing sustainability; and fostering competition in Europe's internal energy market. Commission officials place particular emphasis on the links between energy security, energy efficiency, and an EU-wide reduction in carbon emissions.

In what some consider a reflection both of increasing public pressure to address global climate change and continued member state reluctance to cede national economic and foreign policy making authority, the EU's proposed energy policy focuses largely on enhancing sustainability. Member states have committed in principle to take some steps toward further liberalizing the EU-wide energy market and have broadly endorsed increased foreign policy coordination on securing energy supplies. However, the EU's most far-reaching commitments focus on increasing energy efficiency, decreasing greenhouse gas emissions, and promoting the use of renewable energy and alternative fuels and associated technologies.

18 Both the Commission's Green Paper and more detailed action plan are available at [http://ec.europa.eu/energy/energy_policy/index_en.htm].

Specifically, EU member states have committed to reducing total EU-wide carbon emissions by 20% compared with 1990 levels by 2020. They have also pledged to seek international agreement on a 30% reduction target by 2020 in a post- Kyoto Protocol international carbon emissions reduction treaty.[19] In addition, the EU seeks a 20% increase in Europe-wide energy efficiency by 2020 and has mandated that 20% of all EU energy consumption come from renewable sources and 10% of transport fuel from biofuels by 2020. The Commission hopes that EU heads of state will agree on proposed country-specific targets to achieve their goals during the spring of 2008.

European Commission president Jose Manuel Barroso estimates that achieving these targets could cost or up to $87.7 billion, or 0.5% of EU member states' combined annual GDP. However, he has argued that this approximately $4.50 (3 euros) per week per European citizen represents far less than the cost of inaction.[20]

According to a March 2007 study by the global consulting firm McKinsey and Company, EU member states will need to invest approximately $1.5 trillion (1.1 trillion Euros) in new technologies over the next 14 years in order to achieve their targets. In preparation for the new push toward alternative and renewable energy sources, the European Investment Bank has announced plans to earmark slightly over $1 billion (800 million Euros) for loans for renewable energy projects from 2007- 2010.[21] In addition, the Commission envisions a 50% increase in EU-spending on carbon technology over the next seven years. Although the EU approach appears to be focused largely on developing new and alternative technologies, the McKinsey study suggests it may be more cost efficient for the EU to focus more of its efforts on reducing energy use than on developing and promoting alternative and renewable energy sources.

The Russia-Ukraine and Russia-Belarus oil and gas crises have also ignited calls from the European Commission and several EU member states to increase foreign policy coordination to secure and diversify energy supply. To this end, EU member states have agreed to begin cooperating to form an external policy centered on expanding political partnerships with, and increasing pipeline and energy infrastructure investment in, producer and transit countries. European states have specifically singled out Central Asia and the Caspian and Black Sea areas as focal points for such activities. In addition, the EU has called for the formation of a European energy dialogue with African countries of strategic importance. The EU's flagship project in the Central Asia and the Caspian region is the Nabucco gas pipeline. The proposed pipeline, which is backed by a promise of $7 billion (5 billion euros) in European Commission funding, would bring Central Asian and Caspian gas to Europe, without passing through Russia. However, as discussed below, the potential construction start-date has been pushed back several years, and some observers question the project's viability.

With regard to Russia, some member states hope to institutionalize a common commitment to market principles as outlined in the Energy Charter Treaty through an EU-Russia Energy Dialogue and in a new EU-Russia Partnership and Cooperation Agreement. However, bilateral energy agreements between some member states, notably Germany, Italy,

---

[19] The United Nations Kyoto Protocol, to which the United States is not a party, is set to expire in 2012. European leaders reportedly see 2009 as the deadline for international agreement on a post-Kyoto treaty.

[20] See "20 20 by 2020: Europe's Climate Change Opportunity," speech08/34 by Jose Manuel Barroso, January 23, 2008. Available at [http://ec.europa.eu/commission_barroso/president/index_en.htm].

[21] "EU's 2020 Energy Goals to Cost over 1 Trillion," EU Observer, March 28, 2007.

and Bulgaria, and Russian firms, illuminate continued disagreement within the EU on how best to deal with Moscow. In the face of such discord, the European Commission is seeking to strengthen multilateral mechanisms, including the Energy Charter, to better coordinate global energy policy among consumer, transit, and producer nations.

Many observers consider the European Commission's call to increase competition within and among traditionally protected European energy markets to be its most controversial. The Commission has advanced proposals to reduce the power of state-owned energy companies by forcing them to split up ownership of generation and distribution businesses. In the face of opposition from countries such as Germany, France, and Spain, the Commission could seek backing for a compromise proposal that would allow national energy industries to continue to operate both generation and distribution facilities, but would subject them to oversight from an independent European regulatory body. European leaders are expected to agree on a solution during the first half of 2008.

Although European leaders increasingly pay lip-service to the need for enhanced energy cooperation, the success of an EU energy strategy will likely depend on the ability of member states to overcome differences in addressing three fundamental challenges. First, how to develop strong partnerships with energy producing and transit regions; second, how to use and further develop indigenous and alternative energy sources while seeking to curb overall consumption; and third, how to establish an internal system to provide dependable and secure energy supplies to all of Europe.

Challenge 1: An External Policy for Energy Security

Growing energy demand within the EU's 27 member states is mirrored in regions throughout the world. Growth in China and India has added considerably to global demand, as has rising population growth and economic modernization in Latin America, Africa, and even the energy-rich Middle East. In the face of this strain on limited supplies, Europeans must compete for existing and new energy sources. Projections for European energy consumption indicate that one of the most important energy security challenges facing the EU over the next 20 years will be Europe's ability to diversify the sources and modes of transit of its energy imports.

The bulk of the world's energy resources, located in Russia, the Caspian Sea region, the Middle East and North Africa, are all well within geographic reach of the European Union. In fact, Europe already receives energy supplies from each of these regions. However, Europe's growing dependence on Russia and Russia's apparent willingness to use its energy resources for political purposes have spurred calls from some member states and the United States for a more cohesive EU-wide strategy to further diversify supply. The key for Europe may be to determine the equilibrium point for supply from each region and how to best manage relations with the governments in those regions. By strengthening political relations with these governments, the EU opens additional options for its external energy strategy. According to some, the EU strategy in this regard differs from the stated aim of many U.S. politicians and Administration officials in that Europeans acknowledge they can never gain complete energy independence and therefore seek to better manage their energy dependence rather than achieve outright energy independence.[22]

EU member states have collectively endorsed the Commission's calls to develop a collective international energy policy. Nonetheless, while acknowledging that the EU may at times be in a better position to determine what leverage could be used to advance the collective interests of the Union as a whole, member states have been careful not to sacrifice their individual rights to independently pursue external relations to secure energy supplies. Europe's energy relations with Russia best elucidate the tension between calls for a collective external energy policy and support for individual member state policies. Some of the EU's newer member states in Central and Eastern Europe appear skeptical of Russia's reliability as an energy partner, and therefore call on EU member states work collectively to prevent Russia from exploiting long-term dependencies for political purposes. At the same time, other member states continue to pursue long-term bilateral supply contracts with Russia's state-run energy companies, increasing both their energy and, according to some, their political dependence on Russia.

Russia.[23] Russia is a major player in world energy markets. In 2004, its 1,700 trillion cubic feet (tcf) of natural gas reserves were the largest of any country, making it both the world's largest gas producer and exporter. Russia is also the world's second largest oil exporter. According to the European Commission, EU member states imported 29% of their natural gas and 26% of their oil from Russia in 2004.[24] With gas consumption expected to rise more dramatically than oil consumption in the future, some experts predict that Europe could rely on Russia for more than 40% of its natural gas by 2020.[25] While Russia's resources and proximity to Europe make Euro-Russian collaboration a necessity, Russia's apparent willingness to use its energy wealth to achieve controversial foreign policy objectives has fueled debate within Europe on how best to manage energy relations with Russia.

Most observers contend that Russian president Vladimir Putin views his country's vast energy resources as a tool to regain Russia's stature as a major force in global affairs.[26] Thus, Putin sees energy as an important political force as much as it is the force driving Russia's economic development. Some experts believe that Russia seeks to control as much of Europe's energy infrastructure as possible in return for its delivery of reliable energy supplies. From this perspective, Moscow knows that if the EU is successful in creating a Europe-wide single market for electricity and gas, "[Russia] will be presented with opportunities to become part of the world's largest and most integrated energy market right on its border."[27] According to analyst Daniel Yergin, "Putin believes that energy security is about [Russia's] retaking

---

22 Interviews of German and EU officials, January-March 2007.

23 For additional information on Russian energy policy see CRS Report RL34261, Russian Energy Policy Toward Neighboring Countries, by Steven Woehrel.

24 European Commission Document SEC(2007)12 Op. Cit.

25 Keith Smith, Russian Energy Pressure Fails to Unite Europe, Center for Strategic and International Studies (CSIS), Washington, DC. January 24, 2007.

26 Keith Smith, CSIS, "Russian Energy Policy and Its Challenges to Western Policy Affairs," testimony before the Congress, May 16, 2006.

27 Andrew Monagham, "Russia-EU Relations: An Emerging Energy Security Dilemma," Russia Research Network, London, 2006.

control of the 'commanding heights' of the energy industry and extending that control downstream...."[28]

Energy's political importance is evident in the fact that the two major Russian energy giants, Gazprom and Rosneft, have close ties to the Kremlin and, in particular, to President Putin himself. Rosneft is led by a close associate and former KGB colleague of Putin. Gazprom is run by Alexey Miller, a close Putin ally, and Dimitry Medvedev, Russia's First Deputy Prime Minister and Putin's choice to be his successor as president. Gazprom dominates the Russian gas sector and controls 100% of Russian gas flowing to the EU. Medvedev is expected to win March 2008 presidential elections.

On the investment side, analysts also see Russia playing the political card. The International Energy Agency estimates that the Russian gas sector will require upwards of $10 billion in annual investment to meet future global demands. The EU has urged Russia to provide European energy companies the opportunity to invest in the total range of the energy sector from oil and gas fields to the pipeline system. Thus far, Russia has refused to meet EU demands and in turn, has warned the EU not to attempt to block Gazprom's plans to buy or invest in Europe's energy sector. Brushing aside the EU's policies regarding competition and monopoly practices, as well as the Energy Charter, Gazprom CEO Miller told EU Ambassadors in a not so veiled attempt to exert Russia's energy-driven influence that "attempts to limit Gazprom's activities in the European market ... will not produce good results ... it is no coincidence that competition for energy resources is growing ... and it should not be forgotten that we [Gazprom] are actively seeking new markets such as China...."[29]

Rather than rely on significant outside investment in its energy infrastructure, observers believe Russia intends to satisfy its long-term gas contracts with European nations through its near monopoly on gas from Central Asia (Kazakhstan, Turkmenistan, and Uzbekistan). Russia currently controls the overwhelming majority of oil and gas transportation routes from Central Asia and, according to analysts, intends to exploit this control and its political leverage over central Asian governments and European countries to impede European and U.S. efforts to develop alternative pipelines that bypass Russia. For example, critics of Gazprom activities, such as analyst Vladimir Socor, believe that Gazprom's strategy is to "establish permanent control of the [Hungary/Balkans] markets before Caspian gas can reach them through the proposed Nabucco pipeline...."[30] Specifically, Socor and others point to Gazprom's proposed South Stream gas pipeline, intended to transport Caspian and Central Asian gas to Europe via an almost identical route as the EU's planned Nabucco pipeline. Deals struck between Gazprom and Italian, Bulgarian, and Serbian energy companies in 2007 and January 2008 appear to confirm the view that Gazprom may seek to convince nations that agreed to fund the Nabucco pipeline to withdraw their commitments and rely on the South Stream pipeline instead.[31] These critics also warn that Russia's state-owned energy companies aim to increase their influence in the face of further European diversification by seeking to acquire controlling stakes in natural gas concerns in North Africa.[32]

---

[28] Daniel Yergin, "What Does Energy Security Really Mean?," Wall Street Journal, July 11, 2006.

[29] Gazprom CEO Miller in a speech to EU Ambassadors in Brussels as reported by BBC News, April 4, 2006.

[30] "Gazprom Broadens, Deepens Inroads Into European Union's Internal Markets, Transport Systems," Vladimir Socor, Eurasia Dailey Monitor, March 21, 2006.

[31] See "Russia: Gazprom's Advance in Europe Continues," Radio Free Europe / Radio Liberty, January 25, 2008.

[32] Keith Smith, Russian Energy Pressure Fails to Unite Europe, Op cit.

The European Response. EU member states have established two primary institutional mechanisms with which to collectively address energy relations with Russia: the Energy Charter Treaty and the EU-Russia Energy Dialogue. As outlined above, the Energy Charter Treaty, which Russia has signed but not ratified, would oblige Russia to adopt a legal framework governing investment, transit, and trade in energy resources. Some analysts contend that European leaders should make Russian implementation of the Treaty a requirement of any future EU-Russia Partnership and Cooperation Agreement. Such calls have contributed to delays in current negotiations of such an agreement and have hindered progress on the EU-Russia Energy Dialogue.

Even as EU leadership in Brussels moves forward with its ideas on a common external energy strategy, many question how far individual member states will agree to push Russia (and Gazprom) to adopt the EU's principles of competition, open its energy sector to outside investment, and ratify the Energy Charter. Some believe that without such Russian concessions, Europe will ultimately find its energy security largely under Russian control. Indeed, several member states have pursued bilateral energy deals with Russia that will increase their dependence on Russia for years to come. Both Germany and Italy, the largest importers of Russian gas, have negotiated long-term deals with Russia to lock in future gas supplies. For Germany and a few others, "Russia's role as a key supplier of oil and gas makes Putin a vital strategic partner who cannot be ignored or antagonized."[33] Such deals are not limited to the major energy consumers. Slovenia and Belgium have entered into negotiations with Gazprom to build a pipeline across the former and to enter the gas distribution market in the latter. Hungary's oil and gas company, Mol, has joined with Gazprom to extend Gazprom's Blue Stream pipeline across the Black Sea through the Balkans into Hungary. In January 2008, Bulgaria signed a deal with Gazprom to join the proposed South Stream project.

These examples of individual member states dealing with Russia bilaterally have drawn harsh criticism from other member states, such as Poland and the Baltic states. They have warned their European colleagues not to make energy deals that will give Russia an undue and possibly dangerous amount of political influence over European decision-making. Many of these nations understand that Europe's dependence on Russian energy is likely to last no matter what alternatives are included in an EU energy policy. But they also feel Europe does not gain real security by becoming more dependent on Russia. In fact, the growing presence of Gazprom throughout the European energy market has led many to worry about the EU's ability to develop an energy policy insulated from Gazprom's influence.[34] In a July 2006 speech, Romania's President Basescu went so far as to warn that "Europe's dependence on Russian gas monopoly Gazprom ... could be the biggest threat to the region since the former Soviet Union's army."[35]

Initial efforts to institutionalize an EU-Russia energy dialogue came to fruition in 2000 with the creation of the EU-Russia Energy Dialogue. However, results of the Dialogue have reportedly been mixed at best. Despite continued Russian reluctance to allow European investment in its energy sector and adopt energy market and transportation principles laid out in the Energy Charter, several European nations continue to enable increased Russian

---

[33] William Drozdiak, "Russia: More Awkward, But Still Indispensable," European Affairs, Spring/Summer 2006.

[34] Comments provided through discussions with representatives of several European member states.

[35] Traian Basescu, President of Romania in a speech to the Jamestown Foundation in Washington, D.C. July 2006.

influence in their energy sectors. On the other hand, 2006 negotiations on a renewed Russia-EU partnership agreement were derailed due to Polish and other member state concerns regarding Russian intransigence in several areas, including energy cooperation. In the face of internal discord, the EU appears to have agreed to focus on achieving its energy goals with regard to enhancing sustainability, while Russia has shown few signs of deviating from its recent policies.

The energy situation with Russia is not yet dire. Russia will continue to be Europe's primary energy supplier for the long-term, and healthy Russian-European relations remain a priority on both sides. If a common external EU energy security policy is to emerge, two options may be considered. First, Europe may move to curb its dependence on Russian energy by increasing its diversification to other regions without threatening Russia's own market security in Europe. In doing so, Europe might ask if there is a point at which Russia could decide that the EU's commitment to diversification no longer makes it financially attractive for Russia to continue to invest in new supplies destined for the European market. Second, the EU may attempt to regulate the behavior and practices of Gazprom as it becomes more of a dominant energy player in Europe. Thus far, few European countries have demonstrated restraint in seeking bilateral deals with the Russian monopoly that would do just that. If this continues, Europe could risk having Gazprom interfere more and more in its internal political decision-making. To avoid this, the European Union will likely continue to apply pressure on Gazprom to play by Europe's rules on competition and work to change Gazprom's corporate mentality by allowing European firms to invest in Russia's gas industry.

According to some analysts, however, internal discord on how to approach Russia is preventing the EU from applying this kind of pressure. These critics argue that the EU could collectively pressure Russia by enforcing existing EU competition laws and even using Russia's prospective World Trade Organization (WTO) membership as leverage to open Russia's domestic energy sector to outside investment.[36] On the other hand, countries such as Germany appear reluctant to take any concerted action that may antagonize Russia, citing the need for a healthy strategic partnership, given Euro-Russian mutual dependencies and the importance of Russian cooperation in addressing other issues of global concern, such as the Iranian nuclear program and the future status of Kosovo.[37]

Central Asia and the Caspian / Black Sea Regions.[38] One of the focal points of European energy diversification strategies is Central Asia and the Caspian and Black Sea regions. Indeed, the EU's January 2007 energy policy paper recommends strengthening the EU's so-called Neighborhood Policy with these areas, and European leaders have sought to bolster ties with countries in these regions.[39]

The Caspian Sea in Central Asia is bordered by Azerbaijan, Iran, Kazakhstan, Russia and Turkmenistan. After the collapse of the Soviet Union, the international community took an

---

[36] Keith Smith, Russian Energy Pressure Fails to Unite Europe, Op cit.

[37] Interviews of German officials, January-March 2007.

[38] For more information see CRS Report RS2 1190, Caspian Oil and Gas: Production and Prospects, by Bernard Gelb, and CRS Report RL30294, Central Asia: Regional Developments and Implications for U.S. Interests, by Jim Nichol.

[39] For more information on the EU's political and energy-related agreements with states in the Caspian region and in Central Asia see, Richard Youngs, "Europe's External Energy Policy: Between Geopolitics and the Market," Center for European Policy Studies, November 2007.

active interest in the region because of the potential oil and gas reserves thought to be located in at least six identified hydrocarbon fields beneath the Caspian Sea.

Presently, the Caspian Sea region is a significant, but not major supplier of crude oil to world markets The untapped reserves held by four of these nations might offer Europe an opportunity to move away from increased dependence on Russian energy. Estimates of the Caspian Sea region's proven oil reserves range between 40 and 50 billion barrels. Production levels in 2005 were estimated to be around 2 million barrels per day. The Caspian Sea region's natural gas reserves are estimated at 232 trillion cubic feet (Tcf). Natural gas production in 2004 was approximately 5 Tcf.

Europe's formal interest in the energy resources of the region dates back to 1995 with the creation of the Interstate Oil and Gas Transport to Europe program (Inogate), This EU initiative (currently with 21 member countries) was designed to promote the construction of regional pipeline systems in order to facilitate the transport of oil and gas to Europe.[40] This was followed by another EU proposal, the "Baku Initiative," which was launched in November 2004 with the participation of the European Commission and the Black Sea and Caspian littoral states. The Baku Initiative was designed to facilitate the progressive integration of the energy markets of the region into the EU market as well as the transportation of the extensive Caspian oil and gas resources toward Europe.

At the time Inogate was formed, Russia dominated both oil and gas production and distribution in the region. Since most of the countries involved are landlocked, their oil and gas had to be transported via pipelines. Reflecting Soviet era dictates and infrastructure, nearly all Caspian crude oil traveled north or west via pipeline to or through Russia to European markets. Some oil also went by tanker through the Bosporus straits to Western European markets via the Mediterranean. Natural gas transportation was tied to pipelines traveling mainly north or west through Russia and its monopoly pipeline system — Gazprom. This has provided Russia with the market power to dictate, in part, the price it is willing to pay for the oil or gas, to set transit fees on Caspian energy shipped through its transportation network, and to determine in some cases how much, if any, it is willing to transport. This latter point was evident in 2005 when Russia's oil pipeline company, Transneft, refused to allow oil from Kazakhstan to be shipped through its pipeline system to Lithuania for refining. The Caspian region nations thus have incentives to develop alternatives to routes through Russia to reach European and other markets and provide leverage in negotiating transit fees on shipments that do go through the Russian pipeline system.

Changing the region's energy flow from the existing North-South axis to an East-West axis toward Europe could be integral to Europe's energy strategy. Currently, three big pipeline projects serve to reduce the region's dependence on Russia:

- The Caspian Pipeline Consortium (CPC) project connects Kazakhstan's Caspian Sea area oil deposits with Russia's Black Sea port of Novorossiysk. Oil loaded at Novorossiysk is then taken by tanker to world markets via the congested Bosporus Straits.[41]

---

[40] A more detailed account of the activities of Inogate is available on its website at [http://www.inogate.org].

[41] Additional analysis of the Caspian region can be found at, "Country Analysis Brief: Caspian Sea," Energy Information Administration, U.S. Department of Energy, 2005.

- The Baku-Tbilisi-Ceyhan oil pipeline (BTC), which opened in July 2006, exports oil from Azerbaijan and up to 600,000 bl/d from Kazakhstan along a 1,040-mile route from Baku, Azerbaijan via Georgia to the Turkish Mediterranean port of Ceyhan. This will allow oil to bypass the Bosporus Straits.
- The South Caucasus Gas Pipeline (SCGP), a new gas pipeline venture completed in December 2006, runs parallel to the BTC oil pipeline for most of its route before connecting to the Turkish energy infrastructure and on to Europe via a transit pipeline through Greece.

In addition to these pipelines already in service, several additional projects in Europe could be involved. One option for additional oil transport would be to upgrade the existing oil pipeline which runs from Baku in Azerbaijan to Supsa in Georgia. That line could be extended under the Black Sea or the oil could be loaded onto tankers and shipped to Odessa, Ukraine. The oil could then be pumped through the Odessa-Brody pipeline into Poland. Some, including the Poles, have suggested that the Brody line be extended to northern Poland and possibly into the Baltic states for use at the Mazeikai refinery in Lithuania.

On the gas front, two additional projects offer important options for Europe. As mentioned above, the EU's flagship project is the Nabucco gas pipeline. The other, the Trans-Caspian pipeline, is intended to bring additional gas from the Caspian to Georgia and across the Black Sea to Romania and the Balkans. EU officials had hoped that construction of the Nabucco pipeline — intended to transport up to 31 billion Cubic meters of Central Asian gas through Turkey into Bulgaria and on to Austria — would begin in 2007. However, the project has been repeatedly delayed, and recent advances by Gazprom on signing contracts for its South Stream pipeline have caused some to doubt Nabucco's viability. EU officials remain optimistic, aiming for construction of the pipeline to begin in 2009. In what some ciritics see as an additional attempt by Gazprom to undermine the Nabucco project, Gazprom is reported to have acquired a 50% stake in the planned terminus of the Nabucco pipeline, the Baumgarten transmission center in Austria, in January 2008.[42]

There can be no doubt that the energy resources of the Caspian Sea region can offer Europe a viable alternative source of energy supply. However, the full realization of the energy potential of the region could be impeded by several additional factors.

One issue that continues to raise questions regarding regional stability is the unresolved legal status of the Caspian Sea. Despite a number of efforts, so far only Azerbaijan, Kazakhstan, and Russia among the littoral states have reached agreement on delineating ownership of the Sea's resources or their rights of development. The EU could offer its legal assistance to help resolve outstanding issues.

A second issue is the ability of the EU to work to ensure the long-term political stability of the region. The conflict between Azerbaijan and Armenia over NagornoKarabaugh leaves the BTC and the future SCP pipelines vulnerable to sabotage. Internal political strife involving Georgia and its two breakaway regions also threatens future pipelines through that country. Continued political uncertainty in Ukraine and growing Iranian influence in the southern Caucasus could deter future long-term investment by the private sector. However,

---

42 See Vladimir Socor, "OMV Joins with Gazprom to Undercut Nabucco," Eurasia Daily Monitor, Jamestown Foundation, January 29, 2008.

the January 2007 entry of Romania and Bulgaria into the European Union and the EU's special relationship with Turkey should help keep the Black Sea region settled.

A third issue involves the willingness of the EU to compete with Russia for political and economic influence in the region and to prevent Gazprom from closing off the Caspian market, or at least the Central Asian part of the region, to Europe and its private sector. Russia's higher priced gas exports to Europe depend on Gazprom's ability to control gas from Kazakhstan and Turkmenistan. This dependency is expected to increase over the next 7-10 years until Russia's huge gas fields in the Barents Sea come on line.[43] According to some, although Gazprom was unable to prevent the BTC pipeline from being completed, Gazprom intends to continue to press the countries around the Black and Caspian Sea regions to agree to gas supply and transit arrangements that satisfy the company's goals of channeling lower-cost Central Asian gas to Russian customers and protecting its lucrative European market.[44] Gazprom has already locked up much of Turkmenistan's gas in a 25-year contract and has pursued a similar strategy toward Kazakhstan.

The final issue revolves around whether Europe is the optimal market for Caspian oil and natural gas. Oil demand over the next 10 to 15 years in Europe is expected to grow by little more than 1 million bl/d. Oil exports eastward, on the other hand, could serve Asian markets, where demand for oil is expected to grow by roughly 10 million bl/d over the next 15 years.[45] In fact, China, which opened an oil pipeline to Kazakhstan in 2005, sees Kazakhstan as a major source of oil for the long term.

The Caspian region will continue to be an important source of energy production for the foreseeable future, especially if the estimates of its reserves, particularly its gas reserves, are accurate. Thus, the region can contribute to the diversification of oil and gas supplies to Europe, which will add to Europe's energy security. Taking full advantage of this potential will require a strong commitment on the part of the EU to encourage the private sector to take the financial risks associated with securing a share of the Caspian energy market for Europe, and to set forth an external strategy that is fully prepared to address the dynamics of the entire region. For some, "a credible energy [strategy] needs to demonstrate that the EU means business in the Caspian/Black Sea regions. Brussels must include energy supply and transit as high priorities ... for the region.[46]

Middle East/North Africa. EU efforts to diversify European energy supplies and decrease dependence on Russia have heightened calls within Europe for stronger political and economic engagement in the Middle East and North Africa. However, political instability in the region and strong competition for its energy resources from countries in Asia and North America present challenges to European efforts.

The Persian Gulf countries (Bahrain, Iran, Iraq, Kuwait, Qatar, Saudi Arabia, and the United Arab Emirates) alone hold over half (57%) of the world's oil reserves, and the Middle East Region produces about 31% of the world's oil.[47] In addition, Libya is estimated to hold

---

43 Keith Smith, "Russian Energy Policy and Its Challenges to Western Policy Affairs," Op. Cit.

44 Zeyno Baran,"Energy Supplies to Eurasia and Implications for U.S. Energy Security," testimony before the Senate Foreign Relations Committee, September 2005.

45 CRS Report RS21 190, Op. Cit.

46 "What Role for the Black Sea Region in the European Union's Energy Strategy," Eurasia Daily Monitor, March 3, 2006.

47 For additional information see "Persian Gulf Oil and Gas Exports Fact Sheet," Energy Information Administration, U.S. Department of Energy.

40bb and Algeria 12bb. The Persian Gulf region also holds an estimated 2,400 trillion cubic feet (tcf) of natural gas reserves, representing 45% of the world's total gas. Algeria is estimated to hold (161tcf), and Libya (52tcf).[48]

Europe already depends on the Middle East/North Africa region for close to 30% of its oil imports and approximately 15% of its piped gas. In 2005, Europe imported approximately 3.1 million b/day of oil from the region. The largest portion of that oil comes from Saudi Arabia, followed by Libya and Iran.[49] Europe's primary supplier of natural gas has been Algeria, via two pipelines that enter Europe through Italy and Spain. A smaller amount comes from Libya via pipelines to Italy. Two additional gas pipelines from Algeria to Spain and Italy are under construction.

Perhaps the most important development for Europe in this region has been the growing availability of liquified natural gas (LNG). Today, Europe accounts for approximately 8% of the world's total consumption of LNG, and in 2005, LNG represented 15% of European gas imports — a 21% increase from 2004.[50] Spain, where 65% of gas imports are LNG, leads Europe in LNG imports, followed by Portugal (39%) and France (27%). The principal suppliers of LNG to Europe include Algeria, Egypt, Oman and Qatar. Algeria is the world's third largest exporter of LNG, with almost all of its gas (25b cubic meters) going to Europe. In 2006, the Algerian national oil company, Sonatrach, signed a 20-year LNG supply contract with the Spanish power company Endessa.[51]

LNG has also become a major factor in the development of gas exports from the Persian Gulf. Although nations such as Qatar, Oman and the United Arab Emerites have produced LNG for the Asian market, European energy companies have begun to express more of an interest in purchasing LNG from the Gulf as well. With vast amounts of gas reserves the Gulf states are positioned to meet a portion of Europe's future demand.

European relations with the states of the Persian Gulf and North Africa have steadily improved over the years. EU relations with North Africa were formalized in 1995 with the creation of the Euro-Mediterranean Energy Partnership. The EU has also created the EU-Gulf Cooperation Council (GCC) Dialogue with the states of the Persian Gulf and has initiated a formal dialogue with the nations of OPEC. European energy companies have also become more involved in the Middle East. On the other hand, observers note that these forums have been slow to evolve, and that countries like Algeria have at times been reluctant to agree to European-proposed terms. For example, in September 2007, Algeria revoked a gas contract with Spanish companies reportedly worth upwards of $7 billion (5 billion euros).[52]

The potential for growth in Europe's energy diversification strategy with respect to the Middle East and North Africa is significant. However, European competition with Asia and North America and long-term political instability throughout the region will likely temper the degree to which Europe seeks to increase its reliance on the region. Nevertheless, as with the Caspian region, if the EU is serious about lowering its dependency on any one source, it must turn more and more to the Middle East and North Africa. Parenthetically, Europe's growing interest in energy resources in North Africa has not gone unnoticed by Russia and Gazprom.

---

[48] See Statistical Review of World Energy, British Petroleum (BP), June 2006.

[49] "EU Energy Policy Data," Op. Cit.

[50] Statistics from Eurogas, [http://www.eurogas.org].

[51] As reported by Lloyd's List, May 25, 2006.

[52] See Richard Youngs, Op. Cit.

Just as in the Caspian region, Russia appears to be bolstering its efforts to influence Europe's energy plans. In March 2006, President Putin, along with Gazprom officials, traveled to Algeria to discuss Russian participation in Algeria's future oil and gas projects, including its LNG export markets. Some contend that because Russia intends to make Europe a major market for LNG produced from its Shtockman gas field in the Barents Sea, Russia is seeking to position itself to influence Algeria's future role as a major supplier of energy to Europe.

Norway. Norway, not a member of the EU, is the second-largest exporter of natural gas to the EU, behind Russia. Norwegian exports represented 17% of European gas consumption in 2004. Germany (25%), France (30%), and the United Kingdom (3 0%) are the largest consumers of Norwegian gas exports.[53] As of January 2005, Norway had 73.6 trillion cubic feet (Tcf) of proven natural gas reserves. The North Sea holds the majority of these reserves, but there are also significant quantities in the Norwegian and Barents Seas. Norway is the eighth-largest natural gas producer in the world, producing 2.59 Tcf in 2003.[54] The United States Geological Survey has estimated that almost 25% of the globe's yet to be discovered resources are located in the Arctic region. Norway's recently opened Snohvit gas field along with Russia's field at Shtockman will make the Barents Sea a new European energy region.

According to industry estimates, Norway had 8.5 billion barrels of proven oil reserves as of January 2005, the largest in Western Europe. The bulk of Norway's oil production occurs in the North Sea, with smaller amounts in the Norwegian Sea. In 2005, Norway's oil production averaged 2.95 million bl/d. As North Sea fields continue to mature, Norwegian oil production will likely remain steady for several more years and then begin to decline. There is some hope that new developments in the Barents Sea will offset some of this decline. The largest single recipient of Norway's oil exports is the United Kingdom, which imports around 814,000 bl/d from Norway, or 34% of Norway's total exports. Other significant destinations include the Netherlands and Germany.

Norway's entry into the LNG export market opens a new opportunity for the EU to work with its northern neighbor on energy security issues. Norway's energy giant, Statoil, plans to construct the first large-scale LNG export terminal in Europe, with connections to the Snohvit project. Although the initial LNG production from the Snohvit project has been committed to the United States, follow-on production and future fields in the Barents Sea could be shipped to facilities in Europe. The EU has recognized the growing importance of Norway in Europe's energy security debate and has expressed interest in "facilitating Norway's efforts to develop resources in the high north of Europe."[55] Individual European nations have also recognized Norway's potential future role in providing secure energy. Poland, along with the Baltic states, has already begun discussing with industry the construction of an LNG terminal along the Polish coast to receive LNG from Norway for transport to other parts of Europe.

External Strategy Conclusion. Establishing a diversified network of secure energy suppliers has become one of the foremost challenges facing the nations of Europe and its Union. In one sense, Europe is fortunate to have such large sources of available energy within a relatively small geographical space. However, like other countries, Europe faces the fact that for the foreseeable future, those energy producing nations pose different levels of risk,

---

53 Jonas Store, Minister of Foreign Affairs, Norway, "The High North-Top of the Agenda," Speech at CSIS, Washington, D.C., June 2006; "EU Energy Policy Data," Op. Cit.

54 Energy Information Administration.

55 European Commission Green Paper, Op. Cit.

ranging from outright political instability to more subtle questions of political reliability and long term intentions.

The EU can continue to use its Common Foreign and Security Policy (CFSP) and collective trade policy to promote political stability and security in countries providing the bulk of Europe's energy supply. However, perspectives on energy security policy differ among the 27 member states themselves, and between the states and the European Commission. Long-term bilateral energy agreements such as the Baltic pipeline agreement between Russia and Germany, the South Stream pipeline contracts between Italy and Bulgaria and Gazprom, and LNG contracts signed between Spain and France and Algeria demonstrate that member states continue to view energy security primarily as a national policy issue. Such bilateral agreements may or may not take broader Union security into account. However, they could become more commonplace unless member states agree that a continued reluctance to coordinate may threaten the long-term energy security to the Union, especially if the states gravitate to single energy suppliers. For many, the European Commission's call for a common European energy policy makes sense. Nonetheless, progress towards a common external energy security strategy tied to the EU's CFSP appears to require greater coordination than has been demonstrated heretofore.

Challenge 2: Promoting Indigenous Energy Supply

While efforts to develop a coherent external energy policy remain a top priority for the European Commission and some member states, there is broad European agreement that the EU should also look inward to determine how its dependence can be mitigated by indigenous energy supply. Specifically, the EU has taken steps to increase Europe-wide production and use of alternative and renewable energy sources and to invest in "clean coal" technology. These efforts are aimed both at increasing European energy independence and addressing growing public and political concern with the effects of energy consumption on global climate change. Accordingly, current initiatives are focused on the two sectors that together account for a significant majority of the EU's total carbon-dioxide emissions: power and heat generation (35% of total carbon emissions) and transport (27%).[56] Energy consumption in the transport sector is overwhelmingly fueled by imported oil. About a third of Europe's power generation comes from coal burning plants, and another third from nuclear plants. Figure 2 illustrates the breakdown of power generation by fuel.

The mix of energy supply in EU member states continues to be determined largely by national governments or energy companies. As a result, energy mix varies widely across the EU, influenced by a number of factors ranging from resource cost and availability to political factors, such as legislation curbing nuclear energy production. In France, for instance, nuclear power accounts for over 70% of all electrical generation, while Germany and Spain have enacted laws to phase out the use of nuclear power; in Poland and the Czech Republic coal is the dominant fuel.

[56] The European Commission's Energy and Transport in Figures, Op. Cit.

EU Electricity Generation by Source -2005

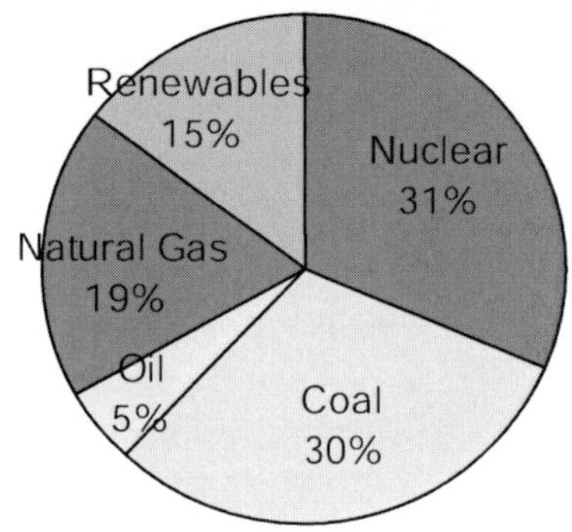

Figure 2. EU Electricity Generation

Source: IEA.

Nonetheless, decisions regarding national energy mix are increasingly influenced by EU and international agreements to reduce greenhouse gas and carbon emissions. Under the 1997 Kyoto Protocol, the EU is obligated to reduce its carbon emissions by 8% of 1990 levels by 2012. In order to achieve this collective target, member states have agreed to individual National Action Plans. The more heavily coal-reliant and less economically developed member states such as Poland have less stringent targets than countries with more developed renewable energy portfolios. Member states are expected to take a similar approach to realizing the EU's most recent goals of reducing emissions by 20% of 1990 levels by 2020.

The EU's ability to reduce its import dependence while mitigating negative environmental effects will depend largely on individual member state decisions regarding energy mix. European efforts are expected to focus on promoting renewable energy and cleaner burning fuels, developing 'clean coal' technology, and increasing energy efficiency and reducing overall consumption. Because nuclear power generation does not directly produce carbon emissions, some experts and government officials advocate an increase in nuclear power generation. Others, on the other hand, cite safety and proliferation concerns in opposing a rise in nuclear power generation.

Coal. Just over one-third of the total electricity generated in Europe is coal fired. As with other energy sources, coal use and production varies among member states. Coal burning accounts for the bulk of electricity production in member states such as Poland (92%), the Czech Republic (65%), Greece (62%) and Germany (just over 50%), but has been almost completely phased out in countries like France, which relies largely on nuclear power. Despite the fact that coal burning accounts for close to 25% of the EU's total carbon dioxide emissions, its abundance — Europe has proven reserves of close to 40 billion tons — leads most analysts to believe that coal will continue to play a significant role in Europe's energy make-up.[57] Given Europe's continued reliance on coal, and acknowledging the "huge possible benefits of a sustainable use of [coal]," the EU has moved to invest in technologies, such as

---

57 European Commission, Sustainable Power Generation from Fossil Fuels: Aiming for Near-Zero Emissions from Coal after 2020, COM(2206)843 Final; European Commission, Coal Market Outlook, 2005; Coal: a Clean Source for the Future? Euractiv.com, June 2006.

carbon capture and sequestration (CCS), that will enable so-called clean coal burning. Current technology is thought to enable efficiencies of well above 60%, but the majority of Europe's older and even most recently built plants have efficiencies ranging only from 30% to 43%.[58]

EU member states have endorsed Commission efforts to foster technological advances in the area of clean coal-burning and to bring 12 sustainable fossil-fuel power plants on line by 2015. Nonetheless, in the long term, the ability of member states to meet their commitments to lowering carbon emissions, the potential for using renewable energy, the price of natural gas, and the cost of installing clean coal burning technologies will likely dictate whether coal can remain a viable alternative energy source for Europe.

Nuclear. Although nuclear power accounts for roughly one-third of Europe's overall electrical generation, pronounced differences in national nuclear energy policies have prevented the EU from developing a common nuclear energy policy. Approximately 175 nuclear reactors are in operation in Europe today. However, while nations such as France, Finland, Sweden and the United Kingdom rely heavily on nuclear power, others oppose it on the grounds that it is dangerous and creates difficult waste disposal problems. Germany and Spain, for instance, have committed to phasing out all of their nuclear reactors over the next several years and replacing those with gas powered facilities. Political pressure to rethink these decisions, or at least extend timetables for the phase out is reportedly growing in both countries, and the United Kingdom, Finland and Lithuania have all decided to add new reactors. Nonetheless, given the substantial costs of putting a nuclear reactor on line and the controversial nature of nuclear waste, it appears unlikely that Europe will see a resurgence of new nuclear reactors in nations where nuclear power does not already play a role. At best, those nations that already utilize nuclear power could be expected to either replace or upgrade existing reactors. On the other hand, advocates of nuclear power generation appear to be gaining favor within Europe given the fact that nuclear power generation emits virtually no greenhouse gas emissions.

One promising alternative may be found in the International Thermonuclear Experimental Reactor (ITER) program. The EU has joined the United States and several other nations in an effort to produce electrical power from nuclear fusion which unlike current nuclear power does not generate dangerous waste. The first facility will be constructed in France but initial results are not expected for at least 15-20 years.

Renewable Energy. Hydro, wind, solar and bio-mass energy currently account for just under 7% of Europe's total energy consumption and 15% of its electricity generation. In March 2007, EU member states agreed to a legally binding target mandating that 20% of total European energy consumption be fueled by renewable energy sources by 2020. In January 2008, the European Commission proposed individual national renewable energy targets intended to realize the EUwide goal. Under the Commission proposal, countries with advanced renewable energy sectors like Austria, Sweden, and Denmark would be expected to achieve significantly more ambitious targets than newer member states in Central and Eastern Europe. Member states would also be given the option to invest in renewable energy projects in other EU countries. Individual heads of states are expected to consider the proposed national targets at their spring 2008 summit.

Although EU-wide support for renewable energy is strong, individual member states' renewable energy portfolios vary. For instance, Austria and Latvia promote hydro power,

---

58 *Ibid.*

while the Czech Republic and Portugal have committed financial support to large solar energy facilities. Germany, Sweden and the UK are home to major wind farms off their coasts. Bio-mass and biofuel programs are becoming more attractive. Given that Europe's oil-dependent transport sector accounts for roughly a quarter of the EU's total carbon emissions, the EU has mandated that biofuels make up a 10% share of all European transport fuel by 2020.

Whether the EU meets its renewable energy targets will likely depend on cost of production and the extent to which member states are willing to subsidize their development on a large scale. As noted earlier, a March 2007 McKinsey and Company report estimates that EU member states will need to invest approximately $1.5 trillion (1.1 trillion Euros) in new technologies over the next 14 years in order to achieve their carbon emissions and accompanying renewable and energy efficiency targets. It appears that the Commission, the European Investment Bank, and individual member states are poised to substantially increase their investment in these sectors, although specific amounts are difficult to estimate. Some member states have announced programs to subsidize and provide low-interest loans to fund research and development on renewable energies, with countries like Germany hoping to create the industrial capacity to supply what German officials believe will become an increasingly lucrative global market for renewable energy.[59]

Challenge 3: Providing Energy Security through an Internal Energy Market

The European Commission has long argued that member states could substantially increase energy supply security and network and cost efficiency by integrating national gas and electricity markets into the EU's single European market.[60] However, despite reforms to liberalize markets in some member states, national energy markets remain largely under state control. As energy security considerations have risen to the forefront of the EU agenda, debate over energy market liberalization has increased. The Commission and some member states contend that market integration and liberalization will increase energy security by forging network connectivity and EU-wide interdependence and diversifying supply sources, while others argue that continued national protection is important to guarantee stable and secure supply and distribution and pto rotect consumers from fluctuations in an unpredictable free market.[61] Commission proposals to develop a Europe-wide internal market for gas and electricity transmission and distribution continue to stoke debate, and analysts do not expect member states to take concerted action in this area before mid-2008.

The Commission launched its efforts to create a competitive EU-wide gas and electricity market began in the mid-to-late 1 990s by issuing a series of Directives focused on four primary objectives: (1) to implement the single market for energy by promoting competition and efficiency in the production and delivery of electricity and gas; (2) to lower prices and give all EU customers the opportunity to choose their energy supplier by 2007; (3) to help

---

59 Interviews of German officials, January-March 2007.

60 The 1986 Single European Act, adopted in 1987, established the so-called single European market allowing for the free movement of goods, services, people, and capital across the EU.

61 Katinka Barsch, Simon Tilford and Aurore Wanlin, The Lisbon Scorecard VII - Will Globalization Leave Europe Stranded?, Centre for European Reform, March 2007.

improve the environment; and (4) to enhance energy security. However, most member states appear to regard energy policy as too important to their own economic development to cede national controls, arguing that in Europe, nationalized industries have, for the most part, provided stability in the energy market. In fact, the dominant position of energy industries in some countries has led some national governments to take measures to protect their industries, even while ostensibly subscribing to the theory of open market competition.[62]

Over the past year, the Commission has increased efforts to promote EU guidelines to determine ownership and access to electricity grids, pipelines, and emergency energy storage facilities. The Commission's most recent and controversial proposal would reduce the power of state-owned energy companies by obligating them to split up ownership of generation and distribution networks (unbundling). The Commission contends that dual ownership blocks competition and allows for price manipulation. However, opposition to unbundling from Germany, France, and Spain has led the Commission to seek to negotiate alternative proposals. These countries are expected to present such proposals in early 2008.[63]

Network Interconnection. As progress toward a liberalized European energy market continues, discussion on energy security is expected to focus on the internal market's ability to deliver energy supplies through interconnected pipelines and electricity grids and to provide infrastructure security and emergency supply.

The European power transmission grid is divided into seven regional "pools" that, according to the Commission, are only weakly connected. Cross-border energy exchanges have increased recently. For example, in July 2006, the French electricity sector purchased additional power from Germany to offset the demand in France brought on by a heat wave. However, in 2005, only about 10% of the currently installed electrical generation capacity of Europe could be delivered across national borders. Although there are examples of successful cross-border cooperation, especially in regions such as the Nordic pool, many contend that a wave of blackouts in 2005 and 2006 were caused by weak links between Europe's power grids, poor coordination between national and regional power markets, and insufficient generation capacity. As a result, the EU has increased efforts to encourage investment in the construction of cross-border electricity grid connection and extending this kind of activity to gas, oil, and new LNG distribution systems.

Assessment

Most EU member states have long held that energy policy should remain the primary responsibility of the states themselves. However, European countries have begun to re-think energy not only as an element of individual national security but as an element of the EU's Common Foreign and Security Policy (CFSP). Furthermore, growing European public and political concern regarding global climate change appears to have spurred European action to mitigate its energy dependence by seeking to increase efficiency and reduce carbon emissions within Europe. Indeed, of the three focal points of current EU energy security policy — external relations, indigenous energy supply, and the internal energy market — efforts to

---

[62] For additional information see CRS Report RS22468, *Europe: Rising Economic Nationalism?*, by Raymond J. Ahearn.

[63] See "'Third way' for energy competition," *European Voice.com*, January 24, 2008.

promote a cleaner and more efficient indigenous energy supply have gained the most member state support.

Some skeptics doubt the ability of EU member states to ultimately come to agreement on a host of energy-related issues, particularly in managing external relations and creating an internal market. There continues to be strong disagreement on how to deal with Russia and on an appropriate diversification strategy, and most decisions regarding foreign supply sources and contract terms will likely remain in the hands of individual member states and their energy sectors. Open and competitive energy markets are desired, but protection of national energy industries still prevails in several key nations, including Germany, France and Spain. Some countries that have reluctantly agreed to open their energy sectors to more competition appear unenthusiastic about turning regulatory decisions over to Commission bureaucrats in Brussels.

On the other hand, the EU is increasing its role in coordinating and financing the development of renewable energy and the storage and use of emergency energy supplies. Although member states and their energy industries appear likely to retain absolute authority in determining which energy mix makes the most sense for individual countries, the EU has set binding EU-wide targets in some areas and may continue this trend. The EU also stands to play a larger role in determining power grid interconnection arrangements and energy infrastructure investment levels. With regard to foreign policy, efforts to advance energy dialogues with Russia and other energy producing and transit regions are being pursued in a more open and coordinated manner between the EU's Office of the High Representative for Common Foreign and Security Policy and the individual member states, and the Commission has outlined specific foreign policy goals with regard to multilateral treaties and an expansion of the EU's Neighborhood Policy. Nonetheless, foreign policy continues, first and foremost, to be determined by national governments.

ENERGY SECURITY IN THE TRANSATLANTIC CONTEXT

Over the past 55 years, relations between the United States and the EU have steadily broadened and deepened so that the two are inextricably linked. Nowhere has transatlantic integration manifested itself more than in the economic sphere. The U.S-EU economic partnership has been described by many as the single most important influence on global economic growth, prosperity and trade.[64] Within the deepening transatlantic economic relationship, energy security policy is becoming a higher priority for both the United States and the EU. Together, the United States and the European Union represent the world's largest energy market. The United States and the EU produce approximately 23% of the world's energy but combine for almost 40% of global energy consumption.[65] Combined, the United States and the EU account for almost 40% of the world's total carbon emissions.[66]

---

[64] For a more detailed explanation of the economic relationship see Hamilton and Quinlan, "Deep Integration: How Transatlantic Markets are Leading Globalization," Johns Hopkins University School of Advanced International Studies, 2004.

[65] According to BP's Statistical Review, the United States' share of global oil consumption is approximately 43% and gas consumption, 23%. The EU accounts for 18% of global oil consumption and 19% of natural gas.

[66] See BP Statistical Review, Op. Cit.

At the 2006 and 2007 U.S.-EU Summits, the parties agreed to increase cooperation on energy security, climate change, and sustainable development issues. Three institutional mechanisms to facilitate this cooperation were established: an annual strategic review of U.S.-EU energy cooperation; a U.S.-EU High Level Dialogue on Climate Change, Clean Energy and Sustainable Development; and a U.S.-EU Energy CEO Forum. None of these forums has convened more than twice, however, and at the April 2007 U.S.-EU Summit in Washington, D.C., the United States Administration reportedly rejected European calls for a commitment to pursue binding international global emissions and energy efficiency targets.[67] Specifically, European officials have urged U.S. support for an international treaty regulating greenhouse gas emissions after 2012, when the U.N. Kyoto Protocol is set to expire, and for an international market-based carbon emissions credit trading system.

European officials appear encouraged by what they perceive as an increasing U.S. willingness to acknowledge climate change as a problem with serious global ramifications, and to link energy and climate change policy. However, they are reportedly frustrated by U.S. reluctance to commit to binding international emissions and energy efficiency targets. The United States is not party to the Kyoto Protocol, and the Administration reportedly views global regulation to address climate change skeptically; instead, U.S. officials advocate transatlantic and international cooperation to develop alternative and renewable energy sources and liberalize international energy markets. To this end, the United States and EU used their 2007 Summit to launch a series of initiatives jointly promoting technological advances in clean coal and carbon capture and storage, biofuels, energy efficiency, and methane recovery. U.S. officials argue that such technological innovation is proving more effective in reducing emissions than global regulation and that such regulation may actually impede the economic growth necessary to sustain further technological advances. To support their claims, Administration officials point out that, despite EU participation in the Kyoto Protocol, carbon dioxide emissions increased at a faster rate in the EU than in the United States from 2000-2004. During the same period, they add, U.S. economic growth outpaced economic growth in the EU.[68]

Just as EU member states have expressed concern regarding a perceived U.S. reluctance to link transatlantic energy security to pursuit of a global climate change treaty, U.S. Administration officials and analysts point to a potential long-term threat to transatlantic relations arising from European dependence on Russian energy and Gazprom's growing influence in large segments of Europe's energy infrastructure.[69] To this end, the Administration has been supportive of efforts to build pipelines and develop other transportation routes from Central Asia and the Caspian Region to Europe that bypass Russia. In addition, U.S. critics of EU policy toward Russia argue that the EU should strengthen its resolve in requiring Russia to ratify the Energy Charter Treaty and to accept standard open market business practices, competition, and foreign investment in its energy sector. Some Europeans, and particularly Germans, on the other hand, appear reluctant to take concerted

---

67 "EU-US Summit To See Trade Move but No Climate Deal," EU Observer, April 26, 2007; "EU's Climate Change Drive Postponed," European Voice, May 2, 2007.

68 Speech by Kurt Volker, Principal Deputy Assistant Secretary of State for European and Eurasian Affairs, at the German Marshall Fund Berlin, February 12, 2007. Available at [http://www.gmfus.org]; "EU and U.S. Head for Climate Change Clash," European Voice, March 29, 2007.

69 Keith Smith, Op. Cit.

action toward Russia that may be viewed as antagonizing a country they view as an essential strategic partner on a variety of important issues beyond energy.[70]

Lastly, transatlantic discussion of energy supply security also includes energy crisis management and infrastructure protection. In this regard, some have called for NATO involvement in energy security issues, including in securing supply sources, distribution routes, and storage facilities. In 2006, Poland circulated a proposal for a so-called "Energy NATO," calling on an increased role for NATO in guaranteeing the protection of member state energy supplies. In a similar vein, in an address at NATO's November 2006 Summit in Riga, Latvia, Senator Lugar proposed the extension of NATO's collective defense clause, Article 5, to cases where a member state's energy security is threatened. Other EU member states, notably Germany and France, have greeted such proposals skeptically, preferring to advocate an enhanced EU role in energy security matters.

However, still others assert that NATO's role in energy security could be complementary to the EU's effort to strengthen market forces and interdependence in the international energy sector by offering assistance for the protection of pipelines or sea lanes during times of political unrest or conflict. NATO Partnership for Peace countries, such as Kazakhstan and Turkmenistan, which are important energy producers are seeking ways to associate themselves more closely with NATO, in part to diminish Russian influence and in part to develop reliable partners in an unstable region. For some, NATO has the ability to help secure the energy infrastructure of such countries.[71]

---

[70] Interviews of German and European officials, January - March 2007.

[71] For additional information see CRS Report RS22409, NATO and Energy Security, by Paul Gallis.

In: Global Energy Security
Editor: Roger T. Attenberg

ISBN: 978-1-60692-087-9
©2009 Nova Science Publishers, Inc.

Chapter 4

ENERGY SECURITY WITH A HIGH EXTERNAL DEPENDENCE: THE STRATEGIES OF JAPAN AND SOUTH KOREA*

Pablo Bustelo

ABSTRACT

Apart from China, there are two other large energy consumers in North-East Asia: Japan and South Korea. These economies, which belong to the OECD, are highly dependent on imports (which account for over 80% of domestic consumption in both cases), especially oil and natural gas. In recent years their energy security has been subject to serious threats. This chapter briefly addresses the position and energy forecasts for the two countries. It proceeds to analyse the strategic responses of Tokyo and Seoul to the deterioration (whether perceived or real) of their energy security, highlighting the strengths and weaknesses of each approach. Finally, the chapter details some of the lessons that other countries which are highly dependent on energy imports might learn from the experience of Japan and South Korea.

INTRODUCTION

Energy security in East Asia is important not only for the economic development of a region which could become the centre of the world in the XXI century, but also for the energy situation in other parts of the world. For example, if the main Asian energy-consuming countries (China, Japan, India and South Korea) find themselves compelled to notably increase their consumption of natural gas, either in the form of liquid natural gas (LNG) or that transported by gas pipeline, the availability and prices of gas in Europe would doubtless be significantly affected.

---

* Excerpted from Munich Personal RePEc Archive, Paper No. 8323, posted 18. April 2008 / 13:06; Online at http://mpra.ub.uni-muenchen.de/8323/

In recent years, experts have understandably focused above all on the case of China, given the Asian Giant's high increase in energy demand. The interest in China has prevailed somewhat at the expense of analysis of the other great East Asian consumers, Japan and South Korea: economies well-known as being of great importance and which are also OECD members. Japan and South Korea are high in the rankings of the world's main energy consumers and importers: suffice, for the time being, to recall that Japan is the world's third-largest consumer of oil and that Japan and South Korea are the world's two biggest importers of coal and LNG.

Given their importance as energy consumers and importers, energy security – especially insofar as it affects self-sufficiency, reliability and security of external supplies, and the prices of purchases abroad – is forming an increasingly essential aspect of their economic development strategy.

The present paper has two objectives. First, it aims to analyse the scope and implications of Japan and South Korea's high external dependence. Secondly, it attempts to analyse the short and midterm perspectives for this dependence and for the two countries' strategies for addressing the situation.

1. Energy Security in North-East Asia: An Overview

Energy security is a particularly important matter in North-East Asia, in view of the fact that this region is home to great world consumers, and given the high dependence with respect to imports, especially of oil and natural gas.[1] Although in recent years experts have stressed the case of China, as a consequence of the high increase in its demand,[2] there are two other countries in the region – Japan and South Korea– in which questions of energy security are at least as serious as in China.

In 2006, Japan occupied the fourth place in world energy consumption (behind the US, China and Russia), it was third in oil consumption (after the US and China), fourth for coal (behind China, the US and India) and sixth for natural gas and nuclear power. For its part, South Korea, of less demographic weighting and per capita income, is also a significant consumer of energy (it is tenth for primary energy, seventh for oil and sixth for nuclear power).

Net energy imports account for over 80% of consumption in Japan and South Korea, a proportion which is higher than Spain's and much higher than those of Germany and France. Furthermore, since Japan and South Korea do not have oil, natural gas and coal resources, they import practically everything that they need (as opposed to China, which does have significant coal and oil production, and which, until now, has even met its natural gas needs with its own production). Japan is the second-biggest importer of oil (behind the US), while South Korea is in fifth place, ahead of France and India. Japan and Korea are two of the world's biggest importers of LNG and coal.

According to the APERC (2007) index on energy insecurity for 2004, Japan and South Korea have a relatively high degree of energy diversification and are making a serious effort

[1] See the general studies on energy security in Asia in Isbell (2006), Pardesi et al. (2006), CRS (2007), Bubalo and Thirlwell (2007), Harris and Naughton (2007), Niquet (2007), Wesley (2007), Wu et al. (2007) and Cole (2008).

[2] In Spain, see working papers including Bustelo (2005) and Palazuelos and García (2007), among others.

to advance in non- carbon based fuels. However, they have a very high dependence on net energy imports, weighted in terms of the weighting of each source over total consumption (although this is approximately on a par with that of the US), a very high dependence on oil imports, weighted in terms of the weighting of oil within primary energy consumption (76% in South Korea and 69% in Japan), and an extremely high dependence with respect to oil imports from the Middle East (80% in Japan and 76% in South Korea).

Table 1.1 displays the values of the index of energy insecurity proposed by Wu and Morrisson (2007), which refer exclusively to oil.[3]

Table 1.1. Index of energy insecurity (oil), 1995, 2005 and 2015p

	1995	2005	2015p
Japan	77.8	76.7	75.5
South Korea	82.5	76.3	75.2
Europe	46.6	46.0	46.2
US	38.9	41.6	42.1
China	12.9	35.8	45.2
India	47.9	56.3	59.3

Note: a higher index score equals higher insecurity. Source: Wu and Morrisson (2007), table 1, p. 208.

The main conclusions in regard to the values for these indices in Japan and South Korea are that these two countries are very dependent on energy imports, oil imports and the supply of oil from the Middle East. The forecasts for 2015 of Wu and Morrisson's synthetic oil insecurity index (2007) would appear to suggest that the insecurity of Japan and South Korea will remain very high (in spite of a slight decrease in both cases), that is to say, much higher than the scores forecast for Europe (which will practically remain unchanged), for the US (which will grow slightly) and for China and India (which will rise considerably).

2. Japan's energy security strategy: the shadow of China

The energy system in Japan has two main features:[4] first, as stated above, the country is a very important consumer of energy;[5] secondly, it is highly dependent on abroad. Net energy imports accounted for 81.2% of its primary energy consumption in 2005, a percentage that is

3 The index is a synthesis of three primary indicators: net oil imports in relation to oil consumption, weighted at 40%; oil consumption in relation to primary energy consumption, weighted at 35%; and imports from the Middle East as a proportion of total oil imports, weighted at 25%. In view of the primary indicators and weightings that it employs, it is obviously an index that may be questioned. See the somewhat more sophisticated international comparison of indicators in APERC (2007), although only for APEC member countries and territories.

4 See a general overview of Japan's energy industry in EIA (2006).

5 In short, in 2006 Japan accounted for 4.8% of the world's energy consumption, but this proportion was significantly lower than its weighting in the gross world product, which was 10.3% in 2005 (and 7.04% in purchasing power parity), according to the new World Bank estimates. See World Bank, The 2005 International Comparison Program – Preliminary Results, Washington DC, 2007.

amongst the world's highest and which has hardly fallen since 1990 (in fact, it has increased slightly since 2001).[6]

In recent years, Japan has had to face up to growing challenges to international energy security (an increase in geopolitical risks in the main producing regions, the strong increase in demand from China and India, energy nationalism in certain countries, etc). This context has led the Japanese authorities to implement firm policies to increase government intervention in the internal and external markets, as well as to increase the country's energy security, factors which could lead to an increase in Japan's rivalry with China and South Korea.

But Japan is not only a large consumer and importer of energy. It is likewise a country which in terms of its energy development could serve to inspire others in many ways.[7] For example, in recent years its oil consumption in absolute terms and its per capita energy consumption have both fallen. Japan's energy intensity is currently lower than that of Germany and France, and substantially lower than that of the US. As Prime Minister Fukuda said at the Davos meeting of the World Economic Forum (in January 2008), energy consumption in industry has remained stable for 30 years, despite the fact that GDP has doubled, while CO_2 emissions per unit of GDP are amongst the lowest in the world (Fukuda, 2008). Moreover, the country is proposing to undertake significant investments in alternative energy sources. These could be promising (including, for example, those in the automobile, electronics and hybrid industries), and could turn Japan into a fundamental incubator of innovative techniques for the future of humanity. In addition, given its high level of development and its significant cooperation with other Asian countries, Japan's capacity for disseminating energy advances amongst neighbouring countries is very high.

2.1. Energy Production and Consumption

Production of primary energy reached 99.7 million tonnes of oil equivalent (Mtoe) in 2005 (for a consumption five times higher), of which 79.4 Mtoe were of nuclear origin. In other words, 75% of energy production in 2005 came from nuclear power, since resources of fossil or mineral fuels (coal, oil and natural gas) are practically non-existent, while there is very little hydraulic or other types of renewable energy. Production of primary energy in Japan is therefore very low, especially in comparison with other advanced countries. These include Germany (134.5 Mtoe), France (136.9 Mtoe) and the UK (204.3 Mtoe), according to IEA statistics (2007a and 2007b).

In 2006, according to the BP data (2007) primary energy consumption (PEC) stood at 520.3 Mtoe, making Japan the fourth biggest global consumer (behind the US, China and Russia), with a share of 4.8% of the world total (Table 2.1).

PEC distribution by energy type shows that oil remains the main source of energy (at 47% of PEC in 2005), followed by coal (22%), natural gas (14%) and nuclear power (11%). According to IEEJ data (Kanekiyo, 2007), since 1990 oil has fallen proportionately, coal and natural gas have risen, and the part corresponding to nuclear power has remained constant. Another of Japan's significant characteristics is that hydroelectric energy and renewable

---

[6] According to the Energy Balances of OECD Countries, it was 81.9% in 2004, a percentage which is similar to that of South Korea (82.1%), higher than that of Spain (77.1%) and, obviously higher than those of Germany (60.9%), France (50.1%) and the US (29.4%).

[7] See, for example, 'Japan's Energy Conservation Obsession', International Herald *Tribune*, 7/I/2007, and 'Japan's Energy Wisdom', International Herald Tribune, 28/III/2007.

energies make up barely 6% of PEC, a proportion that has hardly altered since 1990 (Table 2.2).

Table 2.1. Main consumers of primary energy in 2006 (Mtoe and % of world total)

Ranking	Country	Mtoe	% total world
1	US	2,326	21.4
2	China	1,698	15.6
3	Russia	705	6.5
4	Japan	520	4.8
5	India	423	3.9
6	Germany	329	3.0
7	Canada	322	3.0
8	France	263	2.4
9	United Kingdom	227	2.1
10	South Korea	226	2.1

Source: BP.

Table 2.2. Japan: consumption of primary energy by energy type (as % of the total), 1973, 1990 and 2005

	1973	1990	2005
Oil	77	57	47
Coal	15	17	22
Natural gas	2	10	14
Nuclear	1	10	11
Hydro-geothermic	4	4	4
New energies	1	2	2

Sources: IEA and IEEJ.

Another prevailing feature of the structure by energy types is that Japan (as is the case with South Korea) depends significantly on oil. At 45% in 2006 (47% in Korea), the percentage is higher than in China (2 1%) and India (28%), which continue to depend greatly on coal, but also greater than in Germany (3 8%) or the US (40%), where the relative weighting of natural gas is appreciably higher, according to BP data (2007).

The evolution of PEC in recent years shows relatively high growth in the 1 980s (with an average annual growth rate of 2.0%) and even in the 19990s (1.4%). But since 2000, this expansion has been much slower (0.5% in 2000-2006). Indeed, as we shall see below, oil consumption has fallen since 1996, when it peaked at 5.81 million barrels a day (Mbd). In 2006, consumption fell to 5.16 Mbd. For the purpose of comparison, oil consumption in the US rose from 18.30 Mbd in 1996 to 20.58 Mbd in 2006.

Table 2.3 shows the distribution of the consumption of final energy by sector. It can be seen that the main sectors are heavy industry, at 40%, and commercial-residential, at 33%. The significant fall in the transport industry (and also heavy industry) since the 1980s stands out, as does the increase in the proportion corresponding to the commercial-residential sector.

In comparison with other developed countries, the relative weighting of heavy industry is fairly high (40%, as opposed to an average of 22% for the OECD as a whole), given that

Japan still retains a significant industrial base, which, despite improvements in efficiency, continues to consume large quantities of energy.

Table 2.3. Japan: consumption of final energy by sectors (as a % of the total), 1973, 1980, 2002 and 2005

	1973	1980	2002	2005
Transport	16.4	55.4	27.4	26.5
Commercial/residential	18.1	21.4	31.6	33.1
Heavy industry	65.5	52.9	41.1	40.4

Sources: IEA, APERC and IEEJ.

The IEA forecasts (2007c) suggest that PEC will increase from 530 Mtoe in 2005 to 589 Mtoe in 2015, and to 601 Mtoe in 2030, with average annual growth rates of 1.1% in 2005-2015 (practically the same as for the period 1990-2005), and of barely 0.1% for 2015-2030. Worthy of note is Japan's much lower PEC growth in the past and forecast PEC growth with respect to China and India.

Table 2.4. Forecasts for primary energy consumption and oil, 2015 and 2030, with data for 1990 and 2005 (Mtoe and average annual growth rates in %)

PEC	1990	2005	2015	2030	AAGR (%) 1990-2005	AAGR (%) 2005-2015	AAGR (%) 2015-2030
Japan	444	530	589	601	1.2	1.1	0.1
China	874	1,742	2,851	3,819	4.7	5.0	2.0
India	320	537	770	1,299	3.5	3.7	3.5
OECD	4,518	5,542	6,180	6,800	1.4	1.1	0.6
World	8,755	11,429	14,361	17,721	1.8	2.3	1.4
TPEC	**1990**	**2005**	**2015**	**2030**	**AAGR (%) 1990-2005**	**AAGR (%) 2005-2015**	**AAGR (%) 2015-2030**
Japan	253	249	238	211	-0.1	-0.5	-0.8
China	116	327	543	808	7.2	5.2	2.7
India	63	129	188	328	4.9	3.8	3.8
OECD	1,893	2,247	2,385	2,478	1.1	0.6	0.3
World	3,261	4,000	4,720	5,585	1.4	1.7	1.1

Source: IEA, 2007.

Table 2.4 also reveals that consumption of oil in Japan has fallen since 1990 and that it will continue to do so in the coming decades.

The IEEJ forecasts for PEC distribution by types of energy (see Kanekiyo, 2007a and 2007b) indicate that the weighting of oil will continue to fall, in favour of natural gas and of nuclear power, while the proportion of coal will remain constant (Table 2.5). It is equally noteworthy that the weighting of hydroelectric energy, geothermic energy and the new energies will barely increase at all: from 6% in 2005 to 7% in 2030, after peaking at 9% in 2010.

In terms of consumption of final energy by sectors (Table 2.6), it can be seen that the proportions will remain practically unchanged during the coming decades. Thus, according to

APERC (2006), in 2030 the weighting of heavy industry will be 39%, ahead of the commercial-residential sector (34%) and transport (26%).

Table 2.5. Japan: forecasts for primary energy consumption by energy type (as a % of the total), 2010 and 2030

	2005	2010	2030
Oil	47	44	37
Coal	22	18	18
Natural gas	14	14	18
Nuclear	11	15	20
Hydro-geothermic	4	4	2
New energies	2	5	5

Source: IEEJ.

Table 2.6. Japan: forecasts of consumption of final energy by sectors (Mtoe and % the total), 2010, 2020 and 2030

	2002	%	2010	%	2020	%	2030	%
Industry	141.6	41.1	147.4	39.7	152.4	39.0	156.6	39.4
Transport	94.4	27.4	102.8	27.7	104.9	26.8	104.4	26.3
Res/com	108.8	31.6	121.5	32.7	133.4	34.1	136.5	34.3
Total	344.8	100.0	371.7	100.0	390.7	100.0	397.5	100.0

Source: APERC.

2.2. Energy Imports

Net energy imports rose from 303.2 Mtoe in 1980 to 430.7 Mtoe in 2005. The latter figure represents 81.2% of consumption, a percentage that has increased since 2001 and that only recorded a slight decrease since 1990. These imports do not take into account imports for the nuclear industry. If they are included, dependence with respect to imports would rise to 96% (METI, 2007).

Table 2.7. Japan: Production, consumption and net imports of primary energy, 1980-2005 (thousands of Toe and % of consumption)

	1980	1985	1990	1995	2000	2001	2002	2003	2004	2005
Production	43,281	66,747	76,817	100,414	107,373	106,161	98,670	85,457	96,758	99,792
Consumption	346,526	364,903	445,966	502,449	528,936	520,836	521,652	516,106	533,201	530,463
NetM	303,245	298,156	369,149	402,035	421,563	414,675	422,982	430,649	436,443	430,671
Mn/PEC (%)	87.5	81.7	82.8	80.0	79.7	79.6	81.1	83.4	81.9	81.2

Source: IEA.

In terms of crude imports (4.19 Mbd in 2006), it is important to note two facts: in absolute terms they have fallen since 1997, when they reached a maximum of 4.59 Mbd; and they are largely from the Middle East (87%), a region where the main suppliers are Saudi Arabia, the UAE, Iran and Qatar (Table 2.8).

The percentage of oil imports from the Middle East is very high in Japan, compared with 47% for China, 24% for Europe and 17% for the US. In addition, it is likely that dependence with respect to oil imports from the Middle East will rise in the coming years, in spite of government attempts to develop supplies from other regions (Russia, Central Asia, Africa etc.) and to ensure increased use of natural gas and of nuclear power (Hosoe, 2005).

Table 2.8 also highlights several significant facts: imports from the Middle East are quite diversified by countries; the weighting of Iraq is practically zero; the weighting of Africa is now appreciable (especially for Sudan), and the very recent increase of that of Russia (3.5% in 2007) which will surely increase greatly in the coming years.

Table 2.8. Japan: imports of crude by regions and countries of origin, 2000-2007 (in %)

	2000	2004	2006	2007
The Middle East	87.9	88.9	89.2	86.7
Saudi Arabia	*24.2*	*24.5*	*30.0*	*26.9*
UAE	*25.5*	*25.3*	*25.4*	*24.5*
Iran	*12.8*	*15.0*	*11.5*	*12.1*
Qatar	*9.6*	*9.3*	*10.2*	*10.4*
Iraq	*1.4*	*2.2*	*1.0*	*1.0*
Russia	–	0.7	0.7	3.5
Central Asia	–	–	–	0.2
South-East Asia	sd	5.3	4.4	4.7
Indonesia	*4.8*	*3.5*	*2.8*	*3.0*
Africa	0.7	4.2	4.4	3.6
Angola	–	–	*0.8*	*0.2*
Guinea	–	*0.1*	*0.3*	*0.5*
Sudan	–	*1.8*	*2.6*	*2.5*
Australia	1.4	0.6	0.8	1.0
China	2.2	0.2	0.4	0.1
Total	**100.0**	**100.0**	**100.0**	**100.0**

Sources: IEA and METI.

Table 2.9. Japan: imports of oil derivatives, by regions and countries of origin, 1996-2004 (in %)

	1996	2000	2004
The Middle East	44.8	44.1	46.0
Saudi Arabia	*18.4*	*16.1*	*17.6*
UAE	*10.3*	*15.0*	*13.6*
Kuwait	*10.7*	*9.4*	*8.4*
Qatar	*2.2*	*2.0*	*4.4*
Asia	31.4	35.9	31.2
South Korea	*11.9*	*21.1*	*12.0*
Indonesia	*7.8*	*4.8*	*8.4*
Singapore	*5.6*	*3.2*	*3.7*
India	*1.0*	*1.9*	*2.8*
US	9.1	7.3	8.3
Former USSR	0.8	0.6	1.7
Rest	13.9	12.1	12.8
Total	100.0	100.0	100.0

Source: IEA.

As far as oil derivatives are concerned (table 2.9), according to the IEA data for 2004, the Middle East accounted for 46% of supplies, followed by Asia (31%; including 12% from South Korea and 8% from Indonesia) and the US (8%). These proportions have remained more or less stable in recent years.

The following suppliers stand out for natural gas imports: Indonesia (27%), Malaysia (23%), Australia (15%), Qatar (12%), Brunei (11%) and UAE (9%), according to METI data 2004 (Christoffels, 2007, p. 11). There are plans to increase the proportion imported from the Middle East, Russia and Australia. To date, these have been LNG imports (shipped by sea), although there is currently a debate regarding the possibility of constructing a gas pipeline from Sakhalin, on Russia's Pacific coast (Akasura *et al.*, 2007).

Finally, Australia (33%) and Canada (27%) are the main countries for uranium imports. The proportion from Kazakhstan is still very low (1%), but there are plans to considerably increase it (Masaki, 2007).

Japan is now beginning to import significant quantities of oil and gas from Russia. There are plans for imports of Russian crude from Sakhalin and Siberia to rise considerably, as well as natural gas imports, not only from Sakhalin, but also from the Kovykta field, near Irkutsk, if the problems for the construction of a gas pipeline to the Pacific are overcome (Ahn and Jones, 2008; Masaki, 2008).

The Taishet-Skovorodino-Nakhodka oil pipeline is at design stage. This will link the oil sources to the west of Lake Baikal and the Pacific coast (Lee and Lee, 2006; Itoh, 2007). In 2003, China was initially able to ensure that the oil pipeline would not have this route, but rather that it would link Angarsk with Daqing (China). Nevertheless, Japanese diplomacy won the day and in 2005 Russia announced that it would build the Taishet-Skovorodino section and a first branch from there to Daqing (to supply China), followed by a second branch to Nakhodka (to supply Japan and Korea).

An underwater gas pipeline between Sakhalin and Japan could greatly increase Russia's gas exports, but it has remained at discussion phase until now. As is well known, Russia has two different projects in Sakhalin: Sakhalin 1 –in which ExxonMobil is participating– which exports oil to East Asia[8] and supplies natural gas to the rest of Russia, without having ruled out the possibility of gas exports too; and Sakhalin 2, which, in addition to Gazprom, features the participation of Shell, Mitsubishi and Mitsui, and which exports gas to Japan, Korea and the US.

2.3. Strategies to Guarantee Supply

2.3.1. Ways

Prior to 2006, important steps were taken to diversify energy sources and suppliers, as well as to create a strategic oil reserve and to invest in resources, via exploration and the development of oil and gas fields abroad (Hosoe, 2005; Yokobori, 2006; PAJ, 2007). More recently, the report coordinated by the Japan Forum on International Relations (JFIR), also known as the '27 recommendations', was delivered to Prime Minister Koizumi in May 2006 (see the complete text in JFIR, 2006, and a summary in Toichi, 2006). In June of 2006, the Ministry of Economy, Trade and Industry (METI) published the *New National Energy*

[8] Sakhalin-1 will become increasingly important in Japan's crude supply. See 'Nippon Oil Buys Crude from Sakhalin-1 Under Long Term Contract', Bloomberg, 19/II/2008.

Strategy (NNES) –see the complete text in METI, 2006, and a summary in Hughes, 2006; Masaki, 2006a, and ECCJ, 2007–. Lastly, in March 2007, the Diet of Japan definitively approved the Basic Energy Act, based on the NNES.

The analysis underlying Japan's new energy strategy is based on taking into consideration the growing challenges to the country's energy security that have arisen in recent years and which have been highlighted by many experts (see, for example, Evans, 2006; Christoffels, 2007; and Jain, 2007).

These challenges are both general (those affecting all the big consumer countries) and particular (those specific to Japan).

The general challenges are well known: awareness-raising of the finite nature of fossil fuels; an increase in the geopolitical risks in the main producing areas (terrorism, nationalisation, restrictions on foreign investment etc.);[9] the exhaustion of oil and gas reserves in the US and the EU; the high past and future increase in demand by China and India; global investments that are likely to be insufficient; and price increases, that are likely to prevail in the mid-term.

Japan's particular challenges are diverse. First, the increase in the demand from China[10] has had and will continue to have serious repercussions in Japan, given questions of geographic proximity and historic rivalry between the two countries. Secondly, in recent years there has been a heightening of competition with China and India for the oil and gas resources of the Middle East[11] and Russia, as well as with China and South Korea for resources in Central Asia.[12] In this context, the difficulties involved in solving the territorial disputes in the East China Sea (the Eastern Sea for the Koreans and the Sea of Japan for the Japanese) involving the Diaoyu or Senkaku islands and the small islands of Tokto or Takeshima (Lynch, 2007). Third, the likely increase in dependence with respect to the Middle East (APERC, 2006, p. 4; APERC, 2007, p. 66) is especially worrying given that geopolitical risks have increased significantly in that region, making it potentially very unstable – especially because of the situation in Iraq and Afghanistan and the dispute between the US and certain EU countries and Iran–. Fourth, as is the case in other producing zones, there is growing energy nationalism in Russia.[13] This phenomenon has caused serious concern in Japan: for Taro Aso, the second biggest risk for international energy security is that produced by the fact that 'resource nationalism has resurged throughout the world as a result of the high prices of oil and other minerals' (Aso, 2007, p. 37). Fifth, Japan's foreign policy, aligned with that of the US, has led to setbacks in Iran (Pontius, 2006; Masaki, 2006b) and might accentuate rivalry with China. Sixth, the international regulations on nuclear power (the non-

---

[9] The former Foreign Affairs Minister, Taro Aso, wrote in an article published at the start of 2007 that the main risk to world energy security was the 'geopolitical uncertainty in the Middle East and other energy producing regions' (Aso, 2007, p. 37).

[10] And –although to a lesser extent– also that of other East Asian countries, and even India.

[11] In the next few years it is also likely that there will be an increase in dependency on demand from southern and eastern Asia with respect to the Middle East.

[12] Especially, at least for the time being, of oil and uranium and not so much gas (Masaki, 2007), although the situation may change once the gas pipeline between Turkmenistan and China has been built, in accordance with the two countries' 2007 agreement.

[13] As far as East Asia is concerned, this nationalism was patent in the case of Sakhalin-2, with the initial presence of Mitsui, followed by that of Shell and Mitsubishi, companies which were obliged to sell a majority holding to Gazprom in December 2006. The result is that since then the Russian company has controlled LNG exports to Japan. See Christoffels (2007, p. 44 onwards).

proliferation regime) and climate change (the Kyoto Protocol) are additional challenges for countries having opted for nuclear power and which still depend heavily on fossil fuels. Seventh, the fact that Japan is developing links with nontraditional suppliers, such as those in Central Asia, Africa or Latin America, entails considerable lengthening of the supply chains. Eighth and last, in Japan the high proportion of energy transported through the so-called *chokepoints* (the straits of Hormuz, Malacca and Singapore) is especially worrying, in view of how susceptible they are to being affected by terrorism by a large-scale conflict (for example, between the US and Iran), as Watkins indicates (2006). For Taro Aso, insecurity in the straits is the third biggest challenge to international energy security in general and to that of Japan in particular: he highlights the need for 'further international co-operation for enhancing the security and diversity of energy transportation routes. The straits of Malacca, Singapore and Hormuz are particularly important' (Aso, 2007, p. 37).

Based on the foregoing analysis, the general objectives of the NNES of 2006 are the following three. First, more active governmental intervention in the markets, so as to create strategic links between the energy companies and the governmental institutions –with aim of ensuring greater security in the supply of resources (as a strategy it contrasts with the more liberal one of the 1 990s)–. Secondly, a strengthening of resource diplomacy: economic aid, technical support, soft power, both in producing countries and in consumer countries competing with Japan (for example, aid for diversification and energy-saving in China), in order to reduce the competition in international demand for oil and gas. Third, an increase in emergency response capacity (with an increase in strategic oil and gas reserves).

The specific NNES objectives can be listed as follows:

(a) Diversification of the energy sources towards nuclear power, natural gas and the renewable energies, to which end the government proposes to reduce oil as a proportion of PEC from 47% in 2005 to 40% in 2030 and oil as a proportion of consumption in the transport industry from 100% to 80%, via the introduction of bio-ethanol and of electric automobiles, as well as by increasing the percentage of electricity of nuclear origin in total electricity from 30% to 40% in the same period. Likewise, the government seeks to strengthen Japan's dominant position on the LNG market, which has been somewhat weakened in recent years (Christoffels, 2007, p. 51 and ff.).
(b) Geographical diversification of suppliers: from the Middle East towards Russia, Central Asia, Africa (Libya or Nigeria), Australia, Latin America, Canada, etc, with the Siberia-Pacific coast oil pipeline being of 'strategic importance'.
(c) A substantial reduction in energy intensity, with an increase of 30% in GDP generated per unit of PEC between 2005 and 2030.
(d) An increase in stocks of oil and gas.
(e) Greater direct control on resources abroad. The government aims to increase the level of the so- called *equity oil* (oil controlled by national companies) from 15% of crude imports in 2005 to 40% in 2030.

2.3.2. Requirements and Demands

On an initial viewing, Japan's new energy strategy would appear to be sensible. Nevertheless, the objectives set might also appear to be extremely modest; although it is also true that they will not be easy to achieve.

Diversification towards sources other than oil and coal must be substantially quickened if Japan is to comply with its Kyoto Protocol commitments. According to the Protocol, Japan must reduce its greenhouse gas emissions by 6% in 2008-2012, with respect to 1990. However, it would appear difficult for the country to meet this target, given that in 1990-2005 emission of these gases actually rose by 7%, according to UNFCCC data.[14] Therefore, in order to comply with its commitments in the fight against climate change, Japan must accelerate the pace of restructuring of its energy industry towards nuclear power, hydroelectric energy and the renewable energies.

Moreover, the nuclear power projects –which are significant (there are 55 operating reactors, and a further 13 planned for 2030)– are to a certain extent blocked due to the technical problems they involve and the fact that public opinion is generally opposed to nuclear power. In July 2007, the Kashiwazaki nuclear power station (in Niigata prefecture) – the world's largest– which has seven reactors, was affected by an earthquake which measured 6.8 on the Richter scale, and had to be indefinitely closed. The resistance of local government to the construction of new power stations has put the brakes on the expansion of the nuclear programme, which aims to increase the proportion of electricity of nuclear origin to 40% of total electricity in 2030.

Significant difficulties must be faced in increasing the proportion of hydroelectric energy and, above all, the renewable energies (such as wind power). There is a lack of political will and there are very few initiatives, as shown by a forecast revealing that these forms of energy will hardly increase their weighting in PEC over the coming decades.

In regard to the second overarching objective –geographic diversification of energy imports–, at the outset, it must be stated that its progress is proving extremely slow as far as oil is concerned, as mentioned above (see Table 2.8 above).

Indeed, it may be possible that dependence with respect to the Middle East will actually rise. In the words of an APERC report, 'Despite oil's declining share in the primary energy mix, dependence on the Middle East is expected to rise as imports from Asia –such as Indonesia and Malaysia– decline' (APERC, 2006, p. 42).

The high dependence on the Middle East and the likelihood that this will remain (or even increase), over the coming years are especially serious, particularly in view of the fact that the percentage of oil imports controlled by Japanese companies is very low (15% in 2005), in spite of the repeated attempts to increase investments abroad (Mitchell and Lahn, 2007; Lahn, 2007).

Although links with Russia are both growing and promising (Koyama, 2007) and could lead to a reorientation of imports towards the former, financial and technical problems (the high cost and complexity of the planned oil pipeline between eastern Siberia and the Russian Pacific coast) have to be surmounted, as does the dispute over the Kuriles islands (or Northern Territories for the Japanese), as Buscynski (2006) points out. In addition, optimal exploitation of the Russian resources would require the collaboration of companies from Japan, China and South Korea, as well as the eventual involvement of the great US oil companies, given that exporting oil and LNG to the west coast of the US forms part of Russia's strategic interests (Atsumi, 2007, p. 42). However, the Asian companies have displayed precious little interest in partnership projects to date (Choo, 2006) and it remains to

[14] UNFCCC, *National Greenhouse Gas Inventory Data for the Period 1990-2005*, FCCC/SBI/2007/30, 24/X/2007.

be seen if American companies finally decide to invest massively in exporting hydrocarbons from the Russian Far East.

In terms of the planned reduction in energy intensity, although it is very necessary, past experience would not appear to indicate that it will be easy to achieve. The improvement achieved between 1973 and 1990 did not continue into 1990-2005. Smil (2007) highlights two factors accounting for this change in trends: an increase in energy intensity in heavy industry, especially in metal- machinery and ceramics, and an increase of 45% in per capita consumption of residential electricity, due to increased home usage of electrical and electronic devices. Thus, while energy intensity in the US fell by 12%, in Japan it increased by 3%. The elasticity of PEC with respect to GDP, which was 0.12 in 1973-1980 and 0.51 in 1980-90, increased to 1.19 in 1990-2000 (Kanekiyo, 2007a). It is true that in 2000-04 it was 0.16, but this can be partly explained by the higher GDP growth during this period.

Table 2.10 shows that energy productivity (GDP in terms of kilogrammes of oil equivalent) fell by 1.4% between 1990 and 2005, while it increased by 25% in the US and by 32% in Germany. Per capita consumption of electricity rose much more in Japan (21.8%) than in the US (11.9%) or Germany (10.4%). Table 2.10 also shows that South Korea performed substantially worse than Japan.

Table 2.10. Consumption of primary energy and electricity, energy productivity and per capita GDP, for several countries

	PEC Mtoe 2005	**PCpec Toe/inhab 2005**	**PCelc kwh/inhab 2004**	**Var90-04**	**GDP/kep US$ PPP 2004**	**Var90-04**	**GDPpcPPP US$ 2005**
Japan	530.5	4.15	8,459	21.8	6.4	-1.4	31,267
South Korea	213.8	4.43	7,710	178.3	4.2	-6.3	22,029
US	2,340.3	7.89	14,240	11.9	4.6	25.3	41,890
Germany	344.8	4.18	7,442	10.4	6.2	31.6	29,461
Spain	145.2	3.35	6,412	63.3	6.9	-4.9	27,169

Sources: IEA, Selected Indicators and UNDP, Human Development Indicators.

With regard to strategic oil and gas reserves, it is curious that the Japanese government should insist, more than on increasing said reserves, on the possibility of having recourse to the IEA emergency mechanism and on the importance of the stocks of other consumer countries, such as China and India. In the words of former Minister Aso, 'As uncertain and vulnerable elements in the oil market increase, the role of the IEA in ensuring global energy security takes on new importance. The Agency's emergency response mechanism mobilises the 27 member countries to collectively release supplies from their oil stockpile in the event of unexpected supply disruptions [...]. An important task ahead of us is identifying how to work closely with non-member emerging economies [of the OECD] like China and India in enhancing their oil stockpiling system' (Aso, 2007, p. 37 and 38).

Insofar as control of resources abroad is concerned, it is worthwhile recalling that the proportion of imports executed by Japanese companies is barely 15%. There is an urgent need to increase this proportion, but it may not be easy to do so, in view of the setbacks suffered by Japan in recent years in certain of its investments abroad.[15]

Moreover, the possibility that similar problems could occur in central Asia must not be dismissed. Competition between Asia's national oil companies (ANOCs) is becoming very intense there. For example, it cannot be ruled out that China's rapprochement with Central Asia via Shanghai Cooperation Organisation (OCS) will cause growing difficulties for Japan in terms of its access to these supply sources (Herberg, 2007). However, Prime Minister Koizumi's visits to Central Asia (Kazakhstan and Uzbekistan) in December of 2006, and those of Minister Amari –the head of the METI–, in January of 2007, were apparently successful for Japan, at least insofar as the supply of uranium is concerned. Kazakhstan possesses the world's second largest reserves of this mineral, after Australia.

In conclusion, of the five specific objectives included in Japan's new energy strategy, it would appear that the following three are attainable: increasing efficiency by 30% by 2030 (as long as they change the habits of electricity consumption); reducing the part of oil in PEC to 40%; and increasing the proportion of electricity of nuclear origin to 40% (in the latter two cases, as long as the increase in nuclear power stations is not subject to technical or political setbacks). However, it is not at all obvious that the proportion of equity oil in the imports can be easily increased from 15% to 40%, in view of the heightened competition between Japanese, Chinese, Korean and Indian companies.[16] Lastly, it would appear very difficult to reduce the ratio of oil in the final energy consumption of the transport industry from 100% to 80%, given that such a reduction would require a very decisive commitment to biofuels such as ethanol, electric automobiles, engines based on photovoltaic cells and other systems; and it does not seem likely that this is going to happen without a change in the scope of car manufacturing and demand.

2.4. Conclusions

Japan's energy horizon and the main elements of its new strategy –as detailed in the foregoing sections–, point to several conclusions:

(1) Japan's growing energy insecurity may increase its already considerable frictions with China (Calder, 2007; Khan, 2007; Liao, 2007), unless energy-related questions are treated in isolation within the bilateral relationship (Gupta, 2008) and/or lead to higher cooperation on an international scale and between the two countries (for example in questions of maritime security).

(2) Japan's energy requirements may lead to it establishing increasingly closer links with Russia (as a way of reducing dependence on the Middle East), especially in the cases of the Siberian oil (via the future trans-Siberian oil pipeline) and the gas from Sakhalin and of Kovykta (Ahn, 2007). The importance that Central Asia might

---

[15] In addition to the aforementioned case of Sakhalin-2, we should recall the Azadegan oil field, in Iran. In September 2004, a consortium of Japanese companies led by Inpex agreed to participate in its development, but the conflict between Washington and Teheran concerning the Iranian nuclear programme led to the Japanese consortium postponing its investment. Lastly, in October 2005, Iran reduced the Japanese holding in Azadegan from 75% to 10%. See Watkins (2006) and Christoffels (2007, p. 25 and onwards).

[16] Nevertheless, Toichi (2006) argues that –given that Japan's crude imports will be reduced from 4 to 3 million barrels a day between 2005 and 2030–, it is sufficient to double from 0.6 to 1.2 mbd the part controlled by Japanese companies in order to achieve the target of 40%.

acquire in the future supply of oil and gas to Japan remains to be seen, although it would appear clear that this region (and especially Kazakhstan) will be of the utmost importance in obtaining mineral uranium.

(3) Avoiding frictions with China and promoting imports of resources from the former USSR require a framework of international cooperation (obviously including these two countries, but perhaps India and the US too, as has been stated by observers including Blank, 2006, and Kangas, 2007). Due to the existence of a historical rivalry this is a question on which little progress has been made until relatively recently (Choo, 2006).

(4) Japan needs to be more ambitious and muster a clear political will for the development of renewable energies, in view of the very low weighting that these forms of energy currently have despite the country's technical sophistication.

3. South Korea's Energy Security Strategy: Caught in a Stronghold

The Republic of Korea (henceforth South Korea throughout the present paper) is the world's 10th largest consumer of energy (at 233.4 Mtoe), ahead of Brazil and Italy, and almost on a par with the UK. It is the world's sixth biggest consumer of nuclear power, the seventh for oil, and ninth for coal.[17]

Like Japan, its external dependence is very high. In 2006, its net energy imports reached 188.8 Mtoe, that is to say 80.9% of its consumption. South Korea is a very important importer of coal and of LNG: the second in the world for both, after Japan. It is also the world's ninth largest oil importer.

From an energy security standpoint, South Korea presents a triple dilemma (Calder, 2005): a lack of energy resources, a high dependence with respect to oil, and a high dependence on the Middle East for oil and gas imports.

South Korea only has small anthracite deposits, and limited hydroelectricity resources, meaning that it has to use imports and wood consumption for energy generation.[18] Net imports, without taking nuclear power into account, totalled 96.5% of internal consumption (80.9%, as stated above, if the production of nuclear power is included). In 2006, oil accounted for 44% of primary energy consumption, a proportion which is higher than that for the world average (3 8%). Lastly, in 2006, the rate of dependence on imports of crude from the Middle East was 80.7%. In terms of LNG of imports, almost half of the imports in 2006 came from just two Persian Gulf countries: Qatar (25.6%) and Oman (20.7%), something which is important because two Asian countries (Indonesia and Malaysia) accounted for an additional 42%.

This being said, it must be remembered that, thanks to important development of the petrochemical industry, South Korea is a net exporter of oil derivatives. In 2005, exports totalled 35.1 Mmt, while imports were 19.8 Mmt.

[17] A general overview of South Korea's energy situation can be seen in EIA (2007).

[18] South Korea also has some uranium resources, which remain practically unexplored.

3.1. Energy Production and Consumption

Production of primary energy is very low (barely 44.6 Mtoe in 2006). Most of this production corresponds to nuclear power (37.3 Mtoe), to the wood and others category (4.3 Mtoe), to coal (2.1 Mtoe) and to hydroelectricity (1.3 Mtoe). Although production has risen considerably (it stood at 21.9 Mtoe in 1990), thanks to the efforts to develop nuclear power, hydroelectric energy and the renewable energies, it still accounts for only very small quantities.

Primary energy consumption (PEC) stood at 233.4 Mtoe in 2006, and high growth has prevailed in recent years (it was 93.2 Mtoe in 1990). Growth has been constant, except in 1998, when there was a reduction to 165.5 Mtoe from the 179.6 Mtoe of 1997, as a consequence of the financial crisis of that period. However, in 1999, PEC was already over 180 Mtoe, according to the BP data (2007). In 1980-2006, the annual average growth rate for PEC was 6.6%, almost as high as for GDP.

The main reasons for such high growth have been: the rapid rise in GDP (6.8% in 1970-2005, despite significant slowdown over the last decade), significant expansion of heavy industry (cement, steel and petrochemicals) which is very energy-intensive,[19] the marked progress of motorisation (car numbers rose from 557,000 in 1985 to 15.4 million in 2005), and the automation of the plants of the automobile and electronic component industries.

In terms of the structure of primary consumption by energy sources (Table 3.1), it is worth noting that in 1996-2006 the percentage rate for oil fell from 62.3% to 43.6%, in favour of coal (which increased from 18.5% to 24.3%), natural gas (the weighting of which rose from 6.7% to 13.7%) and, to a lesser extent, of nuclear power (11.8% and 15.9%, respectively). In spite of iversification in oil, the fossil fuels still for 82% of consumption.

Table 3.1. South Korea: consumption of primary energy by energy types, 1995-2006 (thousands of toe and %)

	1995	%	2000	%	2005	%	2006	%
Coal	28,092	18.7	42,911	22.2	54,788	24.0	56,687	24.3
Oil	93,955	62.5	100,280	52.0	101,526	44.4	101,831	43.6
Natural gas	9,213	6.1	18,924	9.8	30,355	13.3	32,004	13.7
Nuclear	16,757	11.1	27,241	14.1	36,695	16.1	37,187	15.9
Hydroelectricity	1,369	0.9	1,402	0.7	1,297	0.6	1,305	0.6
Renewable	1,051	0.7	2,130	1.1	3,961	1.7	4,358	1.9
Total	**150,437**	**100.0**	**192,888**	**100.0**	**228,622**	**100.0**	**233,372**	**100.0**

Source: MOCIE.

With regard to distribution by sectors of the consumption of final energy, heavy industry accounts for the lion's share (56% in 2006), and the weighting of this sector has even risen in recent years. The importance of heavy industry in the consumption of final energy is due to the country's productive specialisation (heavy industry, petrochemicals, naval construction and car manufacturing) and to the development of an important refining industry.

As regards forecasts, APERC (2006) considers that PEC will reach 241.3 Mtoe in 2010, 303.5 Mtoe in 2020, and 351.7 Mtoe in 2030; that is to say, that it will rise by over 50%

[19] For example, ethylene production reached 6.5 million tons in 2005, a six-fold increase since 1985 (WP-EPP, 2006, p. 67).

between 2006 and 2030, although the rate of growth during this period (1.7%) will be much lower than that recorded between 1980 and 2006 (6.6%). In recent years, the rate of growth of PEC (3.2% in 2000-2006) has been considerably lower than this average. The two main reasons for this lower PEC growth since the end of the 1990s are the change in the productive structure towards the services and the information technology industries, and the general efficiency improvements which are expected.[20] Thus, the elasticity of consumption with respect to GDP, which was 1.1 in 1980-2002, will be 0.6 in 2002-2030, according to the forecasts.

Table 3.2. South Korea: Consumption of final energy by sector, 1996 and 2006 (in %)

	1996	2006
Heavy industry	51.4	56.0
Residential and commercial	24.0	20.7
Transport	22.6	21.0
Public and other	2.0	2.2
Total	100.0	100.0

Source: MOCIE.

Table 3.3. South Korea: forecasts for growth in consumption of primary energy (Mtoe and %)

	PEC		AAGR
1980	41.4	1980-2002	7.4
2002	199.7	2002-2010	2.4
2010	241.3	2010-2020	2.3
2020	303.5	2020-2030	1.5
2030	351.7	2002-2030	2.0

Source: APERC, 2006.

Table 3.4. South Korea: consumption forecasts for primary energy by types of energy, 2006, 2010 and 2020 (in %)

	2	2010	2020
Oil	43.6	41.8	38.1
Coal	24.3	25.9	24.0
Natural gas	13.7	15.4	15.7
Nuclear	15.9	14.0	18.0
Hydro and RE	2.5	2.9	4.2
Total	**100.0**	**100.0**	**100.0**

Source: MOCIE and KEEI.

As shown in Table 3.4, the forecasts of the Korea Energy Economics Institute (KEEI) would suggest that, in PEC, the oil ratio will fall from 43.8% in 2006 to 38.1% in 2020, in favour of natural gas, nuclear power, hydroelectric energy and the renewable energies. The proportion corresponding to coal will remain stable at 24%. It is noteworthy that, according to

[20] In addition, naturally, to the effects of the 1997-1998 financial crisis, which meant, for example, that oil consumption reached a maximum of 2.4 mbd in 1997. Consumption fell to 2 mbd in 1998. In 2006, at 2.3 mbd, the figure was even lower than in 1997, according to BP data (2007).

these forecasts, the total for hydroelectric energy and renewables will be barely over 4% of PEC in 2020, and that the fossil fuels will account for three-quarters of PEC in this year.

The forecasts for the distribution by sector of final energy consumption do not show significant changes, as shown in Table 3.5.

Table 3.5. South Korea: forecasts for consumption of final energy by sector, 2006 and 2020 (in %)

	2006	2020
Heavy industry	56.0	55.1
Transport	20.7	20.3
Res/com	21.0	22.1
Public and other	2.2	2.4

Source: MOCIE and KEEI.

3.2. Imports

The net energy imports were 188.8 Mtoe in 2006, equivalent to 80.9% of consumption. South Korea's foreign dependence is, therefore, extreme. This dependence, moreover, has risen from 74.1% in 1985 and 76.5% in 1990, although it has fallen slightly since 2000, as shown in Table 3.6.

Table 3.6. South Korea: production, consumption and net energy imports, 1985-2006 (thousands of toe and %)

	1985	1990	2000	2006
Production	14,597	21,908	33,367	44,582
Consumption	56,296	93,192	192,887	233,372
NetM	41,699	71,284	159,520	188,790
Mn/Cons (%)	74.1	76.5	82.7	80.9

Source: MOCIE.

Table 3.7. South Korea: oil imports by regions and countries of origin, 1995 and 2005

	1995	2005
The Middle East	75.6	80.7
Saudi Arabia	35.2	29.1
UAE	11.0	17.5
Kuwait	3.9	9.2
Iran	10.6	8.3
Qatar	1.7	6.2
Asia	14.5	9.1
Indonesia	4.9	4.5
China	4.9	1.0
Africa	5.9	4.0
Australia	0.8	3.8
Former USSR	0.4	1.0

Source: IEA.

In 2006, South Korea imported 101.8 million tonnes of oil, which meant 100% of its consumption. Furthermore, as shown in Table 3.7, over 80% of these imports were from the Middle East, where Saudi Arabia, the UAE, Kuwait, Iran and Qatar are the main suppliers. In recent years, there has even been an increase in the dependence on the Middle East. Although dependence with respect to the Middle East is high, it is relatively diversified between countries from the region (Saudi Arabia, UAE, Kuwait, Iran and Qatar), as is also the case with Japan. There is still no significant increase in imports from Russia and Central Asia, although there is a slight increase in imports from Australia.

South Korea is a net exporter of oil derivatives, with a total of 15.3 Mmt in 2005 (16.4 Mmt in 2000). The main consuming countries of these derivatives that year were China (28.1%), Japan (22.7%), the US (12.7%) and Indonesia (9.2%).

The main suppliers of coal imports (54.4 Mmt in 2006) are Australia, China and Russia (Table 3.8). The strong increase in China's weighting is worthy of note. It had been zero in 1995.

Table 3.8. South Korea: coal imports by regions and countries of origin, 1995 and 2006 (thousands of mt and %)

	1995	%	2006	%
Australia	9,159.0	35.7	17,637.1	32.4
Canada	2,157.0	8.4	942.7	1.7
China	–	0.0	14,306.9	26.3
US	1,658.0	6.5	62.6	0.1
Russia	–	0.0	4,058.1	7.5
Other	12,681.0	49.4	17,414.7	32.0
Total	**25,665.0**	**100.0**	**54,422.1**	**100.0**

Source: MOCIE.

Given that South Korea is the second biggest world importer of LNG, the countries of origin of its imports are of great importance. All the gas imported to date has been in LNG form, due to the lack of gas pipelines, although several are planned. As Table 3.9 shows, halfway through the last decade over 90% of these imports were from two South-East Asian countries (Indonesia and Malaysia). The weighting of these two countries has fallen since then, due to factors including an increase in their internal demand. Qatar and Oman have become leading suppliers, and in 2006 these two Middle East countries accounted for almost half of the imports.

The forecasts for growth in imports indicate that there will be persistent growth in all energy imports (oil, coal and LNG), although it will be at a lower rate than that of recent years. According to APERC estimates (2006), from a base of 189 Mtoe in 2006, net imports may raise to 239 Mtoe in 2020, and 271 Mtoe in 2030. According to these data, gas imports could reach 72.9 Mtoe in 2030 (they stood at 31.5 in 2006), while oil imports could rise from 101.8 in 2006 to 135.6 Mtoe in 2030. It is predicted that coal will grow little, from 54.6 Mtoe in 2006 to 62.8 Mtoe in 2030.

Table 3.9. South Korea: LNG imports by countries of origin, 1995 and 2006 (thousands of mt and %)

	1996	%	2006	%
Indonesia	6,262.1	65.6	5,060.0	20.1
Malaysia	2,572.2	27.0	5,545.9	22.0
Brunei	707.0	7.4	849.8	3.4
Qatar	–	0.0	6,458.8	25.6
Oman	–	0.0	5,220.9	20.7
Other (1)	56.7	0.6	2,087.9	8.3
Total	9,539.5	100.0	25,221.9	100.0

(1) Australia, Algeria, Nigeria, Spain and Abu Dhabi.
Source: MOCIE.

3.3. Strategies to Guarantee Supply

South Korea's energy policy has been driven forward considerably in recent years. In February 2 006, the national parliament passed the Energy Framework Act and in November of that year the President created a National Energy Committee, with four sub-committees (energy policy, technology, development of resources and conflict management).

South Korea's energy strategy has significant similarities with that of Japan, given that both countries have similar structural features; for example, very low internal production, high oil dependence and a high dependence on the Middle East.[21]

The five main objectives of South Korea's energy security strategy can be summarised as follows: diversification of sources of energy, diversification of suppliers, consolidation of strategic oil reserves, conservation, rationalisation and the quest for efficiency, and development of resources abroad.

First, the aim of resource diversification towards nuclear power and the renewables is the fruit of a desire to reduce external dependence and CO_2 emissions. Opting for LNG, although it might be of interest for reasons of pricing and the proximity of certain suppliers, only partially resolves these problems. Furthermore, the recent growth in LNG consumption has been incredible: having risen from 18.9 Mmt in 2000 to 32 Mmt in 2006.

Nuclear power has to face unfavourable public opinion, such that, as has already been mentioned, a significant increase in its contribution to PEC is not planned. The reverse is true for the renewable energies, which are now a priority objective. In particular, there is an official objective to increase the weighting of renewable energies (not including hydroelectric energy) in PEC from the 2.1% of 2005 to a total of 5% in 2011.

Secondly, South Korea aims to heighten the geographical diversification of its imports, by means of an increase in oil and gas imports from Russia and, to a lesser extent, from Central Asia.[22] For the time being, Africa and Latin America are seen as much lower priorities for the supply strategy. The growth potential for oil imports from Siberia and gas imports from Sakhalin is enormous. In 2005, the countries of the former USSR accounted for barely 1% of the imports of crude. LNG imports from Russia are practically zero, in view of the fact that the four leading suppliers (Qatar, Malaysia, Oman and Indonesia), as stated

21 Note, however, that although the rate of self-sufficiency is similar in the two countries (18% with nuclear power and 4% without it), Japan's oil dependency is somewhat higher (47.4% in 2005) than that of South Korea (43.6% in 2006); the importance of the Middle East is also somewhat higher in Japan (88% as opposed to 81%).

22 See 'South Korea Wins Kurdistan Oil Contract', The Financial Times, 15/II/2008.

above, account for 90% of said imports. As regards the possibility of ensuring that a gas pipeline reaches South Korea, the question is largely addressed as a measure to meet the energy requirements not only of Seoul but also of Pyongyang. This means that if the North Korean nuclear crisis were definitively resolved it could drive Russia and China to favour this option (Paik, 2005). This desire for diversification is very similar to that of Japan, and both countries are competing with China, especially in relation to Central Asia and Russia.

Third, the creation and extension of a strategic oil reserve is a priority not only due to the importance of the same in the event of supply problems, but also to the IEA recommendations to its member countries. South Korea joined in April 2001. The strategic reserve is equal to 90 days of imports, an increase on the 60 days recorded at the start of 2001 (WG-EPP, 2006, p. 104).

Fourth, Seoul is planning for much higher conservation, rationalisation and efficiency, in view of the fact that its energy intensity is very high for a country with Korea's level of development.[23] In 2005, primary energy consumption for each US$1,000 of GDP was 0.34 in South Korea, as opposed to 0.21 in the US, 0.18 in Germany, 0.16 in Italy, 0.14 in the UK and 0.11 in Japan (MOCIE-KEEI, 2007, p. 63). The official target is to reduce energy intensity to 0.30 by 2012, an aim which does not seem very ambitious. Per capita energy consumption is very high and even higher than that of Japan, despite the fact that there is a substantial difference in the level of development. PEC by inhabitant (in Toe per person) in 2005 was 4.43 in South Korea, 4.16 in Japan, 4.18 in Germany, 3.88 in the UK, and 3.16 in Italy; although, obviously, this is lower than that of the US, which recorded 7.89 Toe per inhabitant (MOCIE-KIEE, 2007, p. 63).

Finally, the government is attempting to foment greater activity in the development of resources abroad. However, Korean companies are entering into conflict with their Japanese and Chinese counterparts, especially where oil is concerned (Herberg, 2007; Mitchell and Lahn, 2007; Lahn, 2007). All being said, there is also an official target to increase the national companies' proportion of production abroad as a percentage of total imports from 3.7% in 2005 to 15% in 2013 for oil; and from 5.8% in 2005 to 30% in 2013 for LNG (Bang, 2007, p. 12).

3.4. Conclusions

South Korea's situation of energy insecurity is very similar to that of Japan, although there are naturally a number of differences. The two countries share a dependence which remains high with respect to oil and a very high dependence with respect to energy imports. The following two differences stand out most: dependence on the Middle East is somewhat lower in South Korea than in Japan (80.7% as opposed to 87.9%, respectively, in 2005); conversely, the South Korean situation is notably worse in terms of energy intensity and per capita energy consumption (for example, it is significant that PEC per inhabitant was 4.43 Toe in Korea in 2005, as opposed to 4.15 Toe in Japan, in spite of the difference in per capita income).

The conclusions given above for Japan also apply to South Korea, although with certain nuances.

---

[23] This high intensity is accounted for by inefficiency, but also by factors related to industrial specialisation and the importance of the refining industry.

First, South Korea's energy insecurity means that it is competing with China and with Japan, but without being able to wield either China's demographic and economic weighting or Japan's level of development, despite the fact that the Koreans have been trying hard for many years to equal and even surpass Japan –something which they have achieved in a number of heavy industries–. The size and per capita income difference vis à vis its neighbours, means that, to a certain degree, South Korea is caught in a stronghold.

Secondly, as with Japan, the will to diversify its imports from the Middle East –a region on which it has become increasingly dependent for oil and LNG in recent years– is driving South Korea to develop ties with Russia and, to a lesser degree, with Central Asia. If collaboration with Russia to secure oil from Siberia and Sakhalin-1 and gas from Kovykta and Sakhalin-2 is primordial in Japan's case, it is no less so for South Korea, with the particularity that –since it is a continental country– the eventual links by oil pipeline and gas pipeline would appear to be easier. However, it should be recalled that the situation in North Korea is an added factor which could complicate this relationship, or which could rather act to facilitate it, in view of the fact that a definitive solution to the energy problems of North Korea must involve the creation of supply channels throughout the peninsula.

Third, the need for more regional cooperation on matters of energy security is even more pressing for South Korea than for Japan, due to questions of size and those linked to the necessary improvement of inter-Korean relations. Thus, Seoul has a special interest in promoting energy cooperation in North-East Asia. It was not surprising, for example, that in 2004 it took the initiative in founding a mechanism to regulate ministerial level meetings, in order to develop intergovernmental collaboration on issues of energy cooperation. In 2004, said initiative was consolidated by the constitution of Intergovernmental Collaborative Mechanism on Energy Cooperation in Northeast Asia. Its headquarters are in Ulan Bator, and its member countries are South Korea, Russia, Mongolia and North Korea, although still not Japan and China.

Lastly, it is worth noting that the efforts to promote renewable forms of energy have been somewhat greater in South Korea than in Japan. While Seoul has set an official target to attain a 5% contribution of these energies (not including hydroelectric power) by 2011, Japan has no official target of this type, despite having a higher per capita income, and notwithstanding its technical superiority in certain industries.

4. **Implications of the Strategies of Japan and South Korea for the EU and Spain**

Although the implications are very diverse, the most significant may be listed below.

Firstly, regional cooperation is essential to improving energy security in North-East Asia. It would afford very diverse advantages (Chidester and Kessler, 2007): it would bring down supply prices; it would facilitate improved distribution of costs (since it would include more investors); it would improve the reliability of supplies; it would make it possible –via better inter-connections– to respond to emergencies with better guarantees; and it would ensure that projects involving several countries (for example, oil or gas pipelines from Sakhalin to South Korea, passing through North Korea) became more rational and efficient. However, cooperation is only just getting underway and, to date, it has essentially been limited to the East Asia Summit, which, at its Cebu summit (in January 2007), created a regional cooperation mechanism geared towards pooling efforts in the face of a generalised situation in this part of the world: energy inefficiency, a lack of development of renewable energies, a high dependence on the Middle East and insufficient collaboration between the Asian countries in supply terms (Bustelo, 2007).

The EU has a common energy policy that is of interest to North-East Asia (Gavin and Lee, 2007), given that its long-term strategy features measures including improvement of inter-connections, fomenting saving, reducing the effect of greenhouse gas emissions, and increasing the contribution of renewable energies to primary energy consumption –and that of biofuels in transport sector consumption–.

Within the EU, Spain has been very active in defending favourable positions, including those to promote a European energy market, to diversify supply sources, to stimulate renewable energy sources, and to include political dialogue with the Mediterranean region (in particular Algeria and Morocco) as part of the EU's priorities.

Secondly, competition between the EU and North-East Asia for Central Asia's natural gas will surely be disadvantageous for both regions. In the future, it may be possible that Turkmenistan and Kazakhstan are connected by gas pipeline both with Europe (a trans-Caspian gas pipeline, that would connect with the southern Caucasus and with Nabucco) and with East Asia (via China, which has already signed an agreement with Turkmenistan,[24] or even via India,[25] although this second route would have to overcome serious political difficulties).

For the EU, the connection with Central Asia is a way of increasing access to resources (supply from Russia might prove to be insufficient, given the country's strong internal demand, scant investment in infrastructures, and the likely increase in European demand), diversifying the supply sources, and avoiding the pressures of Russia, which is increasingly inclined to wield its energy influence as the centrepiece of its foreign policy (Cornell and Nilsson, 2008).

If North-East Asia and the EU are simultaneously interested in Central Asia's gas resources, the results, if cooperation is lacking, could be negative for both sides.[26] It might therefore be advisable to commence some sort of dialogue on energy issues involving Central Asia, to take place during ASEM, or at a specially-created ad hoc forum.

Thirdly, the conjunction of Russia's strategic interest in bringing out the value of its eastern energy resources (Simonia, 2006) and the interest of the countries of North-East Asia in diversifying their supplies from the Middle East might lead to high a Asian dependence on Russia. The efforts which the EU has been making for a number of years now to reduce its dependence with respect to Russia could be a point of reference if, as planned, Japan and South Korea diversify their oil supply from the Middle East and diversify their gas imports from South-East Asia.

---

[24] The gas pipeline between Turkmenistan and the Chinese region of Xinjiang, via Uzbekistan and Kazakhstan, is planned for 2009. Spanning 2,000 kilometres, it will have the capacity to transport 30,000 million cubic metres a year. See Pao (2007).

[25] India has great interest in developing a pan-Asian network of natural gas inter-connections and, more specifically, in executing the TAPI gas pipeline (Turkmenistan-Afghanistan-Pakistan-India). See Tapi (2006) for an analysis of India's interests in the regional context.

[26] In more general terms, Paltsev and Reilly (2007) forecast that, if a regional gas market is developed in East Asia – connected with the Middle East, Central Asia and Russia– there could be a significant spiralling effect on gas prices in Europe.

Fourth and last, Spain shares certain common features with Japan and South Korea: a high external dependence (81.2% in 2006, very similar to that of the two Asian countries), diversification from oil and coal towards gas and renewable energy sources,[27] diversification of suppliers, a serious problem of energy inefficiency, high greenhouse gas emissions etc.

In recent years, due to its great capacity for LNG imports, Spain has achieved quite significant success in diversification in terms of sources and geography. Between 2000 and 2006 it considerably reduced the relative weighting of coal and oil in favour of natural gas and the renewable energies, while simultaneously diversifying supplies of natural gas from Algeria towards countries like Nigeria, Qatar, Egypt and Trinidad and Tobago. This success was due to the big increase in its regasification capacity: Spain is currently the world's third-largest importer of LNG (behind Japan and South Korea) and its LNG imports account for practically half those of the EU as a whole. In addition, Spain has managed to create a considerable wind power capacity, although the increase in energy intensity between 1996 and 2004 prevented an increase in renewable energies as a proportion of total energy consumption. In other words, Spain is able to offer countries with similar structural features – such as Japan and Korea– its experience in diversification of sources, especially with regard to the development of certain types of renewable energies –including wind power, of course– but also in production of electricity from waste and biomass.

And Spain itself could also learn from Japan and South Korea in relation to increasing regasification capacity, given that the totality of the two Asian countries' natural gas imports are in LNG format. Spain can learn from Japan in improvement of energy efficiency (conservation and saving): in 2005, its energy intensity was 0.11 Toe for each US$1,000 of GDP (2000), while this figure stood at 0.21 in Spain and 0.34 in South Korea. We should recall that, in Japan, oil consumption, as well as per capita energy consumption, has fallen in recent years[28], and that energy consumption in heavy industry has remained constant for 30 years. Japan is a very advanced country in terms of energy saving, as a consequence of its successful combination of cutting edge techniques (home appliances and low consumption cars, hybrid cars etc.), high petrol and electricity prices, frugality in daily life, the popular obsession with saving, subsidies for solar power and for the installation of fuel cells in homes, etc.

General Conclusions

As we have seen throughout this paper, Japan and South Korea are important consumers of energy, not only in terms of East Asia, but also on a world level; although their weighting has largely been overshadowed in recent times by the extraordinary rise of China. In addition, the two Asian countries have a dependence on foreign suppliers equal to over 80% of consumption, which they have not been able to significantly reduce in recent years. It is within this context that we have to analyse the effects that the recent changes in the world energy horizon (increasingly higher awareness of the finite nature of the fossil fuels, an

[27] Although the weighting of oil is more or less similar in the three countries, Spain is ahead in terms of natural gas (21% of PEC, as opposed to 14% in both Japan and South Korea).

[28] At 4.15 tep, per capita PEC is, lower than in Germany and France, and half that of the US and Canada. See AIE, Key World Energy Statistics 2007, Paris, 2007.

increase in the geopolitical risks in the main producing regions, a large increase in demand from China and India, energy nationalism in certain countries, etc.) have had on the energy security strategies of Tokyo and Seoul.

Both countries are aiming to speed up diversification of sources away from oil and coal (which, jointly, still account for two thirds of primary energy consumption) and towards natural gas, nuclear power and renewable energy. This goal not only responds to a desire to reduce external dependence (which is worsened by the long and vulnerable transportation routes for crude and LNG), but also to the demands for a reduction of greenhouse gas emissions. Moreover, given that in both countries over 80% of oil imports are from the Middle East and that gas imports, in LNG form, are very concentrated in just a few South-East Asian and Middle Eastern countries, the governments have set themselves the target of increasing the weighting of Russia and Central Asia, and, to a lesser extent, that of Australia, Africa and Latin America, in their supplies from abroad. Given the important crude and gas fields that Russia is developing in western Siberia and Sakhalin, and the prospects for better access to the resources of Central Asia (particularly in terms of gas and uranium), it is likely that the importance of the former Soviet republics to the energy supply of Japan and South Korea will increase. But geographical diversification does entail a number of difficulties – especially in the case of oil, a resource for which dependence with respect to the Middle East might even increase–. This is not to say that LNG imports might not also increase, in view of the high internal consumption of countries such as Indonesia and Malaysia, whose resources are, moreover, the subject of fierce regional rivalry. Such a development would be worrying, especially in view of the fact that the crude or gas effectively controlled by Japanese and Korean companies has to date constituted only a small part of imports. Lastly, Japan and South Korea are seeking to reduce energy intensity; this is very high in the latter country (which has extremely high per capita energy consumption), and, although relatively low in the case of Japan, it has nevertheless increased over the last 15 years, as opposed to what has happened during this period in Germany or even in the US.

The past experience and potential of Japan and South Korea in regard to energy security may be of tremendous interest to the EU and to Spain. On the one hand, the EU's common energy policy could, in certain aspects, serve as an inspiration for the development of regional cooperation in this respect in East Asia. This is still in its initial stages and is essential for management of the joint needs of countries like China, Japan and South Korea. From Southern Asia they will be joined by India very shortly. EU experience may also be useful as a mirror into which the two Asian countries could look in regard to dependence on supplies from Russia. In other words, the willingness of Japan and South Korea to reduce their dependence on the Middle East –a region which is potentially very unstable– could throw them headlong into the arms of Russia, and it is not clear that this would be wholly positive. It is obvious that geographical proximity and the importance of the resources from Sakhalin and western Siberia mean that it will be reasonable to increase the weighting of Russian supplies to East Asia, but such a development could lead to excessive dependence on this country, a situation shown by EU experience to not exactly be very advisable.

Another point to take into account is that the lack of cooperation between the two regions could lead to a significant increase in the price of gas in Europe. This might occur if both the EU and Eastern Asia establish gas pipeline links with Central Asia as is likely to be the case (given that the link between Turkmenistan and China is now under construction) and in view of the fact that demand in the Asian countries is likely to significantly increase. In order to avoid this scenario, which would also be harmful for Asian consumers, it would be advisable to establish a mechanism for dialogue and consultations between the EU and the East Asian countries on energy supplies in Central Asia.

Lastly, Spain could learn a lot from Japan and South Korea in terms of capacity for regasification and LNG imports, given the importance of these activities in the two Asian countries. It should not be forgotten that Japan and South Korea are the two leading importers of LNG and that they are quite far ahead of Spain, which is in third place. As is well known, the county's increased capacity for importing LNG has brought great benefits to Spain in its strategy to diversify energy sources and supplies, such that it will probably remain a priority in the coming years. Lastly, the converse is also true, insofar as the two Asian countries could also benefit from the lessons learnt in Spain's case, in terms of driving forward renewable energies in general and wind power in particular.

Bibliographical References

This working paper was written for the Energy Programme, led by Paul Isbell, of the Elcano Royal Institute. It also forms part of a project by the Research Group on the Growth of the World Economy (GICEM), led by Enrique Palazuelos (Madrid Complutense University), on the energy systems of the three great regions (EU, North America and East Asia) which are very dependent on energy imports. In 2007, the project received funding from the Autonomous Community of Madrid, within the framework of the IV Regional Research Plan. The writer thanks Enrique Palazuelos for his friendly and detailed comments, and Federico Steinberg for his editing of an earlier version of the present paper. Obviously, the author assumes exclusive responsibility for any errors and shortfalls which might remain.

References

Ahn, S.-H. (2007), 'Energy Security in Northeast Asia: Putin, Progress and Problems', *ARC Working Paper*, nr 20, Asia Research Center, London School of Economics.

Ahn, S.-H., and M.T. Jones (2008), 'Northeast Asia Kovykta Conundrum: A Decade of Promise and Peril', *Asia Policy*, nr 5, January, p. 105-40.

IEA (2006), *Energy Policies of OECD Countries. The Republic of Korea. 2006 Review*, International Energy Agency, Paris.

IEA (2007a), *Key World Energy Statistics*, International Energy Agency, Paris.

IEA (2007b), *Energy Balances of OECD Countries, 2004-2005*, International Energy Agency, Paris.

IEA (2007c), *World Energy Outlook 2007*, International Energy Agency, Paris.

APERC (2006a), 'Japan', in *APEC Energy Demand and Supply Outlook 2006*, Asia-Pacific Energy Research Centre, Tokyo.

APERC (2006b), 'Korea', in *APEC Energy Demand and Supply Outlook 2006*, Asia-Pacific Energy Research Centre, Tokyo.

APERC (2007), *A Quest for Energy Security in the 21st Century. Resources and Constraints*, Asia Pacific Energy Research Centre, Tokyo.

Asakura, K., N. Yamamoto and M. Hirata (2007), 'The Gas Pipeline Project between Sakhalin and Japan', *Northeast Asia Energy Focus*, vol. 4, nr 1, spring, p. 43-50.

Aso, T. (2007), 'Japan's Foreign Policy and Global Energy Security', *OECD Observer*, nr 261, May, p. 37-38.

Atsumi, M. (2007), 'Japanese Energy Security Revisited', *Asia-Pacific Review*, vol. 14, nr 1, p. 28- 43.

Bang, K.-Y. (2007), 'Energy Policy Directions Towards Sustainable Development in Korea', Korea Energy Economics Institute, Seoul, May.

Blank, S. (2006), 'Can East Asia Dare to Tie its Energy Security to Russia and Kazakhstan?', Conference on Eurasian Pipelines, Columbia University, New York, 30 November-1 December.

BP (2007), *Statistical Review of World Energy. June 2007*, BP, London.

Bubalo, A., and M. Thirlwell (2006), 'New Rules for a New 'Great Game': Northeast Asian Energy Insecurity and the G-20', *Policy Brief*, Lowy Institute for International Policy.

Bustelo, P. (2005), 'China and the Geopolitics of Oil in the Asian Pacific Region', *Working Paper*, nr 38/2005, Real Instituto Elcano, September (published as 'China and Oil in the Asian Pacific Region: Rising Demand for Oil'), *New England Journal of Public Policy*, vol. 21, nr 2, 2007, p. 171-201.

Bustelo, P. (2007), 'The East Asian Summit and Energy Security', *ARI*, nr 10/2007, Real Instituto Elcano, January (also in *RUSI Newsbrief*, vol. 27, nr 5, May 2007, p. 54-55.)

Buszynski, L. (2006), 'Oil and Territory in Putin's Relations with China and Japan', *The Pacific Review*, vol. 19, nr 3, September, p. 287-303.

Calder, K.E. (2005), *Korea's Energy Insecurities. Comparative and Regional Perspectives*, Korea Economic Institute of America, Washington DC.

Calder, K.E. (2007), 'Sino-Japanese Energy Relations: Prospects for Deepening Strategic

Competition', Conference on Japan's Contemporary Challenges, Yale University, March. Chidester, R. and C. Kessler (2007), 'Northeast Asian Energy Cooperation to Improve Regional

Energy Security', *Northeast Asia Energy Focus*, vol. 4, nr 3, autumn, p. 53-60.

Chrisstoffels, J.H. (2007), *Getting to Grips Again with Dependence: Japan's Energy Strategy*, Clingendael International Energy Programme, Clingendael, The Hague.

Choo, J. (2006), 'Energy Cooperation Problems in Northeast Asia: Unfolding the Reality', *East Asia*, vol. 23, nr 3, p. 91-106.

Cole, B.D. (2008), *Sea Lanes and Pipelines. Energy Security in Asia*, Greenwood Press, Westport, Conn.

Cornell, S.E., and N. Nilsson (eds.) (2008), *Europe's Energy Security. Gazprom Dominance and Caspian Supply Alternatives*, Central Asia – Caucasus Institute & Silk Road Studies Program, Washington DC – Stockholm.

CRS (2007), *Rising Energy Competition and Energy Security in East Asia: Issues for US Policy*, CRS Report for Congress, RL32466, Congressional Research Service, May.

ECCJ (?), *Japan Energy Conservation Handbook*, The Energy Conservation Center Japan, Tokyo. EIA (2006), 'Japan', Country Analysis Brief, Energy Information Administration, Washington DC, December.

EIA (2007), 'Korea', Country Analysis Brief, Energy Information Administration, Washington DC, June.

Evans, P.C. (2006), 'Japan', Energy Security Series, The Brookings Foreign Policy Studies, The Brookings Institution, Washington DC.

Fukuda, Y. (2008), 'Fact Sheet on Climate Change', World Economic Forum, Davos, 26 January. Gavin, B., and S. Lee (2007), 'Regional Energy Cooperation in North East Asia: Lessons from Europe', *Asia Europe Journal*, vol. 5, nr 3, p. 401-15.

Gupta, S. (2008), 'An 'Early Summer': Sino-Japanese Cooperation in the East China Sea', Policy Forum Online 08-010A, Nautilus Institute, 4 February.

Halloran, R. (2008), 'Securing Energy Supplies in Asia', *Taipei Times*, 14/II/2008.
Harris, S., and B. Naughton (2007), 'Economic Dimensions of Energy Security in the Asia-Pacific', in Wesley (ed.), *infra*, p. 174-94.
Herberg, M. (2007), 'The Rise of Asia's National Oil Companies', *NBR Special Report*, nr 17, December.
Hosoe, T. (2005), 'Japan's Energy Policy and Energy Security', *Middle East Economic Survey*, vol. 58, nr 3, January.
Hosoya, Y. (ed.) (2007), *Energy Policy Scenarios to 2050 Study. The Report for the Asia Region*, World Energy Council.
Hughes, L. (2006), 'Japan's New National Energy Strategy', *The Diplomat*, 25 September.
Isbell, P. (2006), 'Fire-breathing Dragons: Asia and the Challenge of Energy Security', *Working Paper*, nr 17/2006, Real Instituto Elcano, September.
Isbell, P. (2007), 'Revisiting Energy Security', ARI, nr 123/2007, Real Instituto Elcano, December. Ito, K., *et al.* (2006), *Japan Long-Term Energy Outlook. A Projection up to 2030 under Environmental Constraints and Changing Energy Markets*, IEEJ, June.
Itoh, S. (2007), 'The Pacific Pipeline at a Crossroads: Dream Project or Pipe Dream?', *ERINA Report*, vol. 73, January, p. 42-72.
Jain, P. (2007), 'Japan's Energy Security Policy in an Era of Emerging Competition in the Asia- Pacific', in Wesley (ed.), *infra*, p. 28-41.
JFIR (2006), *The Establishment of an International Energy Security System*, Japan Forum of International Relations, Tokyo.
Kanekiyo, K. (2007a), *Energy Trend of Asia and Japan*, The Institute of Energy Economics, Tokyo. Kanekiyo, K. (2007b), 'Japanese Experience Toward Energy Efficient Economy', China International Energy Forum, Beijing, November.
Khan, H.A. (2007), 'China's Energy Security with Special Reference to Japan', GSIS, University of Denver, June.
Kangas, R. (2007), 'The Changing Face of the Russian Far East: Cooperation and Resource Competition between Japan, Korea and China in Northeast Asia', *Perspectives on Global Development and Technology*, vol. 6, nr 1-3, p. 441-60.
Koyama, K., *et al.* (2007), 'Russian Oil/Gas Development and Its Implications for Japan', IEEJ, Tokyo, September.
Lahn, G. (2007), 'Trends in Asian National Oil Company Investment Abroad: An Update', *Working Paper*, Chatham House, November.
Lee, S.-K., and K.-W. Lee (2006), 'Russia's Oil Supply Potential to the Asia-Pacific Region', *Northeast Asia Energy Focus*, vol. 3, nr 3, autumn, p. 16-19.
Liao, X. (2007), 'The Petroleum Factor in Sino-Japanese Relations: Beyond Energy Cooperation', *International Relations of the Asia-Pacific*, vol. 7, nr 1, p. 23-46.
Lynch, M.C. (2007), 'Petroleum and Territorial Disputes', *Northeast Asia Energy Focus*, vol. 4, nr 3, autumn, p. 39-44.
Masaki, H. (2006a), 'Japan's New Energy Strategy', *Asia Times*, 13 January.
Masaki, H. (2006b), 'Goodbye Iran, Hello Iraq: Japan's and China's Oil Prospects in the Balance', *Japan Focus*, 13 November.
Masaki, H. (2007), 'New Energy Fuels Japan's Diplomacy: From the Middle East to Central Asia', *Japan Focus*, 6 May.
Masaki, H. (2008), 'Japan Covets Russian Gas, Hot Air', *Asia Times*, 14 February.
METI (2006), *New National Energy Strategy*, Ministry of Economy, Trade and Industry, Tokyo. METI (2007), *Energy in Japan 2006,* Agency for Natural Resources and Energy, Ministry of Economy, Trade and Industry, Tokyo.

Mitchell, J., and G. Lahn (2007), *Oil for Asia*, Briefing Paper, Chatham House, March. MOCIE-KEEI (2007), *Energy Info. Korea*, Ministry of Commerce, Industry and Energy, Korea Energy Economics Institute, Seoul.

Niquet, V. (2007), 'Energy Challenges in Asia', *Note of l'IFRI*, Paris, October.

PAJ (2007), *Petroleum Industry in Japan 2007*, Petroleum Association of Japan, Tokyo. Pao, G. (2007), 'China's Pipeline Diplomacy', *Asia Sentinel*, 07/IX/2008.

Paik, K.-W. (2005), 'Pipeline Gas Introduction to the Korean Peninsula', Chatham House, London, January.

Palazuelos, E., and C. García (2007), 'La transición energética en China', *Working Paper*, nr 05/07, Instituto Complutense de Estudios Internacionales.

Paltsev, S., and J. Reilly (2007), 'Energy Scenarios for East Asia: 2005-2025', *Report*, nr 152., MIT

Joint Program on the Science and Policy of Global Change, MIT, Cambridge, Mass.

Pardesi, M.S., *et al.* (2006), *Energy and Security. The Geopolitics of Energy in the Asia-Pacific*,

IDSS, Nanyang Technological University, Singapore.

Pontius, P.R.D. (2006), 'Japan's New National Energy Strategy and Iran: Geo-politics or Geoeconomics?', *IEEJ Discussion Paper*, November.

Simonia, N. (2006), 'Russian East Siberia and the Far East: A Basis for Co-operation with Northeast Asia', *Global Asia*, vol. 1, nr 1, p. 70-79.

Smil, V. (2007), 'Light Behind the Fall: Japan's Electricity Consumption, the Environment, and Economic Growth', *Japan Focus*, 2 April.

Toichi, T. (2006), 'International Energy Security and Japan's Strategy', IEEJ, Tokyo, October.

Tuli, V. (2006), 'Regional Cooperation for Asian Energy Security', *RIS Discussion Papers*, nr 112, RIS, New Delhi.

Watkins, E. (2006), 'Japan's Energy Supplies at Risk', *Terrorism Monitor*, vol. 4, nr 22, 16 November.

Wesley, M. (2007), 'The Geopolitics of Energy Security in Asia', in Wesley (ed.), *infra*, p. 1-14. Wesley, M. (ed.), (2007), *Security Energy in Asia*, Routledge, London.

WG-EPP (2006), *Energy Policy and Statistics in Northeast Asia. Country Report for China, Mongolia, Korea, Russia*, Working Group on Energy Planning and Policy, Intergovernmental Collaborative Mechanism on Energy Cooperation in Northeast Asia.

Wu, K., J.G. Brown and T.A. Siddiqi (2007), 'The Asia-Pacific Energy Dilemma', in Wu and Fesharaki (eds.), *infra*, p. 1-16.

Wu, K., and C. Morrisson (2007), 'Energy Security Index', in Wu and Fesharaki (eds.), *infra*, p. 107-15.

Wu, K., and F. Fesharaki (eds.) (2007), *Asia's Energy Future: Regional Dynamics and Global Implications*, East West Center, Honolulu.

Yokobori, K. (2005), 'Japan', in J.H. Kalicki and D.L. Goldwyn (eds.), *Energy and Security: Toward a New Foreign Policy Strategy*, Woodrow Wilson Press/Johns Hopkins University Press, Baltimore, MA, p. 305-28.

In: Global Energy Security
Editor: Roger T. Attenberg

ISBN: 978-1-60692-087-9
©2009 Nova Science Publishers, Inc.

Chapter 5

ENERGY INDEPENDENCE AND SECURITY ACT OF 2007: A SUMMARY OF MAJOR PROVISIONS*

Fred Sissine

ABSTRACT

The Energy Independence and Security Act (P.L. 110-140, H.R. 6) is an omnibus energy policy law that consists mainly of provisions designed to increase energy efficiency and the availability of renewable energy. This report describes the key provisions of the enacted law, summarizes the legislative action on H.R. 6, and provides a summary of the provisions under each of the titles in the law.

The highlights of key provisions enacted into law are as follows:

- *Corporate Average Fuel Economy (CAFE).* The law sets a target of 35 miles per gallon for the combined fleet of cars and light trucks by model year 2020.
- *Renewable Fuels Standard (RFS).* The law sets a modified standard that starts at 9.0 billion gallons in 2008 and rises to 36 billion gallons by 2022.
- *Energy Efficiency Equipment Standards.* The adopted bill includes a variety of new standards for lighting and for residential and commercial appliance equipment. The equipment includes residential refrigerators, freezers, refrigerator-freezers, metal halide lamps, and commercial walk-in coolers and freezers.
- *Repeal of Oil and Gas Tax Incentives.* The enacted law includes repeal of two tax subsidies in order to offset the estimated cost to implement the CAFE provision.

The two most controversial provisions of H.R. 6 that were not included in the enacted law were the proposed Renewable Energy Portfolio Standard (RPS) and most of the proposed tax provisions, which included repeal of tax subsidies for oil and gas and new incentives for energy efficiency and renewable energy.

---

* Excerpted from Congressional Research Service Report for Congress dated December 21, 2007; Order Code RL34294.

INTRODUCTION

The Energy Independence and Security Act (P.L. 110-140, H.R. 6) is an omnibus energy policy law that consists mainly of provisions designed to increase energy efficiency and the availability of renewable energy. This report describes the key provisions of the enacted law, summarizes the legislative action on H.R. 6, and provides a summary of the provisions under each of the titles in the law.

Many analysts in the CRS Resources, Science, and Industry Division contributed to this report; their names and contact information are located on the back of the summary page.

KEY PROVISIONS

Provisions Included

The three key provisions enacted in P.L. 110-140 are the Corporate Average Fuel Economy (CAFE) Standards, the Renewable Fuel Standard (RFS), and the appliance/lighting efficiency standards.

Corporate Average Fuel Economy (CAFE) Standards. The law sets a target of 35 miles per gallon for the combined fleet of cars and light trucks by model year 2020. Also, a fuel economy program is established for medium- and heavy-duty trucks, and a separate fuel economy standard is created for work trucks. (For more details on issues related to the CAFE provision, see CRS Report RL33982, *Corporate Average Fuel Economy (CAFE): A Comparison of Selected Legislation in the 110th Congress*, by Brent D. Yacobucci and Robert Bamberger, and CRS Report RL3 3413, *Automobile and Light Truck Fuel Economy: The CAFE Standards*, by Brent D. Yacobucci and Robert Bamberger.)

Renewable Fuel Standard (RFS). The law sets a modified standard that starts at 9.0 billion gallons of renewable fuel in 2008 and rises to 36 billion gallons by 2022. Of the latter total, 21 billion gallons is required to be obtained from cellulosic ethanol and other advanced biofuels. (For more details on issues related to the RFS provision, see CRS Report RL34265, *Selected Issues Related to an Expansion of the Renewable Fuel Standard (RFS).*) incandescent reflector lamps, and fluorescent lamps. A required target is set for lighting efficiency, and energy efficiency labeling is required for consumer electronic products. Also, efficiency standards are set by law for external power supplies, residential clothes washers, dishwashers, dehumidifiers, refrigerators, refrigerator- freezers, freezers, electric motors, residential boilers, commercial walk-in coolers, and commercial walk-in freezers. Further, DOE is directed to set standards by rulemaking for furnace fans and battery chargers.

Provisions Excluded

Two controversial provisions of H.R. 6 that were not included in the enacted law were the proposed Renewable Energy Portfolio Standard (RPS) and the proposed repeal of tax subsidies for oil and gas.

Renewable Energy Portfolio Standard (RPS). Under an RPS, retail electricity suppliers (electric utilities) must provide a minimum amount of electricity from renewable energy resources or purchase tradable credits that represent an equivalent amount of renewable energy production. The minimum requirement is often set as a percentage share of a supplier's total retail electricity sales. The second degree amendment to H.R. 6 passed by the House on December 6, 2007, proposed a national RPS target that aimed to reach 15% of total electricity sales by 2020. Up to 4 percentage points of the 15% target could be met with energy efficiency measures. This provision was stripped out by the Senate and was not included in the final version of the bill. (For more details on issues related to the RPS provision, see CRS Report RL341 16, *Renewable Energy Portfolio Standard (RPS): Background and Debate Over a National Requirement*, by Fred Sissine.)

Energy Tax Subsidies. The House-passed second degree amendment to H.R. 6 contained provisions that would have repealed about $22 billion of oil and gas subsidies that were designed to offset the cost of supporting a variety of energy efficiency and renewable energy tax incentives. These proposed incentives would have included a four-year extension of the renewable energy electricity production tax credit. Most of those provisions were stripped out by the Senate and were not included in the final bill. Enough tax revenue offsets were included to cover the estimated cost of the CAFE provision. (For more details about the proposed renewable energy incentives, see CRS Report RL34162, *Renewable Energy Issues in the 110th Congress*, by Fred Sissine. For more details about the proposed repeal of oil and gas subsidies, see CRS Report RL33578, *Energy Tax Policy*, by Salvatore Lazzari.)

Brief Summary of Legislative Action

On January 18, 2007, the House passed the 14-page CLEAN Energy Act (H.R. 6) by a vote of 264-163. The bill proposed to repeal certain oil and natural gas subsidies, thus generating revenue for an Energy Efficiency and Renewables Reserve. The Reserve was designed to reduce foreign oil dependence and serve other purposes. The actual uses of the reserve would have been determined at a later date by further legislation.[1]

On June 21, 2007, the Senate passed its first degree amendments to H.R. 6 as the proposed Renewable Fuels, Consumer Protection, and Energy Efficiency Act of 2007. This action transformed H.R. 6 into a 500-page omnibus energy policy bill, with a primary focus on energy efficiency and renewable energy. The Senate's amendments were drawn primarily from the proposed Energy Savings Act of 2007 (S. 1321).[2] The key provisions of the Senate-passed H.R. 6 were appliance efficiency standards, an increase of the renewable fuel standard to 36 billion gallons by 2022, and an increase of the combined corporate average fuel economy standards to 35 miles per gallon (mpg) by 2020.

On August 4, 2007, the House passed the omnibus energy policy bill, H.R. 3221, which had two divisions and 13 titles. Division A contained provisions of the New Direction

---

1 For more details about the reserve, see CRS Report RS2257 1, *The Strategic Energy Efficiency and Renewables Reserve in the CLEAN Energy Act of 2007* (H.R. 6, by Fred Sissine).

2 S. 1321 was, in turn, derived from several other Senate bills. Additional details about legislation that was incorporated into H.R. 6 are available in CRS Report RL3383 1, *Energy Efficiency and Renewable Energy Legislation in the 110th Congress*, by Fred Sissine, Lynn J. Cunningham, and Mark Gurevitz.

for Energy Independence, National Security, and Consumer Protection Act. An adopted floor amendment (H.Amdt. 748) added a 15% renewable energy portfolio standard (RPS). Division B, the Renewable Energy and Energy Conservation Tax Act of 2007, was also added as a floor amendment and contained the House-approved version of H.R. 2776. It added four titles to H.R. 3221 that included a four-year extension of the renewable electricity production tax credit and other efficiency and renewables incentives.

Because the House omnibus bill (H.R. 3221) and the Senate omnibus bill (H.R. 6) had different bill numbers, the bills could not be taken directly to conference committee.[3] However, after the House completed action on H.R. 3221, informal bipartisan negotiations over the omnibus energy bills began between the House and Senate. Key issues included CAFE, the renewable fuel standard, the RPS provision in H.R. 3221, and a proposed repeal of certain oil and natural gas subsidies to offset costs for new energy efficiency and renewable energy tax incentives.

On December 6, 2007, the House passed by a vote of 235-181 its second-degree amendments to the Senate-passed amendments to H.R. 6. This "second version" of a House omnibus energy bill was derived primarily by trimming and modifying H.R. 3221 and adding major new provisions on CAFE and RFS. The House-passed bill included a proposed increase of the CAFE standard to 35 miles per gallon by 2020 and an increase of the renewable fuel standard to 36 billion gallons per year by 2022. The House bill also included a proposed 15% renewable electricity portfolio standard and $21 billion of new tax incentives for energy efficiency and renewable energy measures. The bill proposed to offset the new tax incentives with a repeal of about $21 billion in tax subsidies for oil and natural gas.

The White House threatened to veto the House-passed bill mainly because of the presence of provisions for an RPS and for the repeal of oil and gas tax subsidies.[4] On December 7, 2007, a Senate cloture vote on the House-passed version of H.R. 6 — with provisions for an RPS and for the repeal of oil and gas subsidies — failed (52-43). After stripping out the RPS and modifying the tax package, a cloture vote on S.Amdt. 3841 failed (59-40).

On December 13, 2007, the Senate adopted S.Amdt. 3850, its second-degree substitute amendment to H.R. 6 by a vote of 8 6-8. The Senate substitute was nearly identical to the House-passed bill, except that the RPS provision and most tax provisions had been taken out. The House approved the Senate bill on December 18, 2007. The President signed the bill into law on December 19, 2007 (P.L. 110-140).

---

3 More details about the House and Senate bills developed up to that point are available in CRS Report RL34 135, Omnibus Energy Efficiency and Renewable Energy Legislation: A Side-by-Side Comparison of Major Provisions in House-Passed H.R. 3221 with Senate- Passed H.R. 6, by Fred Sissine.

4 Office of Management and Budget, Statement of Administration Policy on H.R. 6, Energy Independence and Security Act of 2007, December 6, 2007; [http://www.whitehouse.gov/ omb/legislative/sap/11 0-1/hr6sap-h_2.pdf].

TITLE I: ENERGY SECURITY THROUGH IMPROVED VEHICLE FUEL ECONOMY

Subtitle A, Increased Corporate Average Fuel Economy

This subtitle requires an increase in CAFE standards and a restructuring of the fuel economy program. A single CAFE standard of 35 miles per gallon (mpg) by MY2020 is established, and the distinction between the passenger car and light truck fleet is preserved. The new standards will be based on vehicle attributes and expressed in the form of a mathematical function. Interim standards will be set, beginning with MY20 11. Manufacturers will be required to come within 92% of the standard for a given model year. However, manufacturers can earn credits for exceeding the standards in one vehicle class that can be applied to boost, within limitations, the CAFE of a different vehicle class that is falling short of compliance. Additionally, credits may be sold and bought between manufacturers. CAFE credits for the manufacture of flexible-fueled vehicles (FFV) are retained but phased out by MY2020. Civil penalties assessed for non-compliance will be deposited to the general fund of the U.S. Treasury to support future rulemaking and to provide grants to manufacturers for research and development, and retooling in support of increasing the fuel efficiency of their fleets. The law requires the development of standards for "work trucks" and commercial medium- and heavy-duty on-highway vehicles. (For additional information, see CRS Report RL33413, *Automobile and Light Truck Fuel Economy: The CAFE Standards*, by Brent D. Yacobucci and Robert Bamberger.)

Subtitle B, Improved Vehicle Technology

This subtitle establishes a loan guarantee program for advanced battery development, grant programs for plug-in hybrid vehicles, incentives for purchasing heavy-duty hybrid vehicles for fleets, and credits for various electric vehicles.

Subtitle C, Federal Vehicle Fleets

Federal agencies are prohibited from acquiring any light-duty motor vehicle or medium-duty passenger vehicle that is not "a low greenhouse gas emitting vehicle" as defined in this subtitle. Alternatively, the agency may demonstrate that it has adopted cost-effective policies to reduce its petroleum consumption sufficiently to achieve a comparable reduction in greenhouse gas emissions. By 2015, federal agencies are required to achieve at least a 20% reduction in annual petroleum consumption and a 10% increase in annual alternative fuel consumption. These increases are to be calculated from a 2005 baseline. Interim milestones will be established and agencies will report annually on their progress. The regulations governing this program are required to be issued not later than 18 months after enactment.

Title II: Energy Security Through Increased Production of Biofuels

Subtitle A, Renewable Fuel Standard

This subtitle extends and increases the renewable fuel standard (RFS) set by P.L. 109-58 (§1501). The RFS requires minimum annual levels of renewable fuel in U.S. transportation fuel. The previous standard was 5.4 billion gallons for 2008, rising to 7.5 billion by 2012. The new standard starts at 9.0 billion gallons in 2008 and rises to 36 billion gallons in 2022. Starting in 2016, all of the increase in the RFS target must be met with advanced biofuels, defined as cellulosic ethanol and other biofuels derived from feedstock other than corn starch — with explicit carve-outs for cellulosic biofuels and biomass-based diesel. The EPA Administrator is given authority to temporarily waive part of the biofuels mandate, if it were determined that a significant renewable feedstock disruption or other market circumstance might occur. Renewable fuels produced from new biorefineries will be required to reduce by at least 20% the life cycle greenhouse gas (GHG) emissions relative to life cycle emissions from gasoline and diesel. Fuels produced from biorefineries that displace more than 80% of the fossil-derived processing fuels used to operate a biofuel production facility will qualify for cash awards. Several studies are required on the impacts of an RFS expansion on various sectors of the economy. (For more details on issues related to the RFS proposal, see CRS Report RL34265, *Selected Issues Related to an Expansion of the Renewable Fuel Standard (RFS)*, by Brent D. Yacobucci and Randy Schnepf.)

Subtitle B, Biofuels Research and Development (R&D)

This subtitle promotes research on the expansion of the use of biodiesel and biogas as motor fuels. Grants are authorized for R&D and commercial applications of cellulosic biofuels technologies and for the conversion of existing corn-based ethanol plants to produce cellulosic biofuels. The Secretary of Energy is required to report to Congress on the feasibility of algae as a feedstock for biofuels production. The subtitle also promotes university-based R&D on biofuels.

Subtitle C, Biofuels Infrastructure

This subtitle aims to improve information about federal biofuels research programs, focus research on infrastructure and biorefineries, study potential impacts of increased biofuels use, and increase authorized funding for DOE biofuels research. A funding authorization of $25 million is established to provide grants for biofuels research, development, and demonstration (RD&D) and commercial applications in states that have low rates of ethanol production. A university-based program is authorized to provide grants of up to $2 million for R&D on renewable energy technologies. Priority is given to universities in low-income and rural communities with proximity to trees dying of disease or insect infestation.

DOE is directed to create a grant program to help establish or convert infrastructure to use renewable fuels, including E85 (85% ethanol). The Energy Policy Act of 2005 (EPACT, P.L. 109-5 8) authorization for grants to support cellulosic ethanol production is increased. A grant program is authorized to support production of flexible-fueled vehicles. Studies are also required on the market penetration of flexible-fueled vehicles, the feasibility of constructing dedicated ethanol pipelines, the feasibility of using greater percentages of ethanol in fuel blends, and the adequacy of railroad transportation for delivery of ethanol fuel.

Subtitle D, Environmental Safeguards

Previously, under the Clean Air Act (§21 1 [f]), no new fuels or fuel additives could be introduced into commerce unless granted a waiver by the Environmental Protection Agency (EPA). If EPA did not act within 180 days of receiving a waiver request, the waiver was treated as granted.[5] Section 251 tightens the waiver provision. It amends the Clean Air Act to prohibit the introduction of new fuels or fuel additives unless EPA explicitly grants a waiver. After receiving a waiver request, EPA will now have 270 days to take final action.

TITLE III: ENERGY SAVINGS THROUGH IMPROVED STANDARDS FOR APPLIANCES AND LIGHTING

Subtitle A, Appliance Energy Efficiency

This title sets, by statute, new efficiency standards for external power supplies, residential clothes washers, dishwashers, dehumidifiers, refrigerators, refrigerator-freezers, freezers, electric motors, and residential boilers. DOE is allowed to establish regional variations in standards for heating and air conditioning equipment. DOE is required to complete a rulemaking process for furnace fans by 2013. Federal agencies are directed to purchase devices that limit standby power use. DOE is directed to issue a final rule that sets efficiency standards for battery chargers. Certain energy efficiency measures for walk-in coolers and walk-in freezers are set by law. Also, several procedural changes are now in place to expedite the DOE rulemaking process.

Subtitle B, Lighting Energy Efficiency

Section 321 sets an energy efficiency standard for general service incandescent lamps, provides for consumer education and lamp labeling, and requires market assessments and a consumer awareness program. Section 322 sets energy efficiency standards for incandescent reflector lamps and fluorescent lamps. For federal buildings, Section 323 sets

[5] See 42 U.S.C. 7545(f).

energy efficiency requirements for GSA-leased space and for use of energy efficient lighting fixtures and bulbs in those leased spaces. Section 324 sets energy efficiency standards for metal halide lamp fixtures designed to be operated with lamps rated between 150 watts and 500 watts. Section 325 directs the Consumer Product Safety Commission to set energy efficiency labeling requirements for consumer electronic products.

TITLE IV: ENERGY SAVINGS IN BUILDINGS AND INDUSTRY

Subtitle A, Residential Building Efficiency

Section 411 increases the funding authorization for DOE's Weatherization Program, providing \$3.75 billion over five years. Under Section 412, DOE is directed to conduct a study of the renewable energy system rebate program described in §206(c) of the Energy Policy Act of 2005. The study aims to determine the minimum funding the program would need to be viable. Further, DOE is directed to propose an implementation plan. Section 413 requires DOE to establish energy efficiency standards for manufactured housing.

Subtitle B, High-Performance Commercial Buildings

This subtitle encourages the development of more energy-efficient "green" commercial buildings. Section 421 creates an Office of Commercial High Performance Green Buildings at DOE. Section 422 establishes a zero-energy commercial buildings initiative. A national goal is set to achieve zero-net-energy use for new commercial buildings built after 2025. A further goal is to retrofit all pre- 2025 buildings to zero-net-energy use by 2050. Section 423 requires that DOE establish a national clearinghouse for information and public outreach about high- performance green buildings.

Subtitle C, High-Performance Federal Buildings

Section 431 requires that total energy use in federal buildings, relative to the 2005 level, be reduced 30% by 2015. Section 432 directs that federal energy managers conduct a comprehensive energy and water evaluation for each facility at least once every four years. For new federal buildings and major renovations, Section 433 requires that fossil-fuel energy use — relative to the 2003 level — be reduced 55% by 2010 and be eliminated (100% reduction) by 2030. Section 434 requires that each federal agency ensure that major replacements of installed equipment (such as heating and cooling systems), or renovation or expansion of existing space, employ the most energy efficient designs, systems, equipment, and controls that are life-cycle cost effective. Section 435 prohibits federal agencies from leasing buildings that have not earned an EPA Energy Star label. Section 436 requires GSA to establish an Office of Federal High-Performance Green Buildings to coordinate green building information and activities within GSA and with other federal agencies. The Office must also develop standards for federal facilities, establish green practices, review budget and life-cycle costing issues, and promote demonstration of innovative technologies. Section 437 directs the Government Accountability Office (GAO) to audit the implementation of activities required under this subtitle. The audit must cover budget, life-cycle costing, contracting, best practices, and agency coordination. Section 438 requires federal facility development projects with a footprint exceeding 5,000 square feet to use site planning, design, construction, and maintenance strategies to control storm water runoff. Section 439 directs GSA to review the current use of, and design a strategy for increased use of, cost-effective lighting, ground source heat pumps, and other technologies in GSA facilities. Section 440 authorizes $4 million per year over five years to support work under sections 434-439 and 482. For the purpose of conducting life-cycle cost calculations, Section 441 increases the time period from 25 years, in prior law, to 40 years.

Subtitle D, Industrial Energy Efficiency

Section 451 directs DOE to conduct research on, develop, and demonstrate new processes, technologies, and operating practices and techniques to significantly improve the energy efficiency of equipment and processes used by energy-intensive industries. Section 452 directs EPA to establish a recoverable waste energy inventory program. This program must include an ongoing survey of all major industrial and large commercial combustion sources in the United States. EPA is required to identify the potential for economically feasible waste energy recovery, create a grant program to support waste energy recovery, and strengthen "clean energy centers" that analyze waste energy recovery. Section 453 directs DOE to initiate a voluntary national information program for widely used data centers and data center equipment for which there is significant potential for energy savings.

DOE is also tasked with helping to devise strategies to improve energy efficiency at these data centers.

Subtitle E, Healthy High-Performance Schools

Section 461 creates a grant program for *Healthy High-Performance Schools* that aims to encourage states, local governments, and school systems to build green schools. EPA, in consultation with the Department of Education, is allowed to provide grants to state agencies to provide technical assistance and help with the development of state plans for school building design. Also, EPA is directed to develop model voluntary guidelines for school site selection. In addition to other environmental aspects, the grants and guidelines must have a focus on energy efficiency, natural daylighting, and other energy-related features. Section 462 directs EPA to lead a detailed study of how sustainable building features, such as energy efficiency, affect multiple perceived indoor environmental quality stressors on students in K-12 schools.

Subtitle F, Institutional Entities

Section 471 creates a program of grants and loans to support energy efficiency and energy sustainability projects at public institutions.

Subtitle G, Public and Assisted Housing

Section 481 directs the Department of Housing and Urban Development (HUD) to update energy efficiency standards for all public and assisted housing.

Subtitle H, General Provisions

Section 491 calls for the DOE Office of Commercial High Performance Buildings and the GSA Office of Federal High Performance Buildings to jointly develop guidelines for demonstration projects. In accordance with the guidelines, one federal project must be undertaken annually over a five-year period, supported by a $10 million funding authorization. Also, a total of four projects are to be undertaken at different universities over the five-year period, supported by an additional $10 million funding authorization. Section 492 calls for these two offices to undertake a joint survey of research on green buildings, coordinate efforts to develop a research plan, and identify potential benefits of green buildings for security, natural disasters, and emergency needs of the federal government. Section 493 requires EPA to create a program of competitive grants to local governments for green building demonstration projects. Section 494 directs the Office of Commercial High Performance Buildings and the Office of Federal High Performance Buildings to jointly appoint a Green Building Advisory Committee with representatives from a variety of backgrounds, including federal agencies, state and local governments, building industry experts, security advisors, and environmental health experts. Section 495 calls for DOE to create an advisory committee on energy efficiency finance to help lower costs and increase investment for energy efficiency technologies.

TITLE V: ENERGY SAVINGS IN GOVERNMENT AND PUBLIC INSTITUTIONS

Subtitle A, United States Capitol Complex

Section 501 allows the Architect of the Capitol (AOC) to perform a feasibility study regarding construction of a photovoltaic roof for the Rayburn House Office Building. Under Section 502, the AOC is allowed to construct a fuel tank and pumping system for E85 (85% ethanol) fuel at or within close proximity to the Capitol Grounds Fuel Station. Section 503 requires the AOC, to the maximum extent practicable, to include energy efficiency measures, climate change mitigation measures, and other appropriate environmental measures in the Capitol Complex Master Plan. Under Section 504, the AOC is directed to operate the steam boilers and chiller plant at the Capitol Power Plant in the most energy efficient manner possible to minimize carbon emissions and operating costs. Further, Section 505 requires the AOC to install technologies for the capture and storage or use of carbon dioxide emitted from coal combustion in the Capitol Power Plant.

Subtitle B, Energy Savings Performance Contracting

Section 511 eliminates the advance reporting requirement for Energy Savings Performance Contracts (ESPCs) that have a cancellation ceiling exceeding $10 million. Section 512 increases ESPC funding flexibility by allowing a combination of appropriated funds and private financing. Section 513 restricts federal agencies from limiting the duration of ESPCs to less than 25 years or limiting the total amount of obligations. Further, this section permits the criteria for savings verification to satisfy the requirement for energy audits. Also, it directs federal agencies to modify existing ESPCs to conform with the requirements of this subtitle. Section 514 permanently authorizes ESPCs.

Section 515 extends the definition of energy savings reduction to include increased use of an existing energy source by cogeneration or heat recovery, use of excess electrical or thermal energy generated from onsite renewable sources or cogeneration, and increased energy-efficient use of water resources. Section 516 permits agencies to retain the full amount of energy and water cost savings obtained from utility incentive programs. Section 517 authorizes $750,000 per year over five years for a program to train contract officers in negotiating ESPCs. Section 518 directs the Department of Defense (DOD) and DOE to study the potential use of ESPCs in nonbuilding applications, which is defined to include vehicles and federally owned equipment to generate electricity or transport water.

Subtitle C, Energy Efficiency in Federal Agencies

Under Section 521, GSA is directed to use up to $30 million — subject to appropriation — from FY2007 and prior years' unobligated balances of the Federal Buildings Fund to support the installation of a solar photovoltaic system for the DOE headquarters building in the District of Columbia. Section 522 prohibits, except under certain circumstances, the purchase of incandescent light bulbs for use in Coast Guard office buildings. Section 523 requires 30% of the hot water demand in new federal buildings (and major renovations) to be met with solar hot water equipment, provided it is life-cycle cost-effective. Section 524 encourages federal agencies to minimize standby energy use in purchases of energy-using equipment. Section 525 requires federal procurement to focus on use of Energy Star and Federal Energy Management Program (FEMP)-designated products. Section 526 prohibits federal agencies from procuring synfuel unless its life cycle GHG emissions are less than those for conventional petroleum sources. Section 527 directs each federal agency subject to any requirements under this title to issue an annual report that describes the status of initiatives to improve energy efficiency, reduce energy costs, and reduce GHG emissions. Section 528 requires the Office of Management and Budget (OMB) to submit an annual report to Congress that summarizes the information reported under Section 527, evaluates overall progress toward the goals of Section 527, and recommends additional actions needed to meet those goals. Section 529 directs the Federal Energy Regulatory Commission (FERC) to conduct a national assessment of demand response, including an estimate of nationwide demand response out to a 10- year horizon. Further, FERC is required to prepare a National Action Plan on Demand Response, with cooperation from industry. Annual funding of $10 million per year is authorized over three years.

Subtitle D, Energy Efficiency of Public Institutions

Section 531 increases annual funding authorizations for DOE's state energy programs. Under Section 532, electric and natural gas utilities are required to make energy efficiency a priority resource and to integrate energy efficiency into resource plans and planning processes. Further, the utilities are directed to modify their rates to align their incentives with the delivery of cost-effective energy efficiency and promote energy efficiency investments. Utilities are encouraged to consider several policy options for achieving those goals.

Subtitle E, Energy Efficiency and Conservation Block Grants

This subtitle establishes an energy efficiency block grant program. Section 541 provides definitions of program elements. Section 542 directs DOE to establish an energy efficiency and conservation block grant program to help reduce energy use and emissions at the local and regional level. Section 543 establishes allocation percentages for grants provided under this subtitle. Section 544 enumerates the allowed purposes for the use of funds provided under this subtitle, which includes strategic planning, consultant services, and energy audits. Section 545 provides eligibility requirements for grants under this program, including payment of prevailing wage rates, submission of a strategic plan, and sharing of information. Section 546 sets criteria for minimum allocations of competitive grant funding. Section 547 specifies that DOE may review and evaluate the performance of grant recipients and withhold funds from those it deems have failed to achieve compliance. To support the grant program, Section 548 authorizes $2 billion annually over five years. Additional funding is authorized to cover administrative costs of the program. Section 548 stresses that funding will supplement, not replace, funding provided by DOE under the Weatherization and State Energy programs.

TITLE VI: ACCELERATED RESEARCH AND DEVELOPMENT

Subtitle A, Solar Energy

Section 602 aims to improve the cost and effectiveness of thermal energy storage technologies that could improve the operation of concentrating solar power electric generating plants. Section 603 calls for improved integration of concentrating solar power into regional electricity transmission systems.

Subtitle B, Geothermal Energy

DOE is directed to support programs of R&D, demonstration, and commercial application to expand the use of geothermal energy. Section 613 directs DOE to support programs that (1) develop advanced prospecting tools to locate and develop hidden geothermal resources, and (2) demonstrate advanced exploratory drilling technologies and techniques with industry partners. Section 614 directs DOE to support programs to develop components and systems necessary to develop, produce, monitor, and model the performance of geothermal reservoirs used to produce geothermal energy. In addition, Section 614 directs DOE to support programs that mitigate or prevent environmental damage from geothermal energy development.

Section 615 directs DOE to support enhanced geothermal system development, whereby geothermal reservoir systems are engineered (as opposed to naturally occurring systems) by creating fractures and permeable conduits via reservoir stimulation. DOE would support R&D programs for enhanced geothermal system technologies and for reservoir stimulation and support demonstration projects at a minimum of four sites.

DOE is directed to establish a program of R&D, demonstration, and commercial application for geothermal energy production from oil and gas fields and from geopressured resources.[6] Section 616 directs DOE to implement a grant program for at least three demonstration projects that use geothermal techniques to extract energy from marginal, unproductive, and productive oil and gas fields. Also, DOE is directed to establish a grant program for the recovery of energy from geopressured resources.

Section 618 directs DOE to establish a Center for Geothermal Technology Transfer, via a grant to an institution of higher learning or consortium thereof, that would serve as an information clearinghouse for the geothermal industry, make data available to the public, and coordinate R&D efforts among national and international partners. Section 619 would rename DOE's GeoPowering the West program as "GeoPowering America" and expand its geothermal technology transfer activities to cover the entire United States. Section 620 would award a grant on a competitive basis to an institution of higher education to establish a geothermal-powered energy generation facility on the institution's campus.

Section 624 directs DOE to support international geothermal energy development through collaborative efforts to promote geothermal R&D and deployment of geothermal technologies. Section 625 directs DOE to make grants to eligible entities from "high-cost regions"[7] of the United States for a feasibility study, demonstration, and commercial application of technologies related to geothermal energy.

Subtitle B authorizes $90 million annually for geothermal activities, of which $10 million is designated for activities under Section 616. An additional $5 million is authorized annually for the Intermountain West Geothermal Consortium, and $5 million is authorized annually for Section 624. All of the foregoing authorizations are in effect from 2008 to 2012.

Subtitle C, Marine and Hydrokinetic Renewable Energy Technologies

DOE is directed to create an R&D program focused on technology that produces electricity from waves, tides, currents, and ocean thermal differences (§633). A report to Congress is required. Further, DOE is instructed to award grants to institutions of higher education (or consortia thereof) to establish National Marine Renewable Energy Research, Development, and Demonstration Centers (§634).

---

[6] Geopressured resources are geothermal deposits of hot water or steam found in sedimentary rocks under higher than normal pressures and that are saturated with oil and gas.

Subtitle D, Energy Storage for Transportation and Electric Power

The U.S. Energy Storage Competitiveness Act of 2007 directs DOE to conduct a cost-shared RD&D program to support the ability of the nation to remain globally competitive in energy storage systems for electric drive vehicles, stationary applications, and electricity transmission and distribution. An Energy Storage Advisory Council will be created, with responsibility for preparing a five-year research plan. Also, through competitive bids, DOE will establish four energy storage research centers managed by the Office of Science. DOE is required to conduct energy storage demonstration projects. Also, DOE is to investigate secondary applications of energy storage equipment and to examine technologies and processes for final recycling and disposal of energy storage equipment. After five years of program operation, the law will require a review of the program by the National Academy of Sciences. A total authorization of nearly $3 billion is provided over a 10-year period.

Subtitle E, Miscellaneous Provisions

Section 651 directs DOE to establish an RD&D program to determine ways in which the weight of motor vehicles could be reduced to improve fuel efficiency without compromising passenger safety. This will focus on the development of new materials and on reducing the cost of lightweight materials. An $80 million authorization is provided over a five-year period.

Section 652 directs DOE to report on the state of technology development for "advanced" insulation with an R-value greater than R35 per inch. The report is to include an estimate of potential cost savings by applying such insulation to covered refrigeration units. If sufficient cost savings are projected, DOE will then be directed to conduct a cost-shared demonstration program to show actual cost savings. An $8 million funding authorization is provided for that program.

Section 653 changes the sulfur dioxide (SO_2) criterion for clean coal power plants from a percentage basis (99% of SO_2 removed) to a weight-by-energy basis (no more than 0.04 pounds of SO_2 per million Btu).

Section 654 on the "H-Prize" directs DOE to conduct a competitive program to award cash prizes to advance R&D, demonstration, and commercial application of hydrogen energy technologies. Prizes can be a mix of federal appropriations and funds provided by an entity that DOE chooses to administer the program. The program sunsets in 2018. Prize categories include technology advancements in hydrogen production, storage, distribution, and use; prototypes of hydrogen vehicles and products; and technologies that "transform" distribution or production. DOE is required to report to Congress annually, identifying award recipients, technologies developed, and specific actions undertaken to commercialize the technologies. More than $1 billion is authorized over a 10-year period.

Section 655 directs DOE to create the "Bright Tomorrow" lighting prizes for solid state (LED) lighting developments that achieve targeted levels of energy efficiency and other traits. Two specific categories are a solid state replacement for a 60-watt incandescent light

[7] These "high cost" regions are defined as places where the average cost of retail power exceeds 150% of the national average.

and a replacement for the PAR Type 38 halogen light. Also, a prize is established for a "twenty-first century lamp" that achieves certain output, efficiency, and color targets. After the awards are made, DOE is required to develop guidelines for federal agency purchases of the incandescent and halogen replacements, with the goal of complete replacement within five years.

Section 656 directs DOE to establish a cost-shared Renewable Energy Innovation Manufacturing Partnership Program to make awards to support RD&D on advanced manufacturing processes, materials, and infrastructure for renewable energy technologies. Further goals are to increase domestic renewable energy production and better coordinate federal, state, and private resources through partnerships. Solar, wind, biomass, geothermal, energy storage, and fuel cell systems are eligible forms of equipment.

TITLE VII: CARBON CAPTURE AND SEQUESTRATION

Subtitle A, Carbon Capture and Sequestration Research, Development, and Demonstration

DOE's program for carbon capture and sequestration R&D is expanded and will include large-scale demonstration projects. DOE is directed to engage the National Academy of Sciences (NAS) to conduct a review of the program. DOE is directed to work with the NAS to develop interdisciplinary graduate degree programs with emphasis on geologic sequestration science. A university-based R&D grant program will be established to study carbon capture and sequestration using various types of coal. EPA is directed to assess potential impacts of carbon sequestration on public health and safety and the environment. Further, injection and sequestration activities under this subtitle are subject to the requirements of the Safe Drinking Water Act.

Subtitle B, Carbon Capture and Sequestration Assessment and Framework

Section 711 directs the Department of the Interior (DOI) to develop a methodology for an assessment of the national potential for geologic storage of carbon dioxide. Following publication of the methodology, DOI will be required to complete an assessment of national capacity for carbon dioxide storage in accordance with the methodology.

Section 712 directs DOI to develop a methodology for an assessment of the total capacity of ecosystems to sequester carbon and the ability of ecosystems to reduce emissions of carbon dioxide, methane, and nitrous oxides in ecosystems through management practices. Following publication of the methodology, DOI will be required to complete a national assessment of the quantity of carbon stored in and released from ecosystems, and the annual flux of carbon dioxide, methane, and nitrous oxides in and out of ecosystems.

Section 713 calls for DOI to maintain records, and an inventory, of the quantity of carbon dioxide stored within federal mineral leaseholds.

Section 714 directs DOI to submit a report on a recommended regulatory framework for managing geologic carbon sequestration on public lands. The report must include an assessment of options to ensure that the United States receives fair market value for the use of public land, the proposed procedures for public review and comment, procedures for protecting natural and cultural resources of the public land overlying the geologic sequestration sites, a description of the status of liability issues related to the storage of carbon dioxide in public land, identification of legal and regulatory issues for cases where the United States owns title to the mineral resources but not the overlying land, identification of issues related to carbon dioxide pipeline rights-of-way, and recommendations for additional legislation that may be required for adequate public land management and leasing to accommodate geologic sequestration of carbon dioxide and pipeline rights-of-way.

TITLE VIII: IMPROVED MANAGEMENT OF ENERGY POLICY

Subtitle A, Management Improvements

Section 801 directs DOE to conduct a 10-year national media campaign to educate consumers to save energy and reduce oil use. Competitive bidding is required for contracting the media services. A funding authorization of $5 million per year is provided for five years. An annual report to Congress is required.

Section 802 authorizes the Federal Coordinator for the Alaska Natural Gas Transportation Projects to appoint and terminate personnel and to pay appointed and temporary personnel up to a maximum of the level III rate of the Executive Schedule. The Federal Coordinator is granted authority to establish various payment requirements and to use funds raised without further appropriation. This authority does not affect the authority of the Secretary of the Interior.

Section 803 creates a 50% matching grant program for constructing small renewable energy projects that will have an electrical generation capacity less than 15 megawatts. Eligible applicants include local governments, utilities, and Indian tribes. Such sums as necessary are authorized for the program.

Section 804 requires the Energy Information Administration (EIA) to monitor planned petroleum refinery outages and report to the Secretary of Energy when such outages are affecting the price or availability of petroleum products. The Secretary will then be required to share data with refinery operators and encourage reductions in out-of-service refinery capacity.

Section 805 requires the Administrator of the Energy Information Administration (EIA) to develop a five-year plan for enhancing the scope, quality, and timeliness of the agency's data collection efforts. In addition, it requires closer coordination by EIA with state energy officials and with the Federal Energy Regulatory Commission. The section addresses state-level data in several respects and requires the Administrator to submit to Congress within a year an assessment of state level energy data needs. EIA is directed to revisit certain data series that had been terminated due to budget constraints and to identify data gaps that may have resulted from those terminations. To implement this section, $10 million is authorized for 2008, and additional sums are authorized through 2012.

Section 806 expresses the sense of Congress that there is a national goal to use renewable energy resources from agricultural, forestry, and working lands of the nation to provide at least 25% of the nation's energy use by 2025.

Section 807 directs the Department of the Interior's U.S. Geological Survey to conduct a comprehensive assessment of geothermal energy resources in the United States and report the findings of that assessment to Congress.

Subtitle B, Prohibitions on Market Manipulation and False Information

This subtitle prohibits crude oil and petroleum product wholesalers from using any technique to manipulate the market or provide false information. The law directs the Federal Trade Commission to treat such action as an unfair or deceptive practice, subject to civil penalties of not more than $1 million per incident.

TITLE IX: INTERNATIONAL ENERGY PROGRAMS

This title authorizes assistance to promote clean and efficient energy technologies in foreign countries, and it establishes an International Clean Energy Foundation.

Subtitle A, Assistance to Promote Clean and Efficient Energy Technologies in Foreign Countries

The U.S. Agency for International Development (USAID) is directed to report to Congress on efforts to support policies for clean and efficient energy technologies. The Department of Commerce is directed to increase efforts to export such technologies and report to Congress on the results. Other U.S. agencies with export promotion responsibilities are required to increase efforts to support these technologies. Also, a multi-agency Task Force on International Cooperation for Clean and Efficient Energy Technologies is created to support the implementation of clean energy markets in key developing countries.

Section 917 creates a U.S.-Israel Energy Cooperation partnership to support research, development, and deployment (RD&D) of energy efficiency and renewable energy measures.

Subtitle B, International Clean Energy Foundation

The Foundation is established with the long-term goal of reducing GHG emissions. It is directed to use the funds authorized by this subtitle to make grants to promote projects outside of the United States that serve as models of how to reduce emissions. An annual report to Congress is required.

Subtitle C, Miscellaneous Provisions

Section 931 calls for the Secretary of State to ensure that energy security is integrated into the core mission of the Department of State. Energy advisors are required at key embassies, and the Department is required to report to Congress every two years on its energy-related activities. Section 932 adds the Secretary of Energy to the National Security Council. Section 933 calls for the President to submit to Congress a comprehensive annual report that describes a national energy security strategy for the nation.

Section 934 implements the Convention on Supplementary Compensation for Nuclear Damage that was opened for signature in 1997. The convention has since been signed by the United States and 12 other countries but has not yet entered into force. Each party to the convention will be required to establish a compensation system within its borders for nuclear damages to the public. In the United States, this obligation will be fulfilled by the existing Price-Anderson Act (§170 of the Atomic Energy Act of 1954). The convention will also establish a second tier of damage compensation to be paid by all parties. Section 934 requires the U.S. contribution to the second tier to be paid by suppliers of nuclear equipment and services, under a formula to be developed by DOE. Supporters of the convention contend that it will help U.S. exporters of nuclear technology by establishing a predictable international liability system.

Section 935 has the stated purpose of improving national energy security by promoting anti-corruption initiatives in oil and natural gas rich countries and of improving global energy security by promoting programs such as the Extractive Industries Transparency Initiative (EITI) that aim to increase transparency and accountability into extractive resource payments. The sense of Congress is expressed that global energy security should be furthered by encouraging further participation in EITI by eligible countries and companies and by promoting the effectiveness of the EITI program by ensuring that a robust and candid review mechanism is put in place. The Secretary of State is required to report to Congress on progress made in promoting transparency in extractive industries resource payments. An authorization of $3 million is provided to support U.S. contributions to the Multi-Donor Trust Fund of EITI.

TITLE X: GREEN JOBS

This title authorizes up to $125 million in funding to establish national and state job training programs, administered by the Department of Labor, to help address job shortages that are impairing growth in green industries, such as energy efficient buildings and construction, renewable electric power, energy efficient vehicles, and biofuels development.

TITLE XI: ENERGY TRANSPORTATION AND INFRASTRUCTURE

Subtitle A, Department of Transportation (DOT)

An Office of Climate Change and Environment is established at DOT to plan, coordinate, and implement strategies to reduce transportation-related energy use, mitigate the effects of climate change, and address the impact of climate change on transportation systems and infrastructure.

Subtitle B, Railroads

This subtitle directs DOT, in coordination with EPA, to establish and conduct a pilot grant program to assist railroad carriers in purchasing hybrid locomotives, including hybrid switch locomotives, in order to demonstrate the extent to which such locomotives increase fuel economy, reduce emissions, and lower costs of operation. Also, DOT is directed to create a program of capital grants for the rehabilitation, preservation, or improvement of railroad track (including roadbed, bridges, and related track structures) of class II and class III railroads.

Subtitle C, Marine Transportation

Short sea transportation is defined as commercial waterborne transportation that originates at a port in the United States and ends at another port in the United States or at a port in Canada located in the Great Lakes Saint Lawrence Seaway System. The same definition applies for the case where origination and end points are reversed.[8] This subtitle directs DOT to establish a short sea transportation program and designate short sea transportation projects to be conducted under the program to mitigate landside congestion. Short sea shipping activities are made eligible for support from DOT's capital construction fund. A report to Congress on the short sea transportation program is required.

Subtitle D, Highways

Section 1131 increases the federal share for congestion mitigation and air quality (CMAQ) projects up to 100% of project or program cost. Under Section 1132, DOT is directed to redistribute within each state any unobligated balances of the Highway Trust Fund that are rescinded in FY2008 or FY2009. Section 1133 expresses a sense of Congress that, in constructing new roadways or rehabilitating existing facilities, state and local governments should employ policies designed to accommodate all users, including motorists, pedestrians, cyclists, transit riders, and people of all ages and abilities.

TITLE XII: SMALL BUSINESS ENERGY PROGRAMS

Loans, grants, and debentures are established to help small businesses develop, invest in, and purchase energy efficient buildings, fixtures, equipment, and technology.

[8] This is the definition offered in Section 1122. Also, see the definition provided in Short Sea Shipping: Practices, Opportunities, and Challenges, by Gary A. Lombardo, [http://www.insourceaudit.com/WhitePapers/Short_Sea_Shipping.asp].

Section 1201 empowers the Small Business Administration (SBA) to make "express" loans for certain energy efficiency and renewable energy projects. Section 1202 creates a two-year pilot loan program for purchasing energy efficient technologies under Section 7(a) of the Small Business Act at half the cost that would have otherwise been required. After the pilot program terminates, GAO is required to prepare a report to Congress that describes its energy-saving impact. Section 1203 creates small business energy efficiency, sustainability, and telecommuting programs.

Reports to Congress are required for each of those programs. Section 1204 raises the Small Business Investment Act (SBIA) loan ceilings for certain energy efficiency and renewable energy projects undertaken by small businesses. Section 1205 enables qualified small business investment companies to issue energy-saving debentures.

Section 1206 expands certain SBIA provisions to include investments in energy-saving small businesses. Section 1207 creates a Renewable Fuel Capital Investment (RFCI) pilot program that taps into venture capital to help small firms develop renewable energy sources and new technologies. A funding authorization of $30 million is provided for RFCI over two years. Section 1208 requires SBA to study the RFCI program and issue a report to Congress on its findings.

TITLE XIII: SMART GRID

Section 1301 establishes a federal policy to modernize the electric utility transmission and distribution system to maintain reliability and infrastructure protection. The term "Smart Grid" refers to a distribution system that allows for flow of information from a customer's meter in two directions: both inside the house to thermostats, appliances, and other devices, and from the house back to the utility.[9] Smart Grid is defined to include a variety of operational and energy measures — including smart meters, smart appliances, renewable energy resources, and energy efficiency resources. Section 1302 calls for DOE to report to Congress on the deployment of Smart Grid technologies and any barriers to deployment. Section 1303 directs DOE to establish a Smart Grid Advisory Committee and a Smart Grid Task Force to assist with implementation. Section 1304 directs DOE to conduct Smart Grid RD&D and to develop measurement strategies to assess energy savings and other aspects of implementation. Section 1305 directs the National Institute of Standards and Technology to establish protocols and standards to increase the flexibility of use for Smart Grid equipment and systems. Section 1306 directs DOE to create a program that reimburses 20% of qualifying Smart Grid investments. Section 1307 directs states to encourage utilities to employ Smart Grid technology and allows utilities to recover Smart Grid investments through rates. Section 1308 requires DOE to prepare a report to Congress on the effect of private wire laws on the development of combined heat and power facilities. Section 1309 directs DOE to report to Congress on the potential impacts of Smart Grid deployment on the security of electricity

[9] The Smart Grid could allow appliances to be turned off during periods of high electrical demand and cost and give customers real-time information on constantly changing electric rates. The goal is to use advanced, information-based technologies to increase power grid efficiency, reliability, and flexibility, and reduce the rate at which additional electric utility infrastructure needs to be built.

infrastructure and operating capability. (For additional information, see CRS Report RL34288, *Smart Grid Provisions in H.R. 6, 110th Congress*, by Amy Abel.)

TITLE XIV: POOL AND SPA SAFETY

Section 1401 identifies this title as the "Virginia Graeme Baker Pool and Spa Safety Act." Section 1402 finds that proper use of barriers or fencing could substantially reduce the number of childhood residential swimming and pool drownings. Section 1403 provides several definitions employed throughout this subtitle. Section 1404 sets an industry standard (ASME/ANSI A1 12.19.8) as a national performance standard for swimming pool and spa drain cover equipment. Section 1405 establishes a grant program and requires that at least 50% of the funding be used to assist states in hiring and training enforcement personnel to implement and enforce standards. The remaining funds must be used to educate pool construction and installation companies, pool owners and operators, and pool service companies, about the standard. Also, funding of $2 million per year over two years is authorized for the federal Consumer Product Safety Commission to implement the grant program. Section 1406 specifies minimum state law requirements to qualify for a grant under Section 1405. The requirement includes the enclosure of all outdoor residential pools and spas, installation of devices to prevent entrapment by pool or spa drains, and notification to pool owners about entrapment protection standards. Also, in setting minimum state law requirements, the Commission is directed to consider current or revised national standards for barrier and entrapment equipment, and to ensure that the requirements are consistent with the Commission's existing publications on pool safety guidelines. Section 1407 directs the Commission to conduct a public education program on methods to prevent drowning and entrapment in swimming pools. A funding authorization of $5 million per year is provided over five years. Section 1408 directs the Commission to submit a report to Congress that evaluates the implementation of the state grant program.

TITLE XV: REVENUE PROVISIONS

Section 1500 specifies that, unless expressed otherwise, all tax provisions in this act refer to provisions of the Internal Revenue Code of 1986.

Section 1501 extends Federal Unemployment Tax Act (FUTA) taxes for one year. FUTA imposes a 6.2% gross tax rate on the first $7,000 paid annually by covered employers to each employee. In 1976, Congress passed a temporary surtax of 0.2% of taxable wages to be added to the permanent FUTA tax rate. The temporary surtax was subsequently extended through 2007. The President's FY2008 Budget had proposed extending the FUTA surtax. The Treasury Department stated that "extending the surtax will support the continued solvency of the federal unemployment trust funds and maintain the ability of the unemployment system to adjust to any economic downturns." This section enacts the President's proposal for one year, 2008. This provision is estimated to raise $ 1.446 billion over 10 calendar years.

Under Section 1502, the geological and geophysical costs of a major integrated oil company will be amortized (deducted proportionally) over a seven-year period instead of the current five-year period. (A major integrated oil company is defined as one with an average world production of at least 500,000 barrels per day, with 2005 gross receipts exceeding $1 billion, and which has at least a 15% interest in refinery operations.)

TITLE XVI: EFFECTIVE DATE

Section 1601 specifies that this act and the amendments it makes will take effect one day after enactment.

INDEX

A

B

D

E

F

G

H

I

J

K

L

M

N

O

P

Q

R

S

T

U

V

W

Y

Z